Y. Mizuno, M. B. H. Youdim,
D. B. Calne, R. Horowski,
W. Poewe, P. Riederer (eds.)

Advances in Research
on Neurodegeneration

Volume 3 & 4

SpringerWienNewYork

Prof. Dr. Y. Mizuno
Department of Neurology, Juntendo University, Tokyo, Japan

Prof. Dr. M. B. H. Youdim
Institute of Pharmacology, Technion – Israel Institute of Technology, Faculty of Medicine, Haifa, Israel

Prof. Dr. D. B. Calne
Neurodegenerative Disorders Centre, University of British Columbia, Vancouver, Canada

Dr. R. Horowski
Clinical Research, Schering AG, Berlin, Federal Republic of Germany

Prof. Dr. W. Poewe
Department of Neurology, University of Innsbruck, Austria

Prof. Dr. P. Riederer
Department of Psychiatry, University of Würzburg, Federal Republic of Germany

Graphic design: Ecke Bonk
Printed on acid-free and chlorine-free bleached paper

With 46 Figures

ISBN-13: 978-3-211-82934-9   e-ISBN-13: 978-3-7091-6844-8
DOI: 10.1007/978-3-7091-6844-8

Preface

Neurodegeneration is one of the most important subjects of the investigation now and in the coming 21st century. Alzheimer's disease is the leading cause of dementia in the elderly people and Parkinson's disease is one of the major neurologic disorders with the prevalence between 1 and 2/1000 population in advanced countries. Many others are suffering from intractable neurodegenerative disorders such as amyotrophic lateral sclerosis, Huntington's disease, or spinocerebellar degeneration. No truly effective treatment is available for any of these neurodegenerative disorders except for Parkinson's disease; even in Parkinson's disease, still it is impossible to slow down the disease process with the currently available treatment.

It is urgently needed to develop new effective technique to halt or slow down the disease process in each of those disorders. Recent advance in the molecular biological and molecular genetic technique has brought us great progress in the understanding of etiology and pathogenesis of these disorders, but still it is not known how neurons are going to die in these disorders. To explore the question, mutual cooperation and exchange of ideas between basic scientists and clinical peoples are of utmost importance.

With this idea in mind, International Winter Conference was launched in 1992. This conference has been held annually by inviting relatively small number of distinguished basic as well as clinical investigators and it has been intended to have ample time for discussion after each presentation. Major goal has been to create new ideas on how to elucidate the etiology and pathogenesis of neurodegenerative disorders and how to develop neuroprotective treatment for them. Multiple sclerosis, thought it's not a neurodegenerative disorder in the usual sence, has been incorporated in the scope of this conference, as it is an important neurological disease leading many young persons to disability such as blindness and/or paraplegia.

The first conference was organized at Seefeld, Austria (1992), the second one at Whistler village, Canada (1993), the third one at Eilat, Israel (1994), the fourth one at Kananaskis, Canada (1995), and the fifth one at Spitzingsee, Germany (1996). This supplement includes updated papers presented at Eilat and Kananaskis meetings. I hope this volume will be helpful and informative to those who are interested in neuronal death and neurodegenerative disorders. Finally, we would like to thank Schering for generous financial support of this conference.

Tokyo, June 1997   **Y. Mizuno**

Contents

Contents

Treatment strategies for neurodegenerative diseases based on trophic factors and cell transplantation techniques

B. Hoffer and **L. Olson**

Department of Pharmacology and Psychiatry,
University of Colorado Health Sciences Center, Denver, CO, U.S.A.

Summary. Treatment strategies based on transfer of genes, molecules, or cells to the central nervous system are summarized. When neurons are already degenerated, functional compensation can be effected by grafts of syngeneic or allogenic tissue to the target area. This technique is undergoing clinical trials in Parkinson's disease. Before degeneration has occurred, it may be possible to rescue "stressed" neurons, and stimulate terminal outgrowth using treatment with neurotrophic factors. Such approaches, with an emphasis on the NGF family of neurotrophins and their receptors, are reviewed. Finally, new molecular biology techniques may permit the transfer of genes directly into non-dividing cells of the central nervous system. These three approaches may have a more general applicability, and become important not only in neurodegenerative diseases, but also in other afflictions of the nervous system such as ischemia, stroke and injury.

While axonal regeneration is an effective and clinically important mechanism in the peripheral nervous system, it is a well-established fact that long nerve fiber pathways do not regenerate in the adult mammalian central nervous system. The reason for this difference long remained unknown, but the elegant experiments of Aguayo and collaborators (David and Aguayo, 1981) have demonstrated that adult CNS axons are able to elongate efficiently if given the appropriate environment, such as a section of peripheral nerve. The discrepancy between regenerative capacity in CNS and PNS has at least three possible bases: (1) production of neurotrophic factors by Schwann cells but not oligodendroglial cells, (2) the production of nerve growth inhibitory factors by oligodendroglial but not Schwann cells, or (3), scar formation to a greater extent in the CNS than in the PNS. It now appears as if all three possibilities are true. There are relatively effective regenerative mechanisms for peripheral nerve injury, but the extreme complexity of the central nervous system and precise regulation of central connectivity has led to an absence of such regenerative capacity. As longevity has increased in modern society, the need for reparative intervention for neurodegenerative diseases has increased commensurately.

While research during the last two decades has not yet provided us with methods to effectively stimulate regeneration of long fiber tracts in the central nervous system, it has led to the development of two other principal repair strategies. The first is a cell replacement strategy; neurons that have been lost can sometimes be functionally replaced by other cells such as embryonic neurons which are implanted, not at the original site of such nerve cell bodies, but directly into their axonal target regions. With this strategy one avoids the necessity of regenerating a long axon pathway, obviously with loss of the original circuitry, but there is a gain of new nerve terminals. The second approach is applicable before neurons have died, and focuses on prevention of "stressed" neurons from dying and stimulation of nerve terminals from remaining neurons using neurotrophic factors. Obviously, these two principles can be combined. In the following, we shall discuss some of the recent animal research for treatment with grafts and growth factors, as well as comment upon ongoing clinical trials.

One may look at reparative strategies as various ways of transfering molecules or cells to the brain to obtain long-lasting effects. Thus *molecular and cellular transfer techniques* fundamentally differ from current treatment strategies of neurodegenerative diseases such as Parkinson's disease and Alzheimer's disease. Current treatments attempt to increase dopamine and cholinergic neurotransmission, respectively, by pharmacological means such as administering a dopamine precursor, an acetylcholinesterase inhibitor, or various direct and indirect agonists. These treatments may be somewhat effective as long as the drugs are taken regularly, but the positive effects disappear immediately upon drug withdrawal. Moreover, tolerance and side effects are common problems.

Neurotrophic molecules

From a situation just a few years ago when nerve growth factor, NGF (Levi-Montalcini, 1987), was the only neurotrophic factor known, we now are faced with a very complex but exciting situation in which we realize that brain development and adult neuron maintenance, as well as protective effects, are mediated via a large number of specific neurotrophic factors acting upon an array of specific receptors. It appears as if every "growth factor" that has been searched for in the central nervous system (e.g. NGF family, FGF family, PDGF family, TFG family, IGF family, CNTF) has been found to be expressed either by neurons or glial cells; in many cases, physiological and/or pharmacological actions for these factors have been discovered in the developing and adult brain and spinal cord.

The NGF family of neurotrophins

In order to attempt to understand the role of the neurotrophins in the brain (see Ebendal, 1992), one must identify the site of synthesis and localization of the neurotrophins as well as of their receptors. The four neurotrophins together with their common low-affinity receptor, p75, and the more specific so-called high-affinity receptors *trk*, *trkB* and *trkC*, respectively, encompass eight gene products. Sites of synthesis can be revealed by in situ hybridization or possibly by immunohistochemistry using antibodies against pro-sequences within the protein. The presence of mature protein can only be determined by immunohistochemistry which is particularly problematic in the case of neurotrophins for two reasons. First, these proteins appear not to be stored, and are present in extremely small amounts. Second, the high degree of sequence homology between these proteins makes it necessary to generate antibodies against highly localized amino acid sequences in the molecule in order to find neurotrophin-specific epitopes. The bulk of the neuroanatomical data has thus been obtained from in situ hybridization studies.

Once it was realized that brain-derived neurotrophic factor (BDNF) had a very high amino acid sequence homology with NGF (Leibrock et al., 1989), further members of the NGF family could be searched for using PCR techniques. Today four closely related members, NGF, BDNF, neurotrophin-3 (NT-3) and NT-4 (see Barde, 1990; Thoenen, 1991) have been identified. The mRNA for all four neurotrophins are expressed in the brain. NGF, BDNF, and NT-3 are expressed by specific, partially overlapping, subsets of neurons in the hippocampal formation and other areas of the brain. BDNF appears to possess the most widespread distribution, including many neurons throughout the cerebral cortex, most pyramidal cells and granule cells of the hippocampal formation, as well as neurons in several subcortical areas. NGF has a more restricted expression involving scattered cells along the pyramidal cell layer of the hippocampal formation and large neurons in the hilar region of the dentate gyrus bordering the granular cells. NT-3 has the most restricted distribution of the three factors, being expressed in a specific subset of pyramidal cells in the medial CA1 area, in CA2 and in the granular cells of the hippocampal formation. Localization of neurotrophin proteins has not been very successful so far for NGF and NT-3. However, we have been able to develop BDNF-specific antibodies (Wetmore et al., 1993) which have proven useful localizing BDNF-containing structures in the brain (Wetmore et al., 1994, 1991).

The neurotrophin receptors (Bothwell, 1991) are more abundant in the cells that make them and thus have proven somewhat easier to demonstrate using in situ hybridization as well as immunohistochemistry. There is widespread expression of the low-affinity NGF receptor. The presumed high-affinity receptor for NGF, *trk*, is specifically located on forebrain cholinergic neurons which also express the low-affinity receptor. This fits well with the observed effects of NGF on central cholinergic neurons. *trkB* is much more widespread in the brain, *trkB* mRNA is present in neurons that contain BDNF mRNA, and this pattern supports the hypothesis of widespread local actions of BDNF of a paracrine and perhaps also autocrine nature. The neurotrophin receptors occur in different forms. For instance, there is a truncated form of the *trkB* receptor lacking the intracellular protein kinase domain which is even more widespread and probably also present on glial cells (Wetmore, 1992). There is also a truncated form of the NGF receptor, which is elevated rapidly after CNS injury.

Neurotrophin responses to injury and disturbances

Many different kinds of disturbances such as mechanical lesions, ischemia, hypoglycemia, excitotoxic lesions and seizures cause rapid upregulation of NGF and BDNF, while NT-3 appears to change little or not at all. Interestingly, such upregulation can also be seen in the corresponding neurotrophin receptor mRNA levels. This is in contrast to classical neurotransmitter systems, in which there is an inverse relationship between receptor sensitivity and the amount of transmitter, tending to maintain homeostasis of neurotransmission. For the NGF and BDNF neurotrophins, both neurotrophin synthesis and receptor synthesis increase in parallel and rapidly, thus quickly increasing neurotrophin efficacy in response to neuronal "stress". It has recently been shown that not only neurotrophins in the NGF family, but probably several other neurotrophic compounds such as members of the FGF family are also upregulated rapidly in response to injury. A tentative hypothesis derived from these studies is that many neurotrophic factors may act in a local para- or autocrine mode to protect neurons. Thus pharmacological treatment with such factors might be useful clinically to rescue "stressed" neurons.

Glial cell line-derived neurotrophic factor – a trophic molecule for dopamine neurons

One possible source of trophic support for neurons is the nonneuronal cells of the brain, particularly the various glial elements. Neurotrophic activity has been found in conditioned media from both primary glial cell cultures (Gaul and Lubbert, 1992; Nagata et al., 1993; O'Malley et al., 1992; Rousselet et al., 1988) and established cell lines with glial properties (Engele et al., 1991). Recently, a factor from one such cell line (rat B49; Schubert et al., 1974) having marked trophic effects on the dopamine neurons in ventral mesencephalic cell cultures, was cloned and termed glial cell line-derived neurotrophic factor (GDNF; Lin et al., 1993). GDNF behaves as a disulfide-bonded dimer, each part of the mature protein consisting of 134 amino acid residues, with 93% identity between the human and rat sequences. The protein is synthesized as a precursor and may constitute a new subfamily of the transforming growth factor-β superfamily (Lin et al., 1993).

The potential role of glial cell line-derived neurotrophic factor (GDNF) as a torphic molecule for midbrain dopamine neurons was examined in our laboratory using two different approaches: in situ hybridization and intraocular transplantation (Stromberg et al., 1993). The presence of mRNA for GDNF was noted in striatal and ventral limbic dopaminergic target areas in the developing (E20-P7) rat, but not in the adult rat. Signals were also found in nondopaminergic areas during maturation, such as the cerebellar anlage, spinal cord, and thalamus. Grafts of fetal ventral mesencephalon in the anterior eye chamber were exposed to repeated injections of GDNF, which elicited a marked and dose-dependent increase in transplant volume. A low (0.1 µg/eye) and high (1 µg/eye) dose of GDNF both led to a somewhat larger mean area of dopamine fiber outgrowth into host irides. In the transplants, cell counts of tyrosine hydroxylase (TH)-immunoreactive neurons revealed a doubling of cell numbers in the low-dose group and about four times as many cells in the high-GDNFdose group compared to controls. Moreover, the density of TH-immunoreactive nerve fibers was markedly and significantly higher in transplants treated with the high GDNF dose. Since the volumes of these transplants were also larger, the total amount of both TH-positive cells and TH-positive nerve fibers was many-fold greater in the high-GDNF group than that in the controls. Taken together, these data support the concept that GDNF functions as a dopaminotrophic factor in vivo.

Transfer of cells to the CNS

While terms such as "grafts" and "transplants" are used frequently in various experimental situations and in clinical trials, it is important both from a scientific standpoint, and to avoid lay misconception, that the techniques involving transfer of living cells to the central nervous system are properly differentiated from various transplant strategies used in other parts of the body. Thus, grafting to the brain and the spinal cord only involves cell suspensions or small tissue fragments, and does not involve transplantation of whole parts of brain tissue. Cell transfer can be used in several different ways; the most straightforward is the replacement therapy attempted in Parkinson's disease, in which the lost adult dopamine neurons are replaced by implantation of embryonic dopamine neuroblasts into the target area. This technique has proven effective in animals and clinical results to date are relatively promising (Hoffer et al., 1992; Lindvall et al., 1989; Spencer et al., 1992; Widner et al., 1992). As one alternative in Parkinson's disease, one may transfer cells to the brain that secrete the lost transmitter substance, but are not themselves substantia nigra neurons. One possibility that has been supported by extensive animal experimentation (see Olson, 1988) and is also currently undergoing clinical trials is to use adrenal medullary chromaffin cells (Backlund et al., 1985; Olson et al., 1991). Also of neural crest origin, chromaffin cells are closely related to sympathetic neurons. When removed from the surrounding adrenal cortex and, particularly when supplemented with nerve growth factor, adult adrenal medullary chromaffin cells are able to transform into neuronal phenotypes, alongate long neurites, and synthesize noradrenaline as well as dopamine. In the absence of corticosteroids, PNMT activity is markedly reduced and little adrenalin is synthesized. The main advantages with the adrenal medullary cells as compared to fetal substantia nigra cells include the fact that the patient is his or her own donor, which makes the procedure more easily organized and also makes immunosuppressive treatment unnecessary. In addition, the chances of infection from the donor tissue are markedly reduced.

Animal experimentation has suggested other alternatives to fetal substantia nigra neuroblasts in Parkinson's disease; one approach involves transferring the necessary enzymes for synthesizing L-dopa or dopamine into cell lines which could then be transplanted to serve as local sources of the missing neurotransmitter. If such cell lines are obtained from other sources, it is assumed that immunosuppression will again become necessary; however, if one could transfect primary cell lines, for instance fibroblasts from the patient, then immunosuppression could be avoided. One problem with permanent cell lines is their tendency to continue to divide after transfer to the brain. There are now various ways of controlling unwanted mitotic activity including the insertion of genes coding for products that make cell division temperature-sensitive. Thus, cell lines could be constructed that will only divide at temperatures lower than body temperature, and can thus be maintained in vitro, but will cease to divide after transfer to the brain.

Another use of cell transfer techniques would be to transfect cells with genes for synthesizing neurotrophic factors. When introduced into the brain, such cells would presumably produce and secrete these neurotrophic factors which would be released inside the blood-brain barrier to act at appropriate targets. Again, animal experimentation suggests that this principle is effective (Ernfors et al., 1989; Rosenberg et al., 1988; Strömberg et al., 1990). However, there have been repeated problems with maintaining synthesis of neurotrophic molecules after transfer to the brain. Transgene expression appears significantly reduced 2–3 months after grafting. Further experimentation using various promotors, perhaps inducible promotors, might overcome these problems.

As an alternative to using transfected cells, it may be useful to graft of cells that normally secrete neurotrophic factors. Two examples of such cells are the cells of the striated tubules of the male mouse submandibular gland which secrete extremely high levels of NGF and Schwann cells, which after grafting, increase their secretion of NGF and presumably several other neurotrophic factors. Schwann cells have been effective experimentally in supporting cografts of adrenal medullary tissue, an approach that has also been utilized in clinical trials by combining adrenal medullary tissue pieces with peripheral nerve minces in grafts to the putamen in parkinsonian patients.

Transfer of genes to the CNS

With the rapid advances of molecular biology techniques, it now begins to appear feasible to transfer genes also to non-dividing cells and obtain long-lasting effects. The classical retrovirus approaches have been problematic, because of the requirement for mitotic activity in the target cells. However, newer techniques involving herpes virus or adeno-associated virus have permitted gene insertion into non-dividing CNS neurons in vivo with minimal chance of active viral lytic infection.

Treatment with neurotrophic factors

Based on animal experimentation, we have carried out initial clinical trials with intracerebral infusion of NGF in Parkinson's disease and in Alzheimer's disease. In the case of Parkinson's disease, NGF infusion into putamen has

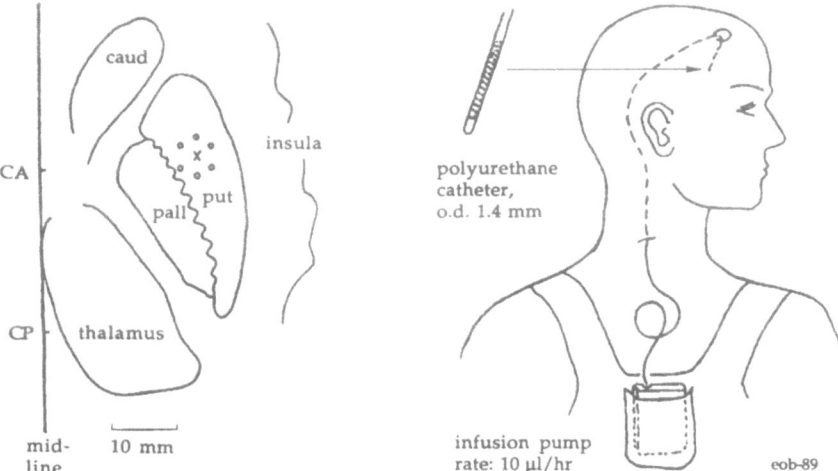

Fig. 1. To the left is depicted a schematic drawing of a horizontal section through the central nuclear complex of the human brain (*CA* anterior commissure, *CP* posterior commissure). The spatial arrangement in the putamen of the six adrenal medullary grafts (circles) and the NGF catheter (cross) is shown. The drawing to the right shows the design used for chronic intraputaminal NGF infusion. The microinfusion pump for long-term NGF administration is percutaneously connected to a catheter, the tip of which is positioned in the putamen

been carried out to support intraputaminal adrenal medullary autografts (Olson et al., 1991). Our ongoing clinical results, with three patients to date, suggests that one can indeed obtain better and more longlasting effects of adrenal medullary implants using temporary NGF support (Fig. 1).

In Alzheimer's disease our clinical trial with NGF infusion is based on the cholinergic hypothesis of cognitive dysfunction seen in Alzheimer's disease. We know that cholinergic neurons are NGF-sensitive and also probably NGF-dependent. We also know that lesions of the cholinergic projections to hippocampus and cortex cerebri will cause memory dysfunction and other cognitive disturbances in animal experiments. Moreover, the cholinergic systems are markedly degenerated in Alzheimer's disease. We have thus initiated clinical trials infusing NGF chronically into the lateral ventricle in this disorder. The first patient received NGF via this route for three months, and we observed several positive changes including an improved cerebral blood flow and increased cerebral nicotine binding (both data obtained by PET), increased power frequency spectrum in the EEG, as well as a few cognitive improvements. These changes lasted as long as the NGF infusion or outlasted the NGF treatment for up to one year (Olson et al., 1992).

Concluding remarks

In this mini-review we have discussed various ways of transfering genes, neurotrophic molecules, or cells into the central nervous system in order to obtain long-lasting improvements in patients with neurodegenerative diseases. Clearly, the clinical trials are in a very early phase for both neurotrophic factor delivery and cell transfer, and have not yet even been initiated for gene delivery. There are many potential future improvements for these technologies. One interesting approach involves binding neurotrophins to molecules that will facilitate their transport across the blood-brain barrier, thus allowing for intravenous injections of neurotrophins (Friden et al., 1991).

Taken together, the approaches outlined above all aim at long-lasting protective and reparative therapeutic actions in the central nervous system. They have general applicability and might become useful treatment strategies not only in neurodegenerative diseases such as Parkinson's disease and Alzheimer's disease, but also in other common afflictions of the central nervous system such as ischemia, stroke and traumatic injury.

Acknowledgements

Supported by the Swedish Medical Research Council, the Swedish Natural Science Research Council, Karolinska Institutets Fonder, M and M Wallenbergs stiftelse and USPHS grants NS09199 and AG04418.

References

Backlund E-O, Granberg PO, Hamberger B, Knutsson E, Martensson A, Sedvall G, Seiger A, Olson L (1985) Transplantation of adrenal medullary tissue to striatum in parkinsonism. First clinical trials. J Neurosurg 62: 169–173

Barde YA (1990) The nerve growth factor family. Prog Growth Factor Res 2: 237–248

Bothwell M (1991) Keeping track of neurotrophin receptors. Cell 65: 915–918

David S, Aguayo A (1981) Axonal elongation into peripheral nervous system "bridges" after central nervous system injury in adult rats. Science 214: 931–993

Ebendal T (1992) Function and evolution in the NGF family and its receptors. J Neurosci Res 32: 461–470

Engele J, Schubert D, Bohn M (1991) Conditioned media derived from glial cell lines promote survival and differentiation of dopaminergic neurons in vitro: role of mesencephalic glia. J Neurosci Res 30: 359–371

Ernfors P, Ebendal T, Olson L, Mouton P, Strömberg I, Persson H (1989) A cell line producing recombinant nerve growth factor evokes growth responses in intrinsic and grafted central cholinergic neurons. Proc Natl Acad Sci USA 86: 4756–4760

Friden PM, Walus LR, Musso GF, Taylor MA, Malfroy B, Starzyk RM (1991) Anti-transferrin receptor antibody and antibody-drug conjugates cross the blood-brain barrier. Proc Natl Acad Sci USA 88: 4771–5

Gaul G, Lubbert H (1992) Cortical astrocytes activated by basic fibroblast growth factor secrete molecules that stimulate differentiation of mesencephalic dopaminergic neurons. Proc R Soc Lond B 249: 57–63

Hoffer BJ, Leenders KL, Young D, Gerhardt G, Zerbe GO, Bygdeman M, Seiger A, Olson L, Stromberg I, Freedman R (1992) Eighteen-month course of two patients with grafts of fetal dopamine neurons for severe Parkinson's disease. Exp Neurol 118: 243–252

Leibrock J, Lottspeich F, Hohn A, Hofer M, Hengerer B, Masiakowski P, Thoenen H, Barde YA (1989) Molecular cloning and expression of brain-derived neurotrophic factor. Nature 341: 149–152

Levi-Montalcini R (1987) The nerve growth factor 35 years later. Science 237: 1154–1162

Lin L-F, Doherty D, Lile J, Bektesh S, Collins F (1993) GDNF: a glial cell line-derived neurotrophic factor for midbrain dopaminergic neurons. Science 260: 1130–1132

Lindvall O, Rehncrona S, Brundin P, Gustavii B, Astedt B, Widner H, Lindholm T, Bjorklund A, Leenders KL, Rothwell JC, Frackowiak R, Marsden C, Johnels B, Steg G, Freedman R, Hoffer B, Seiger Å, Bygdeman M, Strömberg I, Olson L (1989) Human fetal dopamine neurons grafted into the striatum in two patients with severe Parkinson's disease. A detailed account of methodology and a 6-month follow-up. Arch Neurol 46: 615–631

Nagata K, Takei N, Nakajima K, Saito H, Kohsaka S (1993) Microglial conditioned medium promotes survival and development of cultured mesencephalic neurons from embryonic rat brain. J Neurosci Res 34: 357–363

O'Malley E, Sieber B, Black I, Dreyfus C (1992) Mesencephalic type I astrocytes mediate the survival of substantia nigra dopaminergic neurons in culture. Brain Res 582: 65–70

Olson L (1988) Grafting in the mammalian central nervous system: basic science with clinical promise. In: Magistretti P (ed) Discussions in neurosciences. FESN, Geneva

Olson L, Backlund E-O, Ebendal T, Freedman R, Hamberger B, Hansson P, Hoffer B, Lindblom U, Meyerson B, Strömberg I, Sydow O, Seiger Å (1991) Intraputaminal infusion of nerve growth factor to support adrenal medullary autografts in Parkinson's disease: one-year follow-up of first clinical trial. Arch Neurol 48: 373–381

Olson L, Nordberg A, von Holst H, Bäckman L, Ebendal T, Alafuzoff I, Amberla K, Hartvig P, Herlitz A, Lilja A, Lundgvist H, Långström B, Meyerson B, Persson A, Viitanen M, Winblad B, Seiger Å (1992) Nerve growth factor affects ^{11}C-nicotine binding, blood flow, EEG, and verbal episodic memory in an Alzheimer patient. J Neural Transm [PD-Sect] 4: 79–95

Rosenberg MB, Friedmann T, Robertson RC, Tuszynski M, Wolff JA, Breakefield XO, Gage FH (1988) Grafting genetically modified cells to the damaged brain: restorative effects of NGF expression. Science 242: 1575–1578

Rousselet A, Fetler L, Chamak B, Prochiantz A (1988) Rat mesencephalic neurons in culture exhibit different morphological traits in the presence of media conditioned on mesencephalic or striatal astroglia. Dev Biol 129: 495–504

Schubert D, Heinemann S, Carlisle W, Tarikas H, Kimes B, Patrick J, Steinbach H, Culp W, Brandt B (1974) Clonal cell lines from the rat central nervous system. Nature 249: 224–227

Spencer DD, Robbins RJ, Naftolin F, Marek KL, Vollmer T, Leranth C, Roth RH, Price LH, Gjedde A, Bunney BS, Sass K, Elsworth J, Kier E, Makuch R, Hoffer P, Redmond Jr D (1992) Unilateral transplantation of human fetal mesencephalic tissue into the caudate nudeus of patients with Parkinson's disease. N Engl J Med 327: 1541–1548

Strömberg I, Wetmore CJ, Ebendal T, Ernfors P, Persson H, Olson L (1990) Rescue of basal forebrain cholinergic neurons after implantation of genetically modified cells producing recombinant NGF. J Neurosci Res 25: 405–411

Strömberg I, Björklund L, Johansson M, Tomac A, Collins F, Olson L, Hoffer B, Humpel C (1993) Glial cell line-derived neurotrophic factor is expressed in the developing but not adult striatum and stimulates developing dopamine neurons in vivo. Exp Neurol 124: 401–412

Thoenen H (1991) The changing scene of neurotrophic factors. Trends Neurosci 14: 165–170

Wetmore C (1992) Brain-derived neurotrophic factor. Studies on the cellular localization and regulation of BDNF, related neurotrophins and their receptors at the mRNA and protein level. Thesis, Karolinska Institute, Stockholm, Sweden

Wetmore C, Cao YH, Pettersson RF, Olson L (1991) Brain-derived neurotrophic factor: subcellular compartmentalization and interneuronal transfer as visualized with anti-peptide antibodies. Proc Natl Acad Sci USA 88: 9843–9847

Wetmore CJ, Cao Y, Pettersson RF, Olson L (1993) Brain-derived neurotrophic factor (BDNF) peptide antibodies: characterization using a Vaccinia virus expression system. J Histochem Cytochem 41: 521–533

Wetmore C, Bean AJ, Olson L (1994) Regulation of brain-derived neurotrophic factor (BDNF) expression and release from hippocampal neurons is mediated by non-NMDA type glutamate receptors. J Neurosci 14: 1688–1700
Widner H, Tetrud J, Rehncrona S, Snow B, Brundin P, Gustavii B, Bjorklund A, Lindvall O, Langston JW (1992) Bilateral fetal mesencephalic grafting in two patients with parkinsonism induced by 1-methyl-4-phenyl-1, 2, 3, 6-tetrahydropyridine (MPTP). N Engl J Med 327: 1556–1563

Authors' address: Prof. Dr. B. J. Hoffer, Department of Pharmacology and Psychiatry, University of Colorado Health Sciences Center, Campus Box C236, 4200 East Ninth Avenue, Denver, Colorado 80262, U.S.A.

Models of Alzheimer's disease: cellular and molecular aspects

S. Hoyer

Department of Pathochemistry and General Neurochemistry, University of Heidelberg,
Federal Republic of Germany

Summary. Glucose metabolism in the brain is an important process that influences many normal cellular processes, from neurotransmitter synthesis to ATP production. While cortisol and insulin have opposing effects on glucose metabolism, desensitization of the neuronal insulin receptor results in metabolic abnormalities. In the normal aging brain, glucose/energy metabolism is decreased slightly. In the majority of cases, Alzheimer's disease is sporadic and has a late onset. Therefore, age-related variations in cellular metabolism following the principle of self-organized criticality may come into focus with respect to the etiopathogenesis of this neurodegenerative disorder. As a possible primary abnormal event in late-onset sporadic DAT, a prolonged desensitization of the neuronal insulin receptor is assumed to be responsible for cascade-like abnormalities in oxidative energy metabolism and related metabolism with impacts on amyloid formation.

Introduction

Alzheimer's disease is a heterogenous neurodegenerative disorder. Only a minority of cases are due to genetic abnormalities on chromosomes 14, 19, and 21, and these are mostly cases with early onset (genetic abnormalities on chromosomes 14 and 21); in the majority of all cases, Alzheimer's disease is sporadic and with late onset (Farrer et al., 1990; Goate et al., 1991; Hardy et al., 1991; Murrell et al., 1991; Pericak-Vance et al., 1991; Schellenberg et al., 1992; St. George-Hyslop et al., 1990; Tanzi et al., 1991, 1992). Clear evidence has been provided that the prevalence of late-onset sporadic dementia of the Alzheimer type (SDAT) increases exponentially beyond the age of 65 years (Gottfries, 1985; Evans et al., 1989; Ott et al., 1995). Aging leads to inherent changes in fundamental metabolic principles at the cellular and molecular levels including free radical formation, clearing and repair, and energy availability to name some of the functionally most important processes. Whereas one such aberrations alone would not have the capacity to cause decisive damage to cell functions a multiplicity of slight variations may be able to provoke cellular damage in the course of time. Considered in this way, age may be considered as a risk factor for SDAT (Storandt et al., 1988; Hoyer, 1992a) (Fig. 1).

S. Hoyer

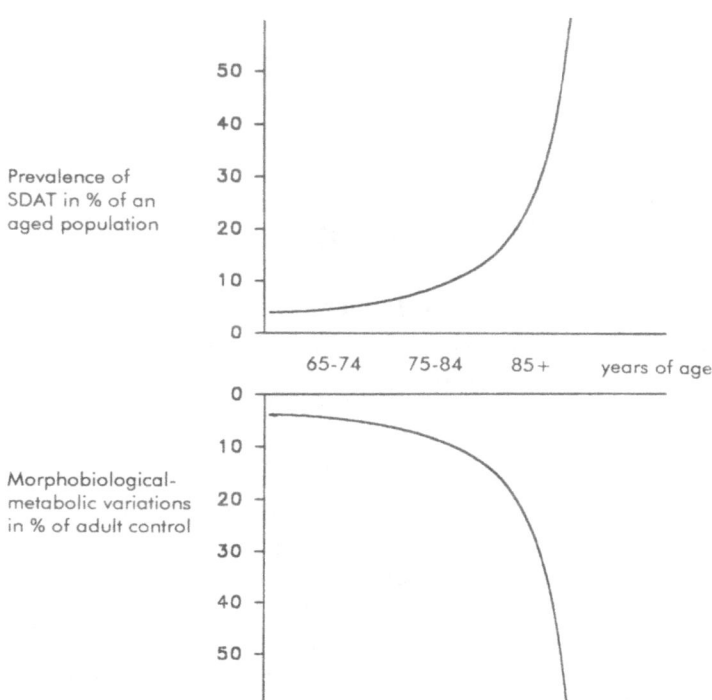

Fig. 1. Average percent changes of prevalence rates and morphobiological-metabolic variations in mammalian brain in old and very old age [after: Evans DA, et al (1989) JAMA 262: 2552–2556; Hoyer S (1990) Aging 2: 245–258]

Normal cellular work depends on the availability of sufficient amounts of biological energy. In the healthy, mature, nonstarved mammalian brain, energy in the form of ATP is formed exclusively from the oxidation of glucose (Erecinska and Silver, 1989; Hoyer, 1992b). Numerous metabolic processes require ATP; these include:

– maintenance of synaptic transmission (Huganir and Greengard, 1990; Kadokaro et al., 1985; Tissari et al., 1969)
– protein synthesis (Buttgereit and Brand, 1995)
– protein folding and maturation (Braakman et al., 1992; Gething and Sambrook, 1992; Hartl, 1996)
– protein phosphorylation (Ehrlich et al., 1986; Greengard, 1978)
– protein degradation (Chiang et al., 1989; Okada et al., 1991) and
– extracellular transmission (Burnstock, 1990; Edwards et al., 1992).

Glucose breakdown also yields the high-energy compound acetyl CoA, a constituent of acetylcholine synthesis, which is accomplished at a rate in proportion to that of pyruvate oxidation (Gibson et al., 1975; Sims et al., 1981). Acetylcholine synthesis was found to be related to difference between cytoplasmic and the mitochondrial NAD/NADH potentials. Any impairment in glucose oxidation altered these redox states and led to a perturbed formation of acetylcholine without changing the ATP concentration or the energy charge potential (Gibson and Blass, 1976). Otherwise, hippocampal acetylcholine release can be augmented by an increase of circulating glucose concentration during mental activity (Ragozzino et al., 1996).

From these results, it may be deduced that any impairment in the control of cerebral glucose metabolism in general, and in glucose oxidation in particular, precedes the reduction of both acetylcholine synthesis and ATP formation in the brain. These abnormalities, along with variations in related metabolism, and the consequences of this will be discussed in more detail with respect to aging and SDAT.

Glucose/energy metabolism during normal aging under resting conditions

In physically and mentally healthy senescent subjects, a 23% reduction was found in the cerebral metabolic rate of glucose without any change in cerebral oxygen utilization (Dastur et al., 1963; Dastur, 1985). When ATP formation is calculated on the basis of these data, a slight fall by around 5% becomes obvious with aging. This also holds true for experimental animals: the availability of energy was reduced by 5% in parietotemporal cerebral cortex of 104-week-old rats compared with 52-week-old rats (Dutschke et al., 1994), but dropped by 15% in very old (130 weeks of age) rats (Hoyer, 1985). This in in keeping with the reduction in acetylcholine synthesis. Otherwise the Ca^+-dependent release of acetylcholine was reduced to as much as one third in senescent compared with adult mouse brain (Gibson and Peterson, 1981). From these data it may be deduced that the cholinergic function is more severely compromised than is the glucose/energy metabolism in the resting aging brain. Thus, transmission failure is more pronounced than energy failure in this period of life, in other words a specialized (luxury) function is more affected than vital function (Oldstone et al., 1977).

Glucose/energy metabolism during normal aging under stress conditions

Recovery of the energy pool in the cerebral cortex after hypoglycemia was found to be more markedly compromised in aged than in adult animals (Benzi et al., 1984). Cerebral ischemia caused a more pronounced fall in the adenosine nucleotide level in aged than in adult animals (Hoyer and Krier, 1986), and the delayed decrease in energy-rich phosphates in the postischemic recirculation period was more severe in aged than in young adult animals (Hoyer and Betz, 1988).

Mental activation increased the cerebral cortex energy pool (sum of adenine nucleotides, and sum of ATP + creatine phosphate) by 25% and 13% in adult and by 21% and 9% in aged animals indicating that an age-dependent decline in the size of the energy pool could not be prevented by short-term mental activation. The diminution of the energy pool with aging was even more marked after mental activation than after mental rest. Mental activation induced an increase in ATP turnover up to 24 mths of age in male Wistar rats

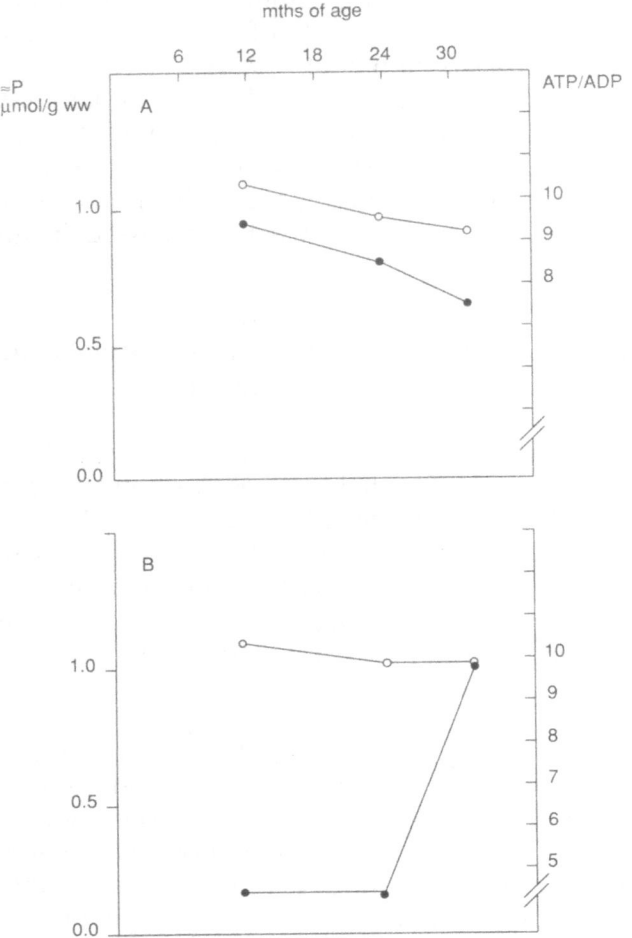

Fig. 2. Course of available energy (≈ P) (o—o) and energy turnover (expressed as ATP/ADP ratio) (●—●) in cerebral parietotemporal cortex during aging. **A** resting state; **B** mental activity

(aged animals) (Dutschke et al., 1994). However, at the age of 30 mths (very old animals), ATP turnover fell drastically although ATP formation was found to be unchanged (Fig. 2). These changes are thought to be due to a reduction of the activity of adenine nucleotide translocase (Nohl and Krämer, 1980; Kim et al., 1988), and may indicate a distinct age-related abnormality in mitochondrial function that becomes particularly evident under stress conditions (Hoyer, 1994).

Control of neuronal glucose metabolism

In nonnervous tissue, glucose metabolism is largely controlled by insulin and its signal transduction via the insulin receptor. Generally the regulation of the celluar signal transduction pathway is mediated by amplification and desensitization of reponses to the respective agonist. Amplification of the insulin signal is forwarded by activation of tyrosine kinase, and desensitization by activation of serine kinase (Häring, 1991). Serine phosphorylation by cAMP-dependent protein kinase counteracts the effect of tyrosine phosphorylation (Roth and Beaudoin, 1987). Catecholamines induced a reduction in tyrosine kinase activity, too (Häring et al., 1986). Insulin release from pancreatic islets was found to be reduced, and insulin receptors were lost as a function of cell age (Wilson and Peterson, 1986; Castro et al., 1993). There is an increasing body of evidence indicating that insulin acts in a similar way in the brain as in nonnervous tissue (Hoyer et al., 1993), although the origin of cerebral insulin is still an issue of controversial debate. Insulin receptors are widely distributed in the brain the highest densities being in olfactory and limbic structures (Hill et al., 1986; Unger et al., 1991). The diverse variations found in cerebral glucose metabolism during aging (Hoyer, 1992a) may point to a dysregulation of the neuronal insulin signal transduction pathway. So ^{125}I-insulin binding was found to be decreased in olfactory bulb, but not in frontal cortex, hippocampus and hypothalamus in aged rats (Tchilian et al., 1990).

The data of noradrenaline concentration in the aging brain reported so far are not uniform. Some studies demonstrated both pre- and postsynaptic reduction in the efficiency in the noradrenergic system with only slight changes in the noradrenaline concentration in the aging brain (Walker and Walker, 1973; Schmidt and Thornberry, 1987; Jones and Olpe, 1983; McIntosh and Westfall, 1987). Otherwise, the presynaptic noradrenaline concentration was found to increase in cerebral cortex with aging (Ida et al., 1982; Harik and McCracken, 1986), and the stress-induced release of noradrenaline was prolonged with aging (Perego et al., 1993). So a functional hyperactivity of noradrenaline in the aging brain is likely what may influence also the function of the neuronal insulin receptor (Häring, 1991).

Recent studies have provided evidence that glucocorticoids counteract the effect of insulin in the brain (Plaschke et al., 1996). After stress, the circulating glucocorticoid concentrations was remained elevated longer in aged animals whereas there was a decline to basal levels in younger animals (Sapolsky et al., 1984, 1986). Prolonged glucocorticoid exposure contributed to an age-related loss of hippocampal neurons (Sapolsky et al., 1985). It is tempting to assume that this effect is due to a glucocorticoid-mediated decrease on tyrosine phosphorylation of the neuronal insulin receptor as was demonstrated for skeletal muscle (Giorgino et al., 1993).

The principle of criticality

In the physical sciences, the term criticality is used to describe a self-organized metalabile steady state. Criticality can progress to supercriticality which is able to induce a catastrophic reaction, subsequently approaching another metalabile steady state. Snow-slips, for example, can be explained in this manner (Bak et al., 1988; Held et al., 1990). Although the principle of criticality has not yet been proven for biological systems, in all probability it is applicable here, too. The principle of coupled synchronization has been demonstrated to exist in biological systems, and this principle may be assumed to be very closely related to criticality (Mirollo and Strogatz, 1990).

As was shown earlier (Hoyer, 1992a), and has been discussed above, numerous smaller but permanent aberrations in glucose/energy metabolism, and in related metabolism occur in the brain from adulthood to senescence. This pattern may reflext a distinct metalabile steady state in aged neurons. Any small additional abnormality, even one that is ineffective in itself, may change the former steady state and may cause a catastrophic reaction representing another new but detrimental steady state. In this context, age may be considered as a risk factor for neuronal damage and, thus, for sporadic DAT.

Brain glucose/energy metabolism in SDAT

The particular vulnerability of the cerebral glucose metabolism has become obvious in SDAT. In incipient late-onset SDAT, cerebral glucose utilization was found to be diminished in all areas of the cerebral cortex, but was accentuated in frontal and parietotemporal cerebral cortices (Mielke et al., 1992). The reduction in the global cerebral metabolic rate (CMR) of glucose ranged from 19% (mild) to more than 40% (severe) (Hoyer et al., 1991; Kumar et al., 1991), indicating that severity of dementia may parallel the diminution in the CMR of glucose. In contrast, the CMR of oxygen was diminished by around 20% only (Frackowiak et al., 1981; Hoyer et al., 1991). The disproportion between the utilization rates of these two parameters in the incipient state of the disorder in arterial normoglycemia points to a disturbance in the control of the cerebral glucose metabolism as the cause rather than the consequence of neuronal loss for this abnormality. The same metabolic pattern became obvious when the brain is undersupplied with glucose in arterial hypoglycemia (Norberg and Siesjö, 1976).

The metabolic abnormalities in cerebral glucose breakdown may be attributed mainly to disturbances in glycolytic glucose catabolism and pyruvate oxidation (see Hoyer, 1992b) causing a fall in the formation rate of acetylcholine to 47% of control (Sims et al., 1981). ATP production from glucose decreased by around 50%, but this was mitigated to 20% by the oxidative utilization of endogenous substrates such as amino acids and fatty acids (Hoyer, 1992b). Abnormalities in oxidative metabolism and the energy state in the brain of mildly demented Alzheimer patients have been reported recently (Pettegrew et al., 1994) and confirm our former findings. Thus, in the brain affected by incipient late-onset SDAT, a perturbed control of glucose breakdown (Hoyer, 1988) may severely damage cellular homeostasis and subsequently cause a drastic reduction of cellular work with substantial impacts on cellular function. Maturation and processing of the amyloid precursor protein may be involved in particular, and may be assumed to have a detrimental role in the pathogenesis of SDAT but "downstream" of the origin of this disorder (Hoyer, 1993).

Working hypothesis

It is assumed that the perturbed control of the cerebral glucose metabolism in SDAT is due to an abnormality in the insulin signal transduction subsequent to prolonged insulin receptor desensitization. Clearly, there are as yet few findings in SDAT supporting this hypothesis beside the disturbance in glycolytic glucose metabolism and pyruvate oxidation: SDAT patients had an abnormal glucose tolerance tests similar in nature to those observed in non-insulin dependent diabetes mellitus (Bucht et al., 1983); in cerebrospinal fluid, insulin levels were found to be enhanced after oral glucose load in SDAT (Fujisawa et al., 1991); in microvessels from SDAT brain, a disease-related increase in cAMP was observed (Grammas et al., 1994). As discussed above, cAMP-dependent kinases inhibit the activity of the tyrosine kinase of the insulin receptor.

To further test the hypothesis, streptozotocin which is known to inhibit insulin receptor signaling (Kadowaki et al., 1984; Giorgino et al., 1992) was administered intracerebroventricularly to inhibit tyrosine phosphorylation at cerebral insulin receptors only. This procedure led to cascade-like abnormalities in cerebral glucose catabolism and energy formation, in membrane phospholipid and monoaminergic catecholamine metabolism along with disturbance in learning and memory capacities (Mayer et al., 1990; Hoyer et al., 1994).

References

Bak P, Tang C, Wiesenfeld K (1988) Self-organized criticality. Phys Rev A 38: 364–374

Benzi G, Pastoris O, Villa RF, Giuffrida AM (1984) Effect of aging on cerebral cortex energy metabolism in hypoglycemia and posthypoglycemic recovery. Neurobiol Aging 5: 205–212

Braakman I, Helenius J, Helenius A (1992) Role of ATP and disulphide bonds during protein folding in the endoplasmic reticulum. Nature 356: 260–262

Bucht G, Adolfsson R, Lithner F, Winblad B (1993) Changes in blood glucose and insulin secretion in patients with senile dementia of Alzheimer type. Acta Med Scand 213: 387–392

Burnstock G (1990) Purinergic mechanisms. Ann NY Acad Sci 603: 1–17

Buttgereit F, Brand MD (1995) A hierarchy of ATP-consuming processes in mammalian cells. Biochem J 312: 163–167

Castro M, Pedrosa D, Osuma JI (1993) Impaired insulin release in aging rats: metabolic and ionic events. Experientia 49: 850–853

Chiang HL, Terlecky SR, Plant CP, Dice JF (1989) A role for a 70-kilodalton heat shock protein in lysosomal degradation of intracellular proteins. Science 246: 382–385

Dastur DK (1985) Cerebral blood flow and metabolism in normal human aging, pathological aging, and senile dementia. J Cereb Blood Flow Metab 5: 1–9

Dastur DK, Lane MH, Hansen DB, Kety SS, Butler RN, Perlin S, Sokoloff L (1963) Effects of aging on cerebral circulation and metabolism in man. In: Birren JE, Butler RN, Greenhouse S, Sokoloff L, Yarrow MR (eds) Human aging – a biological and behavioural study. U.S. Department of Health, Education and Welfare, National Institute of Mental Health, Bethesda (DHEW Publication No 986)

Dutschke K, Nitsch RM, Hoyer S (1994) Short-term mental activation accelerates the age-related decline in brain tissue levels of high energy phosphates. Arch Gerontol Geriatr 19: 43–51

Edwards FA, Gibb AJ, Colquhoun D (1992) ATP receptor-mediated synaptic currents in the central nervous system. Nature 359: 144–147

Ehrlich YH, Davis TB, Bocke E, Kornecki E, Lenox RH (1986) Ectoprotein kinase activity on the external surface of neural cells. Nature 302: 67–70

Erecinska M, Silver IA (1989) ATP and brain function. J Cereb Blood Flow Metab 9: 2–19

Evans DA, Funkenstein HH, Albert MS, Scherr PA, Cook NR, Chown MJ, Hebert LE, Hennekens CH, Taylor JO (1989) Prevalence of Alzheimer's disease in a community population of older persons: higher than previously reported. J Am Med Assoc 262: 2551–2556

Farrer LA, Myers RA, Cupples LA, St. George-Hyslop PH, Bird TD, Rossor MN, Mullan MJ, Polinsky R, Nee L, Heston L, van Broeckhoven C, Martin JJ, Crapper McLachlan D, Growdon JH (1990) Transmission and age-at-onset patterns in familial Alzheimer's disease: evidence for heterogeneity. Neurology 40: 395–403

Frackowiak RS, Pozzilli C, Legg NJ, DuBoulay GH, Marshall J, Lenzi GL, Jones T (1981) Regional cerebral oxygen supply and utilization in dementia. A clinical and physiological study with oxygen-15 and positron tomography. Brain 104: 743–778

Fujisawa Y, Sasaki K, Akiyama K (1991) Increased insulin levels after OGTT load in peripheral blood and cerebrospinal fluid of patients with dementia of Alzheimer type. Biol Psychiatry 30: 1219–1228

Gething MJ, Sambrook J (1992) Protein folding in the cell. Nature 355: 33–45

Gibson GE, Blass JP (1976) Impaired synthesis of acetylcholine in brain accompanying mild hypoxia and hypoglycemia. J Neurochem 27: 37–42

Gibson GE, Peterson C (1981) Aging decreases oxidative metabolism and the release and synthesis of acetylcholine. J Neurochem 37: 978–984

Gibson GE, Jope R, Blass JP (1975) Decreased synthesis of acetylcholine accompanying impaired oxidation of pyruvic acid in rat brain minces. Biochem J 148: 17–23

Giorgino F, Chen JH, Smith RJ (1992) Changes in tyrosine phosphorylation of insulin receptors and 170,000 molecular weight non receptor protein in vivo in skeletal muscle of streptozotocin-induced diabetic rats: effects of insulin and glucose. Endocrinology 130: 1433–1444

Giorgino F, Almahfouz A, Goodyear LJ, Smith RJ (1993) Glucocorticoid regulation of insulin receptor and substrate IRS-1 tyrosine phosphorylation in rat skeletal muscle in vivo. J Clin Invest 91: 2020–2030

Goate A, Chartier Harlin MC, Mullan M, Brown J, Crawford F, Fidani L, Giuffra L, Hanyes A, Irving N, James L, et al (1991) Segregation of a missense mutation in the amyloid precursor protein gene with familial Alzheimer's disease. Nature 349: 704–706

Gottfries CG (1985) Alzheimer's disease and senile dementia: biochemical characteristics and aspects of treatment. Psychopharmacology 86: 245–252

Grammas P, Roher AE, Ball MJ (1994) Increased accumulation of cAMP in cerebral microvessels in Alzheimer's disease. Neurobiol Aging 15: 113–116

Greengard P (1978) Phosphorylated proteins as physiological effectors. Science 199: 146–152

Hardy J, Mullan M, Chartier Harlin MC, Brown J, Goate A, Rossor M, Collinge J, Roberts G, Luthert P, Lantos P, et al (1991) Molecular classification of Alzheimer's disease. Lancet 337: 1342–1343

Harik SI, McCracken KA (1986) Age-related increase in presynaptic noradrenergic markers of the rat cerebral cortex. Brain Res 381: 125–130

Hartl FU (1996) Molecular chaperones in cellular protein folding. Nature 381: 571–580

Häring HU (1991) The insulin receptor: signalling mechanism and contribution to the pathogenesis of insulin resistence. Diabetologia 34: 848–861

Häring HU, Kirsch D, Obermaier B, Ermel B, Machicao F (1986) Decreased tyrosine kinase activity of insulin receptor isolated from rat adipocytes rendered insulin-resistant by catecholamine treatment in vitro. Biochem J 234: 59–66

Held GA, Solina DH, Keane DT, Haag WJ, Horn PM, Grinstein G (1990) Experimental study of critical-mass fluctuations in an evolving sandpile. Phys Rev Lett 65: 1120–1123

Hill JM, Lesniak MA, Pert CB, Roth J (1986) Autoradiographic localization of insulin receptors in rat brain: prominence in olfactory and limbic areas. Neuroscience 17: 1127–1138

Hoyer S (1985) The effect of age on glucose and energy metabolism in brain cortex of rats. Arch Gerontol Geriatr 4: 193–203

Hoyer S (1988) Glucose and related brain metabolism in dementia of Alzheimer type and its morphological significance. Age 11: 158–166

Hoyer S (1990) Brain glucose and energy metabolism during normal aging. Aging 2: 245–258

Hoyer S (1992a) The biology of the ageing brain. Oxidative and related metabolism. Eur J Gerontol 1: 157– 165

Hoyer S (1992b) Oxidative energy metabolism in Alzheimer brain. Studies in early-onset and late-onset cases. Mol Chem Neuropathol 16: 207–224

Hoyer S (1993) Editor's note for debate. Sporadic dementia of Alzheimer type: role of amyloid in etiology is challenged. J Neural Transm [P-D Sect] 6: 159–165

Hoyer S (1994) Age as risk factor for sporadic dementia of the Alzheimer type? Ann NY Acad Sci 719: 248–256

Hoyer S, Krier C (1986) Ischemia and the aging brain. Studies on glucose and energy metabolism in rat cerebral cortex. Neurobiol Aging 7: 23–29

Hoyer S, Betz K (1988) Abnormalities in glucose and energy metabolism are more severe in the hippocampus than in the cerebral cortex in postischemic recovery in aged rats. Neurosci Lett 94: 167–172

Hoyer S, Nitsch R, Oesterreich K (1991) Predominant abnormality in cerebral glucose utilization in late-onset dementia of the Alzheimer type: a cross-sectional comparison against advanced late-onset and incipient early-onset cases. J Neural Transm [P-D Sect] 3: 1–14

Hoyer S, Prem L, Sorbi S, Amaducci L (1993) Stimulation of glycolytic key enzymes in cerebral cortex by insulin. NeuroReport 4: 991–993

Hoyer S, Müller D, Plaschke K (1994) Desensitization of brain insulin receptor. Effect on glucose/energy and related metabolism. J Neural Transm [Suppl] 44: 259–268

Huganir RL, Greengard P (1990) Regulation of neurotransmitter receptor desensitization by protein phosphorylation. Neuron 5: 555–567

Ida Y, Tanaka M, Kohno Y, Nakagawa R, Iimori K, Tsuda A, Hoaki Y, Nagasaki N (1982) Effects of age and stress on regional noradrenaline metabolism in the rat brain. Neurobiol Aging 3: 233–236

Jones RSG, Olpe HR (1983) Altered sensitivity of forebrain neurones to iontophoretically applied noradrenaline in aging rats. Neurobiol Aging 4: 97–99

Kadokaro M, Crane AM, Sokoloff L (1985) Differential effects of electric stimulation of sciatic nerve on metabolic activity in spinal cord and dorsal root ganglion in the rat. Proc Natl Acad Sci USA 82: 6010–6013

Kadowaki T, Kasuga M, Akanuma Y, Ezaki O, Takaku F (1984) Decrease autophosporylation of the insulin-receptor-kinase in streptozotocin diabetic rats. J Biol Chem 259: 14208–14216

Kim JH, Shrago E, Elson CE (1988) Age-related changes in respiration coupled to phosphorylation. II. Cardiac mitochondria. Mech Ageing Dev 46: 279–290

Kumar A, Schapiro MB, Grady C, Haxby JV, Wagner E, Salerno JA, Friedland RP, Rapoport SI (1991) High-resolution PET studies in Alzheimer's disease. Neuropsychopharmacology 4: 35–46

Mayer G, Nitsch R, Hoyer S (1990) Effects of changes in peripheral and cerebral glucose metabolism on locomotor activity, learning and memory in adult male rats. Brain Res 532: 95–100

McIntosh HH, Westfall TC (1987) Influence of aging on catecholamine levels, accumulation, and release in F-344 rats. Neurobiol Aging 8: 233–239

Mielke R, Herholz K, Grond M, Kessler J, Heiss WD (1992) Differences of regional cerebral glucose metabolsm between presenile and senile dementia of Alzheimer type. Neurobiol Aging 13: 93–98

Mirollo RE, Strogatz SH (1990) Synchronization of pulse-coupled biological oscillators. SIAM J Appl Math 50: 1656–1662

Murrell J, Farlow M, Ghetti B, Benson M (1991) A mutation in the amyloid precursor protein associated with hereditary Alzheimer's disease. Science 254: 97–99

Nohl H, Krämer R (1980) Molecular basis of age-dependent changes in the activity of adenine nucleotide translocase. Mech Ageing Dev 14: 137–144

Norberg K, Siesjö BK (1976) Oxidative metabolism of the cerebral cortex of the rat in insulin-induced hypoglycemia. J Neurochem 26: 345–352

Okada M, Ishikawa M, Mizushima Y (1991) Identification of a ubiquitin- and ATP-dependent protein degradation pathway in rat cerebral cortex. Biochim Biophys Acta 1073: 514–520

Oldstone MBA, Holmstoen J, Welsh jr RM (1977) Alterations of acetylcholine enzymes in neuroblastoma cells persistently infected with lymphocytic choriomenigits virus. J Cell Physiol 91: 459–472

Ott A, Breteler MMB, van Harskamp F, Clauss JJ, van der Cammen TJM, Grobbee DE, Hofman A (1995) Prevalence of Alzheimer's disease and vascular dementia: association with education. The Rotterdam Study. Br Med J 310: 970–973

Perego C, Vetrugno CC, DeSimoni MG, Algeri S (1993) Aging prolongs the stress-induced release of noradrenaline in rat hypothalamus. Neurosci Lett 157: 127–130

Pericak-Vance MA, Bebout JL, Gaskell jr PC, Yamaoka LH, Hung WY, Alberts MJ, Walker AP, Bartlett RJ, Haynes CA, Welsh KA, Earl NL, Heyman A, Clark CM, Roses AD (1991) Linkage studies in familial Alzheimer disease: evidence for chromosome 19 linkage. Am J Hum Genet 48: 1034–1050

Pettegrew JW, Pawchalingam K, Klunk WE, McClure RJ, Muenz LR (1994) Alterations of cerebral metabolism in probable Alzheimer's disease: a preliminary study. Neurobiol Aging 15: 117–132

Plaschke K, Müller D, Hoyer S (1996) Effects of adrenalectomy and corticosterone substitution on glucose and energy metabolism in rat brain. J Neural Transm 103: 89–100

Raggozzino ME, Unick KE, Gold PE (1996) Hippocampal acetylcholine release during memory testing in rats: augmentation by glucose. Proc Natl Acad Sci USA 93: 4693–4698

Roth RA, Beaudoin J (1987) Phosphorylation of purified insulin receptor by cAMP kinase. Diabetes 36: 123–126

Sapolsky RM, Krey LC, McEwen BS (1984) Glucocorticoid-sensitive hippocampal neurons are involved in terminating the adrenocortical stress response. Proc Natl Acad Sci USA 81: 6174–6177

Sapolsky RM, Krey LC, McEwen BS (1985) Prolonged glucocorticoid exposure reduces hippocampal neuron number: implications for aging. J Neurosci 5: 1222–1227

Sapolsky RM, Krey LC, McEwen BS (1986) The neuroendocrinology of stress and aging: the glucocorticoid cascade hypothesis. Endocr Rev 7: 284–301

Schellenberg GD, Bird TD, Wijsman EM, Orr HT, Anderson L, Nemens E, White JA, Bonnycastle L, Weber JL, Alonso ME, Potter H, Heston LL, Martin GM (1992) Genetic linkage evidence for a familial Alzheimer's disease locus on chromosome 14. Science 258: 668–671

Schmidt MJ, Thornberry JF, (1978) Cyclic AMP and cyclic GMP accumulation in vitro in brain regions of young, old and aged rats. Brain Res 139: 169–177

Sims NR, Bowen DM, Davison AN (1981) (^{14}C) acetylcholine synthesis and (^{14}C) carbon dioxide production from (U-^{14}C) glucose by tissue prisms from human neocortex. Biochem J 196: 867–876

St. George-Hyslop PH, Haines JL, Farrer LA, Polinsky R, van Broeckhoven C, Goate A, Crapper McLachlan DR, Orr H, Bruni AC, Sorbi S, et al (1990) Genetic linkage studies suggest that Alzheimer's disease is not a single homogenous disorder. Nature 347: 194–197

Storandt M, Bäckman L, Baltes MM, Blass JP, Braak H, Gutzmann H, Hauw JJ, Hoyer S, Jorm AF, Kauss J, Kliegl R, Mountjoy CQ (1988) Relationships of normal aging and dementing diseases in later life. In: Henderson AS, Henderson JH (eds) Etiology of dementia of Alzheimer's type. Wiley, Chichester

Tanzi RE, St. George-Hyslop P, Gusella JF (1991) Moleculargenetics of Alzheimer disease amyloid. J Biol Chem 266: 20579–20582

Tanzi RE, Vaula G, Romano DM, Mortilla M, Huang TL, Tupler RG, Wasco W, Hyman BT, Haines JL, Jenkins BJ, et al (1992) Assessment of amyloid beta-protein precursor gene mutations in a large set of familial and sporadic Alzheimer disease cases. Am J Hum Genet 51: 273–282

Tchilian EZ, Zhelezarov IE, Petkov VV, Hadjiivanova CI (1990) ^{125}I-insulin binding is decreased in olfactory bulbs of aged rats. Neuropeptides 17: 193–196

Tissari AH, Schönhöfer PS, Bogdanski F, Brodie BB (1969) Mechanism of biogenic amine transport. II. Relationship between sodium and the mechanism of ouabain blockade and the accumulation of serotonin and norepinephrine by synaptosomes. Mol Pharmacol 5: 539–604

Unger JW, Livingston JN, Moss AM (1991) Insulin receptors in the central nervous system: localization, signaling mechanisms and functional aspects. Prog Neurobiol 36: 343–362

Walker JB, Walker JP (1973) Properties of adenylate cyclase from senescent rat brain. Brain Res 54: 391–396

Wilson C, Peterson SW (1986) Insulin receptor processing as a function of erythrocytes age. A kinetic model for down-regulation. J Biol Chem 261: 2123–2128

Author's address: Prof. Dr. S. Hoyer, Department of Pathochemistry and General Neurochemistry, University of Heidelberg, Im Neuenheimer Feld 220/221, D-69120 Heidelberg, Federal Republic of Germany.

The relationship of Alzheimer-type pathology to dementia in Parkinson's disease

K. Jendroska

Department of Neurology, University Clinic Charité, Berlin,
Federal Republic of Germany

Summary.   Lewy body degeneration of the subcortical nuclei other than the substantia nigra is common in PD and may represent the substrate for a higher vulnerability to dementia in patients with PD. Cortical pathologies of Alzheimer and Lewy body type seem to be the major determinants of dementia. The prevalence of Alzheimer's disease is not increased in PD, but "early" cortical Alzheimer lesions (usually sub-clinical in normal controls) are frequently associated with dementia in PD. Furthermore, dementia in PD is heterogeneous and should always prompt the clinician to search for treatable causes.

Introduction

Dementia is a frequent complication in the later stages of Parkinson's disease (PD). It has been estimated that the lifetime risk of PD patients to develop dementia is two-fold increased. Degeneration of subcortical nuclei other than the substantia nigra is a common feature of PD, but in most cases dementia is caused by additional pathologies of the cerebral cortex, especially co-incidental changes of Alzheimer's disease and/or cortical Lewy body degeneration. This review focuses on the variety of morphological brain changes implicated in the heterogeneous syndrome of dementia in PD.

The prevalence of dementia in PD

The prevalence of dementia in PD has been the subject of several clinical studies this century. Early estimates varied considerably, ranging from 0% (Patrick and Levy, 1922) to 64% (Lewy, 1923) and 81% (Martin et al., 1973). However, with the advent of more precise clinical and pathological definitions of PD and separation of clinically similar but pathologically distinct parkinsonian disorders (Steele-Richardson-Olszewski syndrome, multiple system atrophy among others) and the application of universally agreed definitions of dementia (DSM-III, DSM-IIIR) estimates of dementia in PD now usually range between 15 and 25% (Brown and Marsden, 1984; Lees, 1985; Gibb, 1989).

Age is a well-recognized risk factor for dementia in PD patients as well as in the general population. In the latter, dementia is rarely found earlier than in the 7th decade (< 1%) but reaches about 10% in the 8th decade and increases up to 40% of people aged over 90 years (Jorm et al., 1988). Similarly dementia is infrequent in the young with PD; in three independent studies (Quinn et al., 1987; Lima et al., 1987; Gibb and Lees, 1988) of a total of 123 cases with early onset PD (onset before age 46, mean duration of disease 14.5 years) only one case (0.8%) was demented, whereas in another report 12.5% of PD patients in the sixth decade were demented and up to 69% of cases in the 9th decade (Mayeux et al., 1992). It has therefore been concluded, that the lifetime risk for dementia in PD is approximately two-fold increased (Gibb, 1989). The low prevalence of dementia in young onset cases indicates that in itself PD is not the main determinant of dementia, however, PD does seem to increase the patients vulnerability to the effects of pathological aging.

Subcortical pathology and neurotransmitter deficiencies in PD

Degeneration of subcortical nuclei other than the substantia nigra is a well recognized feature of PD. In fact, Lewy bodies were first described in the substantia innominata, not in the substantia nigra (Lewy, 1913). Degeneration of Nucleus basalis of Meynert (NBM), locus ceruleus (LC) and dorsal raphe nuclei (DRN) lead to subsequent deficiencies of acetylcholine, noradrenaline and serotonin (Agid et al., 1990). Depletion of up to 60% of NBM neurons in PD occurs without co-incidental cortical AD lesions (Nakano and Jirano, 1984).

The clinical effects of these neurotransmitter deficiencies are, however, controversial. Some authors have associated dementia in PD with cell loss in the NBM (Arendt et al., 1983; Whitehouse et al., 1983; Ball, 1984; Gaspar and Gray, 1984; Jellinger, 1989) and/or in the LC (Chan-Palay and Asan, 1989; Mann and Yates, 1983; Gaspar and Gray, 1984), whereas others did not support a relationship of dementia with pathologies of NBM (Heilig et al., 1985; Candy et al., 1983; Pendlebury and Perl, 1984; Tagliavini et al., 1984) or LC (Boller, 1985; Heilig et al., 1985). Loss of DRN neurons and subsequent reductions in serotonin levels seem to be related to depression rather than dementia (Paulus and Jellinger, 1991). In a recent morphometric study (Duyckaerts et al., 1993) severe cell loss in SN, NBM and LC in PD was demonstrated, but there was no significant difference between demented and non-demented cases. Furthermore, earlier results that had related dementia specifically to severe cell loss in the medial SN (Paulus and Jellinger, 1991) could not be duplicated; this relationship had been suggested since the medial SN projects to the caudate nucleus, and limbic and frontal cortical areas (meso-cortico-limbic pathways) and disturbances in this circuit may result in frontal lobe dysfunction (Brown and Marsden, 1990).

It therefore seems that in PD cellular and subsequent neuro-transmitter depletions in subcortical nuclei may not cause dementia, but rather increase vulnerability to mental disturbances.

Cortical Alzheimer pathology in PD

Senile (neuritic) plaques (SP) and neurofibrillary tangles (NFT) are the hallmarks of Alzheimer's disease, first described in a demented case by Alzheimer in 1905. A mild degree of these pathological changes can often be found in brains of elderly people with no history of dementia (Tomlinson et al., 1968). It is still undecided whether such changes represent a variation of normal aging (Critchley, 1931) or are to be considered early preclinical Alzheimer's disease (Davies et al., 1988). In neurohistopathology, age-related plaque scores are applied to describe the severity of Alzheimer pathology. These scores describe a threshold of pathology at which dementia is usually evident and consequently a final diagnosis of AD is based on them (Fig. 1). These scores may refer to SP counts alone (Mirra et al., 1991) or on counts of SP and NFT (Khachaturian, 1985).

The histopathological diagnostic scores have been criticized because clinical and pathological findings do not always correspond: several cases with severe brain pathology, but no dementia have been described (Crystal et al., 1988). This gives rise to the question of whether SP and NFT truly represent the pathological substrate of dementia or are merely an epiphenomenon of disease. An improved clinico-pathological relationship was found when cortical synapse counts and cognitive state were compared (Terry et al., 1991). However, synaptic degeneration is not a valuable disease-specific marker; synapse losses similar to Alzheimer's disease were also demonstrated in demented cases of Huntington's disease, PD and in vascular dementia (Zhan et al., 1992) and may be common to a variety of dementing disease processes.

Several factors may enhance or diminish the impact of Alzheimer pathology in an individual patient (Fig. 1). It is well-known that a high degree of formal education seems to provide some protection against dementia (Zhang et al., 1990; Bonaiuto et al., 1990). The believe is that better education leads to a higher "functional brain reserve" enabling the individual to longer compensate the adverse effects of Alzheimer changes. In contrast, co-existent brain pathology may increase vulnerability to Alzheimer changes. A combination of vascular pathology and some degree of Alzheimer pathology causes dementia in approximately 25% of the demented elderly (Tomlinson et al., 1970). As we show later, co-existent PD tends to precipitate dementia when Alzheimer pathology is at an early stage, when most control subjects still appear normal.

The relationship between cortical Alzheimer pathology and dementia in PD has been studied by a number of investigators over the past twenty years (Alvord et al., 1974; Hakim and Mathieson, 1979; Boller et al., 1980; Ditter and Mirra, 1987; Leverenz and Sumi, 1986). Some authors claim that increased dementia in PD is due to a higher prevalence of AD in PD and that the two diseases are somehow linked etiologically (Boller et al., 1980). Others criticise these studies on methodological grounds and suggest that increased dementia in PD may be due to higher vulnerability to the effects of "early" Alzheimer lesions (Quinn et al., 1987).

Braak and Braak (1990) found severe neurofibrillary degeneration of the allocortical pre-α neurons of the entorhinal region in demented cases of PD

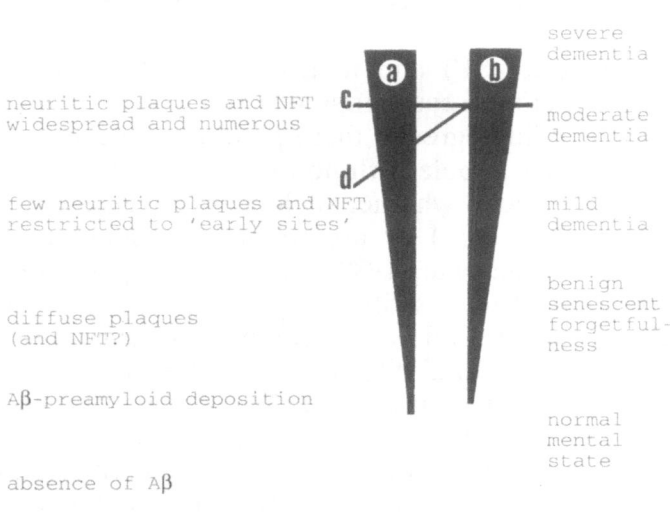

Fig. 1. Idealized model of cortical Alzheimer changes (a) and their putative clinical correlates (b). A final diagnosis of Alzheimer's disease is reached when a score of histopathological changes is present in a demented case (c). However, with additional pathological changes (as in vascular or Parkinson's disease) dementia may occur at an earlier stage of the disease process (d)

with mostly mild-to-moderate neocortical Alzheimer changes. Since the axons of pre-α neurons constitute the perforans pathway which transmits information from the neocortex to the hippocampus a destruction of these neurons may contribute to dementia. NFT in pre-α neurons are a very early lesion in the Alzheimer process; in a prospective examination of non-parkinsonian patients with pre-α tangles no cognitive defects were evident (Dickson et al., 1993).

Relationship of amyloid β-peptide with dementia in PD

The amyloid β-peptide (Aβ) is the key component of amyloid in senile plaques and walls of blood vessels (congophilic angiopathy) of Alzheimer's disease (Glenner and Wong, 1984; Masters et al., 1985). The central role of Aβ in the pathogenesis of AD was underlined by the finding of several mutations of the amyloid β-precursor protein (βAPP) in patients with early-onset familial AD (Goate et al., 1991; Murrell et al., 1991; Chartier-Harlin et al., 1991; Hendriks et al., 1992; Mullan et al., 1992). A double mutation of βAPP at codon 670 and 671 causes abundant Aβ deposition in senile plaques and preamyloid, congophilic angiopathy and NFT (Mullan et al., 1992).

It is not clear, what causes deposition of Aβ in sporadic and familial forms of AD not related to βAPP mutations. In humans, deposition of Aβ can result from to severe head trauma (Roberts et al., 1991) and brain ischemia (Cochran et al., 1991; Jendroska et al., 1995). In experimental animals, βAPP mRNAs were also reported to increase after excitotoxic injury (Solà et al., 1993) supporting a role for βAPP as a stress-response protein involved in wound healing and repair. This putative function is also suggested by the finding of βAPP (Nexin-II) in platelet granules which are released after platelet activation (van Nostrand et al., 1989). Following this line, βAPP expression and subsequent Aβ deposition may serve as an indicator for a yet unknown underlying pathological process.

On the other hand the neurotoxicity of Aβ itself is well established in vitro (Yankner et al., 1989), and experiments using foot-shock active avoidance training as a memory task in mice showed that Aβ in vivo has a specific amnesic effect (Flood et al., 1994).

In order to investigate the relationship between Aβ deposition and dementia in PD, we recently carried out whole hemispheral histoblot-immunostains of post-mortem brains of 50 PD cases, 23 of whom were demented (Jendroska et al., 1996). For this study, half-brains were frozen for histoblot analysis, which in our experience is the most sensitive immunohistological staining technique for Aβ (Taraboulos et al., 1992), and half brains were formalin fixed for routine histology, including ubiquitin-immunohistochemistry for cortical Lewy bodies. We asked whether Aβ-deposition was increased in PD, whether there exists a lowered threshold for dementia related to Aβ and whether dementia not related to Aβ occurs in PD. The distribution of Aβ-immuno-positivity is identical in PD and controls, and corresponds to the distribution of SP in AD. Therefore we adapted the staging criteria for plaque distribution in AD as published by Braak and Braak (1991) to stage our histoblots (Fig. 2): Stage 0: no Aβ present; Stage I: Aβ-immunopositive plaques in the medial temporal lobe, i.e. the parahippocampal and fusiform gyri; Stage II: Aβ-immunopositive plaques in the entire temporal lobe; Stage III: Aβ-immuno-positive plaques in all areas of cerebral cortex. The stages refer to coronal sections of the brain at the level of the lateral geniculate body, including the precentral gyrus which is affected late in the course of AD.

There was no increase of Aβ deposition in PD compared with controls and the number of PD cases with fully developed Alzheimer's disease was in the normal range. All cases with PD and Aβ-immunoreactivity in the entire cerebral cortex (Stage III) were demented, including four cases with fully developed AD

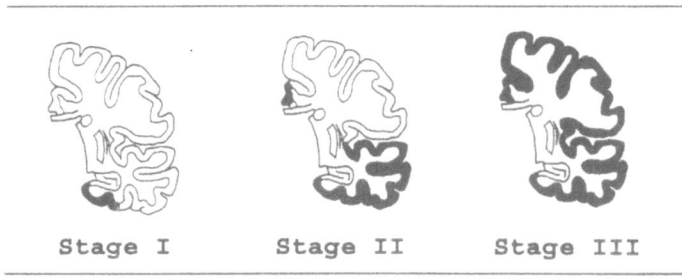

Stage I   Stage II   Stage III

Fig. 2. Progression of plaque pathology in the Alzheimer's disease process

Table 1. Lewy body densities and Aβ deposition

	No dementia	Dementia
Stage III		9 cases (incl. 4 AD) ⌀ = 1.16 (0.09–2.15)
Stage II	7 cases ⌀ = 0.51 (0.07–1.14)	1 case (0.83)
Stage I	4 cases ⌀ = 0.28 (0.11–0.47)	2 cases ⌀ = 0.31 (0.09; 0.52)
Stage 0	16 cases ⌀ = 0.26 (0–0.54)	10 cases ⌀ = 0.43 (0.04–1.82)

⌀ = mean of Lewy body densities in LB/mm^2; ( ) = range of LB density

and five cases which did not yet fulfil pathological criteria for AD (Table 1). In contrast, 5/19 control cases of Stage III had no dementia (data not shown). A causal relationship of dementia with Aβ was suggested in 9/23 PD cases, whereas in the 14 remaining demented cases there was little or no Aβ. Other causes of dementia were normal pressure hydrocephalus and cerebral vascular disease. Furthermore, in four cases (including three with co-incidental AD) the averaged Lewy body densities (cingulate gyrus and parahippocampus examined) exceeded those seen in non-demented PD cases (Table 1).

Since three of four cases of PD with co-incidental AD also had high Lewy body counts, we asked whether Aβ-deposition and Lewy body degeneration were related. We therefore correlated the amount of Aβ-immunoreactivity (expressed as % of cortical area covered with Aβ) with the density of Lewy bodies (in mm^2). Although there was a tendency towards for higher Lewy body counts in cases with more Aβ, this did not reach statistical significance; there were several cases with substantial Aβ deposition and few Lewy bodies, while others presented with high Lewy body densities in the absence of Aβ (Table 1).

Our data confirm that in a subgroup of demented PD cases there is an association between dementia and "early" AD pathology. However, the presence of PD and AD together appears to be purely co-incidental; we found no indication of an etiological relationship between the two diseases. Dementia was heterogeneous in our series of PD and included potentially treatable conditions such as normal pressure hydrocephalus. In most demented cases the pathological lesions were moderate and in few no cause for dementia could be convincingly established by routine neurohistopathology. In these cases, dementia may have resulted from a combination of the various cortical and subcortical lesions and biochemical deficiencies usually present in PD.

Acknowledgement

This work was supported by the Schering Research Foundation.

References

Agid Y, Ruberg M, Raisman R, et al (1990) The biochemistry of Parkinson's disease. In: Stern G (ed) Parkinson's disease. Chapman and Hall Medical, London

Alvord EC, Forno LS, Kusske JA, et al (1974) The pathology of Parkinsonism: a comparison of degeneration in cerebral cortex and brainstem. Adv Neurol 5: 175–193

Alzheimer A (1907) Über eine eigenartige Erkrankung der Hirnrinde. Allg Z Psychiatr 64: 146–148

Arendt T, Bigl V, Arendt A, Tennstedt A (1983) Loss of neurons in the nucleus basalis of Meynert in Alzheimer's disease, paralysis agitans, and Korsakoff's disease. Acta Neuropathol 61: 101–108

Ball MJ (1984) The morphological basis of dementia in Parkinson's disease. Can J Neurol Sci 11: 180–184

Boller F (1985) Parkinson's disease and Alzheimer's disease. Are they associated? In: Hutton JT, Kenny AD (eds) Senile dementia of the Alzheimer type, vol 18. Liss, New York

Boller F, Mitzutani T, Roessman U, et al (1980) Parkinson disease, dementia and Alzheimer disease: clinicopathological correlations. Ann Neurol 7: 329–335

Bonaiuto S, Rocca W, Lippi A, et al (1990) Impact of education and occupation on the prevalence of Alzheimer's disease (AD) and multi-infarct-dementia (MID) in Appignano, Macerata Province, Italy. Neurology 40 [Suppl 1]: 346 (Abstract)

Braak H, Braak E (1990) Cognitive impairment in Parkinson's disease: amyloid plaques, neurofibrillary tangles and neuropil threads in the cerebral cortex. J Neural Transm [P-D Sect] 2: 45–57

Braak H, Braak E (1991) Neuropathological staging of Alzheimer-related changes. Acta Neuropathol 82: 239–259

Brown RG, Marsden CD (1984) How common is dementia in Parkinson's disease? Lancet 2: 1262–1265

Brown RG, Marsden CD (1990) Cognitive function in Parkinson's disease: from description to theory. TINS 13: 21–29

Candy JM, Perry RH, Perry EK, et al (1983) Pathological changes in the nucleus basalis of Meynert in Alzheimer's and Parkinson's disease. J Neurol Sci 59: 277–289

Chan-Palay V, Asan E (1989) Alterations in catecholamin neurons in senile dementia of the Alzheimer type and in Parkinson's disease with and without dementia and depression. J Comp Neurol 287: 373–392

Chartier-Harlin MC, Crawford F, Houlden H, et al (1991) Earlyonset Alzheimer's disease caused by mutations at codon 717 of the β-amyloid precursor protein gene. Nature 353: 844–846

Cochran E, Bacci B, Chen Y, et al (1991) Amyloid precursor protein and ubiquitin immunoreactivity in dystrophic axons is not unique to Alzheimer's disease. Am J Pathol 139: 485–489

Critchley M (1931) The neurology of old age. Lancet ii: 1119–1127

Crystal H, Dickson D, Fuld P (1988) Clinico-pathologic studies in dementia: nondemented subjects with pathologically confirmed Alzheimer's disease. Neurology 38: 1682–1687

Davies L, Wolska B, Hilbich H, et al (1988) A4 amyloid protein deposition and the diagnosis of Alzheimer's disease: prevalence in aged brains determined by immunocytochemistry compared with conventional neuropathological techniques. Neurology 38: 1688–1693

Dickson DW, Singer G, Davies P, et al (1993) Regional immunocytochemical studies of brains of prospectively studied demented and nondemented normal elderly humans. In: Nicolini M, Zatta PF, Corain B (eds) Alzheimer's disease and related disorders. Advances in the biosciences, vol 87. Pergamon Press, Oxford

Ditter SM, Mirra SS (1987) Neuropathologic and clinical features of Parkinson's disease in Alzheimer's disease patients. Neurology 37: 754–760

Duyckaerts C, Gaspar P, Costa C, et al (1993) Dementia in Parkinson's disease: morphometric data. Adv Neurol 60: 442–455

Flood JF, Roberts E, Sherman MA, et al (1994) Topography of a binding site for small amnestic peptides deduced from structureactivity studies: relation to amnestic effect of amyloid protein. Proc Natl Acad Sci USA 91: 380–384

Gaspar P, Gray F (1984) Dementia in idiopathic Parkinson's disease. A neuropathological study of 32 cases. Acta Neuropathol 64: 43–52

Gibb WRG (1989) Dementia and Parkinson's disease. Br J Psychiatry 154: 492–497

Gibb WRG, Lees AJ (1988) A comparison of clinical and pathological features of young and old-onset Parkinson's disease. Neurology 38: 1402–1406

Glenner GG, Wong CW (1984) Initial report on the purification and characterisation of a novel cerebro-vascular amyloid protein. Biochem Biophys Res Commun 120: 885–890

Goate A, Chartier-Harlin MC, Mullan M, et al (1991) Segregation of a missense mutation in the amyloid precursor protein gene with familial Alzheimer's disease. Nature 349: 704–706

Hakim AM, Mathieson G (1979) Dementia in Parkinson's disease. A neuropathological study. Neurology 29: 1209–1214

Heilig CW, Knopman DS, Mastri AR, Frey W (1985) Dementia without Alzheimer pathology. Neurology 35: 762–765

Hendriks L, van Duijn CM, Cras P, et al (1992) Presenile dementia and cerebral haemorrhage linked to a mutation at codon 692 of the β-amyloid precursor protein gene. Nature Genet 1: 218–221

Jellinger K (1989) Pathology of Parkinson's syndrome. In: Calne DB (ed) Handbook of experimental pharmacology, vol 88. Springer, Berlin Heidelberg New York Tokyo

Jendroska K, Poewe W, Daniel SE, et al (1995) Ischemic stress induces deposition of amyloid β-immunoreactivity in human brain. Acta Neuropathol 5: 461–466

Jendroska K, Poewe W, Lees AJ, Daniel SE (1996) Amyloid β-peptide and the dementia of Parkinson's disease. Mov Disord 11: 647–653

Jorm AF, Henderson AS, Jacomb PA (1988) Projected increases in the number of dementia cases for 29 developed countries: application of a new method for making projections. Acta Psychiatr Scand 78: 493–500

Khachaturian ZS (1985) Diagnosis of Alzheimer's disease. Arch Neurol 42: 1097–1105

Lees AJ (1985) Parkinson's disease and dementia. Lancet i: 43–44

Leverenz S, Sumi SM (1986) Parkinson's disease in patients with Alzheimer's disease. Arch Neurol 43: 662–664

Lewy FH (1913) Zur pathologischen Anatomie der Paralysis agitans. Dtsch Z Nervenheilk 50: 50–55

Lewy FH (1923) Die Lehre vom Tonus und der Bewegung. Zugleich eine systematische Untersuchung zur Klinik, Physiologie, Pathologie und Pathogenese der Paralysis agitans. Springer, Berlin

Lima B, Neves B, Nora M (1987) Juvenile parkinsonism: clinical and metabolic characteristics. J Neurol Neurosurg Psychiatry 50: 345–348

Mann DMA, Yates PO (1983) Pathological basis for neurotransmitter changes in Parkinson's disease. Neuropathol Appl Neurobiol 9: 3–19

Martin WE, Loewenson RB, Resch JA, Baker AB (1973) Parkinson's disease. A clinical analysis of 100 patients. Neurology 23: 783–790

Masters CL, Simms G, Weinmann NA, et al (1985) Amyloid plaque core protein in Alzheimer's disease and Down's syndrome. Proc Natl Acad Sci USA 82: 4245–4259

Mayeux R, Denaro J, Hemenegilde N, et al (1992) A populationbased investigation of Parkinson's disease with and without dementia. Arch Neurol 49: 492–497

Mirra SS, Heyman A, McKeel D, et al (1991) The consortium to establish a registry for Alzheimer's disease (CERAD), part II. Standardization of the neuropathological assessment of Alzheimer's disease. Neurology 41: 479–86

Mullan M, Crawford F, Axelman K, et al (1992) A pathogenic mutation for probable Alzheimer's disease in the APP gene at the N-terminus of β-amyloid. Nature Genet 1: 345–47

Murrell J, Farlow M, Ghetti B, Benson M (1991) A mutation in the amyloid precursor protein associated with hereditary Alzheimer's disease. Science 254: 97–99

Nakano I, Jirano A (1984) Parkinson's disease: neuron loss in the nucleus basalis without concomitant Alzheimer's disease. Ann Neurol 15: 415–418

Patrick HT, Levy DM (1922) Parkinson's disease: a clinical study of 106 cases. Arch Neurol Psychiatry 7: 711–720

Paulus W, Jellinger K (1991) The neuropathological basis of different clinical subgroups of Parkinson's disease. Neuropathol Exp Neurol 50: 743–755

Pendlebury WW, Perl DP (1984) Nucleus basalis of Meynert. Severe cell loss in Parkinson's disease without dementia. Ann Neurol 16: 129 (Abstract)

Quinn NP, Rossor MN, Marsden CD (1986) Dementia and Parkinson's disease – pathological and neurochemical considerations. Br Med Bull 42: 86–90

Quinn NP, Rossor MN, Marsden CD (1987) Young onset Parkinson's disease. Mov Disord 2: 73–91

Roberts GW, Gentleman SM, Lynch A, et al (1991) βA4-amyloid protein deposition in brain after head trauma. Lancet 338: 1422–1423

Solà C, Garcia-Ladona FJ, Mengod G, et al (1993) Increased levels of the Kunitz protease inhibitor-containing βAPP mRNAs in rat brain following neurotoxic damage. Mol Brain Res 17: 41–52

Tagliavini F, Pilleri G, Bouras C, Constantinidis J (1984) The basal nucleus of Meynert in idiopathic Parkinson's disease. Acta Neurol Scand 69: 20–28

Taraboulos A, Jendroska K, DeArmond SJ, et al (1992) Mapping of prion proteins in brain. Proc Natl Acad Sci USA 89: 7620–7624

Terry RD, Masliah E, Salmon DP, et al (1991) Physical basis of cognitive alterations in Alzheimer's disease: synapse loss is the major correlate of cognitive impairment. Ann Neurol 30: 572–580

Tomlinson BE, Blessed G, Roth M (1968) Observations on the brains of non-demented old people. J Neurol Sci 7: 331–356

Tomlinson BE, Blessed G, Roth M (1970) Observations on the brains of demented old people. J Neurol Sci 11: 205–242

Van Nostrand WE, Wagner SL, Suzuki M, et al (1989) Protease nexin-II, a potent antichymotrypsin, shows identity to amyloid β-protein precursor. Nature 341: 546–549

Whitehouse PJ, Hedreen JC, White CL, Price DL (1983) Basal forebrain neurons in the dementia of Parkinson's disease. Ann Neurol 13: 243–248

Yankner BA, Dawes LR, Fisher S, et al (1989) Neurotoxicity of a fragment of the amyloid precursor associated with Alzheimer's disease. Science 245: 417–420

Zhan SS, Beyreuther K, Schmitt HP (1992) Synaptophysin expressivity of the cortical neuropil in primary neurodegenerative and vascular dementia. Neurobiol Aging 13: 41 (Abstract)

Zhang M, Katzman R, Salmon D, et al (1990) The prevalence of dementia and Alzheimer's disease in Shanghai, China: impact of age, gender and education. Ann Neurol 27: 428–437

Author's address: K. Jendroska, M.D., Department of Neurology, University Clinic Charité, Schumannstrasse 20/21, D-10117 Berlin, Federal Republic of Germany.

Models to study the role of neurotrophic factors in neurodegeneration

D. Lindholm

Department of Developmental Neuroscience, BMC 587, Uppsala University,
Uppsala, Sweden

Summary. The physiological functions of neurotrophic factors, such as nerve growth factor (NGF), in supporting the survival and differentiation of specific neurons during early development has in many cases been well established. Recent studies have shown that neurotrophic factors can also protect vulnerable neurons against a variety of mechanical and chemical injuries. The role and the effects of neurotrophic factors in various neurological diseases are however less known. Neurodegenerative diseases such as Parkinson and Alzheimer's diseases as well as amyotrophic lateral sclerosis (ALS) are characterized by an impaired function and ultimate loss of specific populations of neurons. The study of the ethiology and molecular biology of these diseases has for a long time been hampered by the lack of good animal models mimicked part of the human disease in experimental animals. Here we will discuss some of the current approaches taken in these studies as well as address the important question of the possible beneficial effect of neurotrophic factors in alleviating the symptoms and possibly retarding the course of neurodegenerative diseases.

Introduction

Neurotrophic factors are important molecules regulating the survival and differentiation of developing neurons (Thoenen and Barde, 1980; Barde, 1989). In the adult animals neurotrophic factors like nerve growth factor (NGF) maintain the chemical phenotype of neurons as shown by effects on neurotransmitter and neuropeptide expression (Levi-Montalcini, 1987). Likewise data are accumulating showing that the terminal arborization of neurons are directly influenced by locally produced neurotrophic factors. Much of our knowledge of the action of neurotrophic factors comes from work in the peripheral nervous system but many of the molecules are also present in brain (Thoenen, 1991; Lindholm et al., 1993). In addition, it has became clear that the receptors for the various factors reside on central neurons also making them targets for neurotrophic factors. Indeed recent data has shown that neurotrophic molecules can rescue injured or degenerating central neurons (Snider and Johnson, 1990). This neuroprotective action observed does not always occur in the strict physiological context of the various neurotrophic factors but involves pharmacological actions of these molecules which are beneficial for damaged neurons. The characterization of the physiological actions of neurotrophic factors in brain involves also to study whether neurotrophic mechanisms are involvedin some neurodegenerative diseases affecting central neurons. A major support for this reasoning comes from studies showing that neurons undergo atrophy and degenerate not only during the course of a disease but also during aging and that the affected neurons are still responsive to neurotrophic factors. In addition, some of the degenerative changes observed eg on cholinergic neurons in Alzheimer ,s disease and aging can partly be mimicked in experimental animal models (Sofroniew and Cooper, 1993). Some of these aspects will be further considered here in addition to discussing some useful approaches and models to study the role of neurotrophins in neurodegeneration especially with regard to the known beneficial effects of NGF on injured septal cholinergic and of CNTF on damaged motoneurons.

Degenerative changes in Alzheimer's disease and the septohippocampal pathway

Senile dementia of the Alzheimer type is a devastating disease affecting an increasing number of elderly people in the Western world.There is no treatment so far for this disease neither is there any good animal mode for the disease process characterized by a loss of cognitive functions and memory decline. Although many different types of neurons are degenerating in Alzheimer's disease an affliction of the cholinergic neurons in the basal forebrain occurs rather early during the course of the disease (Whitehouse et al., 1982). The nucleus basalis of Meynert in humans shows signs of neurodegeneration in Alzheimer's disease and the atrophy involves predominantly the cholinergic neurons normally projecting to hippocampus and neocortex (Allen et al., 1988). Thus it appears that there are some neuropathological changes in septum common to Alzheimer's disease brain and rodents undergoing a fimbria fornix lesion. In the rat axotomy of the septohippocampal pathway leads also to degeneration of the forebrain cholinergic neurons with a decrease in the the enzyme choline acetyl transferase (ChAT) which is responsible for the synthesis of the neurotransmitter acetylcholine.

A major contribution to the study of neurotrophic factors in the central nervous system was the finding that NGF injected into brain can be taken up and retrogradely transported by the septal cholinergic neurons (Schwab et al., 1979; Seiler and Schwab, 1984). Subsequently the presence of endogenous NGF in the target areas of the forebrain cholinergic neurons such as in the hippocampus has been amply demonstrated using a sensitive ELISA assay for the protein and RNA hybridization to determine NGF mRNA levels (Ebendal, 1989; Thoenen, 1991). NGF was also found to increase the activity of

ChAT in cultured septal neurons suggesting a direct trophic role for NGF in this system (Gnahn et al., 1983). The action of NGF on the septal neurons seems to be rather specific since e.g. neither motoneurons (see below) nor dopaminergic neurons of the substantia nigra respond to NGF (Schwab et al., 1979). Most importantly, Hefti (1986) was the first to demonstrate that exogenous NGF is able to rescue degenerating forebrain cholinergic neurons following a fimbria fornix lesion. This study and others thereafter have shown that there are many different ways by which NGF can be delivered into the brain in order to exert its protective function (Kordower et al., 1993). Thus both single injectons as well as minipumps delivering NGF over a long time period have been used in this model in addition to grafts or transplantation of cells eg fibroblasts genetically engineeered to produce and release relatively large amounts of NGF (Gage et al., 1991). More recently a carrier mediated transport to cross the blood-brain barrier have been attempted using NGF coupled to an antibody against the transferrin receptor which is abundantly expressed on blood capillaries (Friden et al., 1993). In addition, T cells having access to the brain parenchyma and enginered to release NGF might be an interesting alternative to consider (R. Kramer et al., unpublished). The feasibility of these methods for NGF delivery and neuroprotection are currently being studied or have already been proven using the fimbria fornix lesion model. However, the search for short peptides or molecules which could act on the NGF receptor and mimick the effects of NGF on neurons has unfortunately been less succesful. The three-dimensional structure of NGF has recently been resolved and it might be that the whole intact molecule is needed to interact with the trkA receptor to exert the full biological activity of NGF. The effects of NGF on lesioned septal neurons are to be considered pharmacological since the doses employed are rather high even when taking into account poor tissue penetration, losses of the protein due to degradation or catabolism. However, this fact does not make the treatment less valuable since NGF act here more as a drug in mimicking its physiological actions. An alternative approach to the implantation of cannulas or engineered cells for NGF delivery might be the enhancement of NGF production within brain tissue. Thus pharmacological agents could be administrated to enhance synthesis of endogenous NGF at physiologically relevant sites where NGF is normally produced eg within the postsynaptic neuron (see Thoenen et al., 1994).

Factors controlling the levels of neurotrophin factors in brain are currently under intense investigation and neuronal activity seems to play an important role in BDNF and NGF regulation (Lindholm et al., 1993). This approach would also avoid possible difficulties arising from the flooding of the system with too much of a neurotrophin which in itself might have adverse effects. NGF infusions have indeed been shown to be able to cause hypophagia or hyperalgesia. Although these side effects can probably be well tolerated they have anyway been taken into account in planing eg clinical trials (Olson, 1993). A more serious effects of NGF given in vivo might be the hyperinnervation of responsive neurons seen in some experiments (Olson, 1993). Likewise high doses of NGF have been reported to impair the performance of rats in the

Morris water maze test (Markowska et al., unpublished). Whether this decrease in learning behaviour in these rats were due to the possible establishment of aberrant projection or sprouting of the cholinergic neurons due to too much NGF is not clear at the moment.

Is NGF then the only neurotrophic factor for the septal cholinergic neurons? The neurotrophins constitue a family of structurally related molecules which beside NGF include brain-derived neurotrophic facor (BDNF), neurotrophin-3 (NT-3) and neurotrophin-4 (NT-4) (Glass and Yancopoulos, 1993). Recent data indicate that BDNF also increases ChAT in the septal cholinergic neurons in culture whilst NT-3 had no effect. Likewise both trk A and trk B representing the high-affinity receptors for NGF and BDNF respectively are found on septal neurons (Thoenen, 1991; Glass and Yancopoulos, 1993). In addition to the neurotrophins other growth factors also act on these neurons. Thus it was recently shown that fibroblast growth factor-5 (FGF-5) elevates ChAT albeit to a lower degree than NGF (Lindholm et al., 1994). FGF-5 is present in the rat hippocampus and is produced by neurons mainly in the dentate gyrus region. Interestingly, in contrast to some of the other members of the FGF family, FGF-5 is a secreted molecule having a signal sequence at its aminoterminal end. This fact makes FGF-5 an interesting candidate as a target-derived neurotrophic factor for the septal cholinergic neurons. In addition, FGF-5 is also produced in different cortical areas and could e.g. give trophic support for different cortical neurons including probably the cholinergic interneurons.

Beside NGF, ciliary neurotrophic factor (CNTF) and FGF-2 which both lack a classsical signal sequence for secretion have been shown to rescue septal cholinergic neurons after axotomy (Hagg et al., 1992; Anderson et al., 1988). Since CNTF and FGF-2 are normally probably not released the effects of these molecules on lesioned septal neurons reflects the ubiquitous presence of their receptors on a variety of central neurons. Both molecules are probably also released after injury and brain trauma and they could thus exert their protective actions on the damaged neurons. As discussed above there seems to be a certain degree of redandancy in the action of neurotrophic factors since eg many of them are expressed in the same brain area such as in the hippocampus. The reason for this could be that the neurons require multiple factors in order to ensure optimal biological function or that the molecules could possible act in concert influencing different aspects of neuronal metabolism. It is also conceivable that some factors cross-regulate each others as shown recently for the BDNF up-regulation of NT-3 in the rat cerebellum (Leingärtner et al., 1994). These aspects are all important to consider since it is eg not fully understood whether the different neurotrophic factors acting on septal cholinergic neurons address a different population of neurons or whether they partly overlap. Answers to these question are theoretically interesting but have even more important implications for the roles ascribed to the growth factors in neurodegeneration. It might be that not only one but several factors are deficient in the different disease complicating the picture and possible treatment strategies. The establishment of novel animal models to study growth factors in brain will help us to better understand these parameters and to what extent the different molecules play a role in neurodegeneration.

What is the relationship of the above-mentioned results obtained in animals or using cell cultures to the pathology of Alzheimer's disease? A deficiency of NGF is clearly not the cause of this disease as can be also deduced from the rather similar levels of NGF mRNA and protein in control and Alzheimer brains (Olson, 1993). In addition, although NGF under some conditions has been shown to regulate expression of amyloid β-protein, which is concentrated in senile plaques, these results were found to be cell line specific and it is not clear how they would relate to the disease process itself. Recently BDNF mRNA levels were found to be lower in patients suffering from Alzheimer's disease (Phillips et al., 1991) but this change might also reflect loss of some of the target neurons within the hippocampus. Likewise it is also possible that the decrease in the levels of BDNF and NT-3 mRNA in Alzheimer's disease is a consequence of cholinergic cell loss since the septal afferent input to the hippocampus exerts a control over particularly BDNF levels within the target hippocampus (Berzaghi et al., 1993). Thus changes in neurotrophin production in brain in this disease is probably not the cause but rather a result of ongoing cell degeneration. Data on normal aging also indicate that the levels of NGF remain unchanged albeit a substantial decrease in NGF levels was observed in one strain of rats, the Fischer rat (Lärkfors et al., 1987). In general, although the levels of the neurotrophins remain essentially constant in brain during aging and in Alzheimer's disease the responsiveness of the neurons to the factors might nevertheless be altered. This could possible result from an intrinsic decrease in the capacity of the neurons to react to the factor or be due to reduced uptake or retrograde transport of the molecule from the nerve endings to the cell soma. Some evidence for an agedependent decrease in the capacity of septal cholinergic neurons to accumulate radiolabelled NGF was recently obtained by Cooper et al. (1994). There was a decrease in the amount of retrogradely transported NGF in the septum of aged rats which was accompanied by a lower ChAT staining of the cholinergic neurons.

In addition some of the septal neurons in aged brains were more affected than others and especially those neurons which were unlabelled with NGF were significantly smaller and appeared severely atrophied. These studies suggest that in aged rat brain there is a reduction in the capacity to sustain receptor-mediated uptake and transport of NGF to the septum. The molecular basis for such an age-related reduction in NGF transport is unknown but it was recently shown that the level of trkA mRNA, the high-affinity NGF receptor, decrease in the septum of aging animals. Whether similar molecular changes as found in aged brains accompany also Alzheimer's disease is not clear at the moment. However, in keeping with the recent finding's discussed above Strada et al. (1992) reported that there is a considerable loss of NGF binding in forebrain of patient with Alzheimer's disease. This loss seemed to even precede ensuing cholinergic neurone degeneration. However further work is needed to confirm this point and to study which genes beside trkA might change in their expression during aging with a concomitant decrease in the trophic support of the neurons. It is also known from animal work that the expression of NGF receptors can be increased by NGF itself (Higgins et al.,

1989) which would constitute an important feed-back loop of trophic interaction for NGF responsive neurons. As mentioned above the septal cholinergic neurons atrophy and degenerate in the rat model following a fimbria fornix lesion and they can be rescued by NGF (Hefti, 1986). In analogy to the beneficial effects of NGF on cholinergic neurons in this lesion model it has been proposed that NGF could improve some of the symptoms of Alzheimer's disease as well (Olson, 1993).

Therapeutical approaches in humans however have to be carefully planned, weighing risks against expected beneficial effects and also take into account ethical issues. After a thorough analysis of these aspects the first patient suffering from Alzheimer's disease was taking into a trial on NGF in Sweden. The patient was a 69 old woman suffering from the disease for a long time with worsening of her condition. She received mouse NGF for a period of 3 months via a cannula implanted into the lateral ventricle in the brain. The outcome of this first clinical trial with NGF has recently been published and the results look encouraging (Olson, 1993). Specifically no obvious adverse effects were noted and there were no immunological complications. However some unexpected effects of the NGF treatment were also observed such an increase in cerebral blood flow of the patient. The reason for this is at the moment not clear The EEG of the patient largely normalized and she performed better in some of the cognitive test following the NGF treatment. Likewise position emisssion tomography showed an increase in nicotine binding sites indicating a shift to normal values (Olson, 1993). However since the improvement in some of the parameters were transitory further studies using a larger group of patients are needed to investigate the possible beneficial effects of NGF treatments in patients suffering from Alzheimer's disease.

Effect of neurotrophic factors on injured motoneurons

Amyotrophic lateral sclerosis (ALS) is a progressive neurodegenerative disease affecting cranial and spinal motoneurons. ALS occurs in a sporadic (about 90% of cases) and a familial form (5–10% of patients) of disease. Recent data has demonstrated an alteration in the enzyme Cu/Zn super oxide dismutase in some but not in all familial cases of ALS suggesting a possible involvement of free radicals in the disease process (Rosen et al., 1993). However, the ethiology and pathogenesis of most cases of ALS are unknown.

In searching for models for motoneuron disease, research workers have used the transection of the rat facial nerve as an experimental paradigm. Axotomy of the facial nerve at birth leads to a nearly complete degeneration of the nerve cell bodies in the facial nucleus. CNTF (see above) which has a survival effect on a variety of neurons in culture (Thoenen et al., 1993) including chick spinal motoneurons (Arakawa et al., 1990; Hughes et al., 1993b) was found to prevent facial motoneuron degeneration also in vivo (Sendtner et al., 1990). Thus in the facial nerve lesion model CNTF applied to the proximal nerve stump protected virtually all motoneurons against degenerative changes caused by the axotomy (Sendtner et al., 1990). This study and those made in vitro identified CNTF as a neurotrophic factor for rat and chick motoneurons. Further support for this view came from studies using the pmn, mouse mutant strain which shows progressive motoneuron degeneration of unknown cause starting from the hind legs. As shown by Sendtner et al. (1992a) CNTF given systematically was able to prevent to a considerable degree the degeneration and functional loss of motoneurons in the pmn homozygote mice. This result pointed to the possibility that CNTF might be useful in the treatment of human motoneuron degenerative diseases such as ALS. Indeed the first clinical trials using CNTF in ALS are underway and the results are soon to become available.

Beside CNTF other growth factors have been shown also to have survival promoting effects on cultured embryonic chick (Arakawa et al., 1990) and rat motoneurons (Hughes et al., 1993b). In the chick system the activity of FGF-2 was about half of that of CNTF whilst the neurotrophins had no effect. However, BDNF and to a smaller extent NT-3 promoted survival of rat motoneurons in culture. Like CNTF, BDNF also prevented degeneration of rat motoneurons in vivo after axotomy (Koliatsos et al., 1992; Sendtner et al., 1992b; Yan et al., 1992). In addition to these factors, FGF-5 (Hughes et al., 1993a) and insulin growth factor-1 (IGF-1) is also active on cultured rat motoneurons. Clinical trial using IGF-1 have recently been initiated in ALS patients and are also intended to be done in some of the peripheral neuropathies (Lewis et al., 1993).

Taken together, studies on motoneurons in culture and on various models in vivo indicate that different factors have a trophic action on these neurons and are thus potentially interesting in considering treatments of motoneuron diseases. The evaluation of the efficacy and the benefits of the various neurotrophic factors in motoneuron degeneration have to await the results of the on-going studies performed both in experimental animals and in clinical trials. It might be that in some of the future clinical approaches a combination therapy is preferred since some of the factors increase motoneuron survival in an additive manner at least in culture.

Mouse genetic models of neurodegeneration

There are a number of mouse mutants exhibiting changes in the morphology and function of the brain. As described above some of the mutant strains such as the pmn one have already been employed to study the role of neurotrophic factors in neurodegeneration. There are also some other mouse strains, e.g. the mnd and the wobbler mouse, which show signs of motoneuron degeneration and thus are interesting models for ALS. Likewise, as shown recently the myelinated axons in the nerve of transgenic mice overexpressing the neurofilament heavy chain show morphological changes resembling ALS (Cote et al., 1993). Research on Alzheimer's disease has still to await the development of accurate animal models e.g. transgenic animals expressing some of the mutant forms of amyloid β-protein which is found in some of the familial cases of Alzheimer's disease. The transgenic approach with overexpression of APP in various parts of brain might give useful information of the role of this protein in the disease process. There are nowadays also a variety of cell-specific promoters available by which the expression can be tailed to occur in specific brain cells or even in a subtype of neurons. An equally interesting approach to the study of neurodegeneration is the creation of mice carrying a null mutation for any of the neurotrophic factors or their receptors. The first animals in which a neurotrophic factor was disrupted in this way was the CNTF knock-out mouse (Masu et al., 1993). The analysis of the phenotype of these mice showed a progressive degeneration of the spinal motoneurons starting after the second month of life. The motor function measured as grip strength was reduced only later in accordance with a greater loss of motoneurons in older age. This study clearly shows the importance of CNTF for maintenance of the normal function of the motoneurons.

Beside CNTF, successful gene-targeting for disruption of some of the neurotrophins and their receptors have recently been accomplished (Klein et al., 1993; Ernfors et al., 1994). All of these mice show a complex but distinct picture of phenotypic changes in brain and in the peripheral nervous system and they even show behaviour alterations which all need to be studied in more detail in the near future. It is to be expected that the creation of these types of animal models in addition to others will help us to understand better the normal physiology of neurotrophic factors in brain. In addition they might also prove useful to analyse the possible pharmacological effects and the therapeutic potentials of neurotrophic factors in some of the neurodegenerative diseases affecting central neurons.

References

Allen SJ, Dawbarn D, Wilcock GK (1988) Morphometric immunochemical analysis of neurons in the nucleus basalis of Meynert in Alzheimer's disease. Brain Res 454: 275–281

Anderson KJ, Dam D, Lee S, Cotman CW (1988) Basic FGF prevents death of lesioned cholinergic neurons in vivo. Nature 332: 360–361

Arakawa Y, Sendtner M, Thoenen H (1990) Survival effect of CNTF on chick embryonic motoneurons in culture: comparison with other neurotrophic factors and cytokines. J Neurosci 10: 3507–3515

Barde Y-A (1989) Trophic factors and neuronal survival. Neuron 2: 1525–1534

Berzaghi M da Penha, Cooper J, Castren E, Zafra F, Sofroniew M, Thoenen H, Lindholm D (1993) Cholinergic regulation of BDNF and NGF but not NT-3 mRNA levels in the developing rat hippocampus. J Neurosci 13: 3818–3826

Cooper JD, Lindholm D, Sofroniew MV (1994) Reduced transport of NGF by cholinergic neurons and down-regulated trkA expression in the medial septum of aged rats. Neurosci Lett 62: 625–629

Cote F, Collard J-F, Julien J-P (1993) Progresive neuronopathy in transgenic mice expressing the human neurofilamnet heavy gene: a mouse model for ALS. Cell 73: 35–46

Ebendal T (1989) NGF in CNS: experimental data and clinical implications. Prog Growth Factor Res 1: 143–159

Ernfors P, Lee K-F, Jaenisch R (1994) Mice lacking BDNF develop with sensory deficits. Nature 368: 147–150

Friden P, Walus L, Watson P, Doctrow S, Kozarich J, Bäckman C, Bergman H, Hoffer B, Bloom F, Granholm A (1993) Blood-brain barrier penetration and in vivo activity of an NGF conjugate. Science 259: 373–377

Gage F, Kawaja MD, Fischer LJ (1991) Genetically modified cells: application for intracerebral grafting. Trends Neurosci 14: 328–333

Glass DJ, Yancopoulos GD (1993) The neurotrophins and their receptors. Trends Cell Biol 3: 262–268

Gnahn H, Hefti F, Heumann R, Schwab M, Thoenen H (1983) NGF-mediated increase of choline acetyltransferase (ChAT) in the neornatal rat forebrain: evidence for a physiological role of NGF in the brain. Dev Brain Res 9: 45–52

Hagg T, Quon D, Higaki J, Varon S (1992) CNTF prevents neuronal degeneration and promotes low affinity NGF receptor expression in the adult rat CNS. Neuron 8: 145–158

Hefti F (1986) NGF promotes survival of septal cholinergic neurons after fimbrial transections. J Neurosci 6: 2155–2162

Higgins GA, Koh S, Chen KS, Gage FH (1989) NGF induction of NGF receptor gene expression and cholinergic neuronal hypertrophy within the basal forebrain of the adult rat. Neuron 3: 247–266

Hughes RA, Sendtner M, Goldfarb M, Lindholm D, Thoenen H (1993a) Evidence that FGF-5 is a major muscle derived survival factor for cultured spinal motoneurons. Neuron 10: 369–377

Hughes RA, Sendtner M, Thoenen H (1993b) Members of several gene families influence survival of rat motoneurons in vitro and in vivo. J Neurosci Res 36: 663–671

Klein R, Smeyne RJ, Wurst W, Long LK, Auerbach BA, Joyner AL, Barbacid M (1993) Targeted disruption of the trkB neurotrophin receptor gene results in nervous system lesions and neonatal death. Cell 75: 113–122

Koliatsos V, Clatterbuck RE, Winslow JW, Cayouette MH, Price DL (1993) Evidence that BDNF is a trophic factor for motor neurons in vivo. Neuron 10: 359–367

Kordower JH, Mufson EJ, Granholm A-C, Hofer B, Friden P (1993) Delivery of trophic factors to the primate brain. Exp Neurol 124: 21–30

Leingärtner A, Heisenberg C-P, Kolbeck R, Thoenen H, Lindholm D (1994) BDNF increases NT-3 expression in cerebral granule neurons. J Biol Chem 269: 828–830

Levi-Montalcini R (1987) The nerve growth factor: thirty-five years later. EMBO J 6: 1145–1154

Lewis ME, Neff NT, Contreras PC, Stong DB, Oppenheim RW, Grebow PE, Vaught JL (1993) IGF-1: potential for treatment of motor neuronal disorders. Exp Neurol 124: 73–88

Lindholm D, Castren E, Da Penha Berzaghi M, Thoenen H (1993) Effects of neurotransmitters and hormones on neuronal production of neurotrophins. Semin Neurosci 5: 279–283

Lindholm D, Hartikka J, Da Penha Berzaghi M, Castren E, Tzimagiorgis G, Hughes RA, Thoenen H (1994) FGF-5 promotes differentiation of cultured rat septal cholinergic and raphe serotonergic neurons: comparison with the effects of neurotrophins. Eur J Neurosci 6: 244–252

Lärkfors L, Ebendal T, Whittemore S, Hoffer B, Olson L (1987) Decreased level of NGF and its messenger RNA in the aged rat brain. Mol Brain Res 3: 55–60

Masu Y, Wolf E, Holtmann B, Sendtner M, Brem G, Thoenen H (1993) Disruption of the CNTF gene results in progressive motoneuropathy. Nature 365: 27–32

Olson L (1993) NGF and the treatment of Alzheimer s disease. Exp Neurol 124: 5–15

Phillips HS, Hains JM, Armanini M, Laramee GR, Johnson SA, Winslow JW (1991) BDNF mRNA is decreased in the hippocampus of individuals with Alzheimer's disease. Neuron 7: 695–702

Rosen DR, Siddique T, Patterson D, Figlewicz DA, Sapp P, Hentati A, Donalson D, Goto J, O'Regan JP, Deng H-X, Rahmani Z, Krizus A, McKenna-Yasek D, Cayabyab A, Gaston SM, Berger R, Tanzi RE, Halperin JJ, Herzfeldt B, Van den Berg R, Huang W-Y, Bird T, Deng G, Mulder DW, Smyth C, Laing NG, Soriano E, Periacak-Vance MA, Haines J, Rouleau GA, Gusella JS, Horvitz HR, Brown RH (1993) Mutation in Cu/Zn superoxide dismutase gene are associated with familial amyotrophic lateral sclerosis. Nature 362: 59–62

Schwab M, Otten U, Agid Y, Thoenen H (1979) NGF in the rat CNS: absence of specific retrograde axonal transport and tyrosine hydroxylase induction in locus coeruleus and substantia nigra. Brain Res 168: 473–482

Seiler M, Schwab M (1984) Specific retrograde transport of NGF from neocortex to nucleus basalis in the rat. Brain Res 300: 33–39

Sendtner M, Kreutzberg GW, Thoenen H (1990) CNTF prevents the degeneration of motor neurons after axotomy. Nature 345: 440–441

Sendtner M, Schmalbruch H, Stöckli KA, Carroll P, Kreutzberg GW, Thoenen H (1992a) CNTF prevents degeneration of motor neurons in mouse mutant progressive motor neuropathy. Nature 358: 502–504

Sendtner M, Holtmann B, Kolbeck R, Thoenen H, Barde Y-A (1992b) BDNF prevents the death of motoneurons in newborn rats after nerve section. Nature 360: 757–759

Snider WD, Johnson EM (1990) Neurotrophic molecules. Ann Neurol 26: 489–506

Sofroniew MV, Cooper JD (1993) Neurotrophic mechanisms and neuronal degeneration. Semin Neurosci 5: 249–257

Strada O, Hirsch E, Javoy-Agid F, Lehericy S, Ruberg M, Hauw J, Agid Y (1992) Does loss of NGF receptors precede loss of cholinergic neurons in Alzheimer's disease? An autoradiography study in the human striatum and basal forebrain. J Neurosci 12: 4766–4776

Thoenen H (1991) The changing scene of neurotrophic factors. Trends Neurosci 14: 165–169

Thoenen H, Barde Y-A (1980) Physiology of nerve growth factor. Physiol Rev 60: 1284–1335

Thoenen H, Hughes R, Sendtner M (1993) Trophic support of motoneurons: physiological, pathophysiological and therapeutic implications. Exp Neurol 124: 47–55

Thoenen H, Castren E, Berzaghi M, Blöchl A, Lindholm D (1994) Neurotrophic factors: possibilities and limitations in the treatment of neurodegenerative disorders. Int Acad Biomed Drug Res 7: 197–203

Whitehouse PJ, Price DL, Struble RG, Clark AW, Coyle JT, DeLong MR (1982) Alzheimer's disease and senile dementia, loss of neurons in the basal forebrain. Science 215: 1237–1230

Yan Q, Elliott J, Snider WD (1992) BDNF rescues spinal motor neurons from axotomy-induced cell death. Nature 360: 753–755

Author's address: Dr. D. Lindholm, Department of Developmental Neuroscience, BMC 587, Uppsala University, S-75123 Uppsala, Sweden.

Induction of experimental autoimmune encephalomyelitis by CD4[+] T cells specific for an astrocyte protein, S100β

K. Kojima[1], H. Wekerle[1], H. Lassmann[2], T. Berger[2] and Ch. Linington[1]

[1]Department of Neuroimmunlogy, Max-Planck-Institute for Psychiatry, Martinsried,
Federal Republic of Germany
[2]Neurological Institute, University of Vienna, Austria

Summary. S100β protein is a calcium binding protein that is not only expressed by astrocytes in the CNS, but also in many other tissues including the eye, thymus, spleen and lymph nodes. Despite this tissue distribution, which was expected to induce a firm state of self-tolerance to S100β, the Lewis rat mounts a strong T cell response to this autoantigen. The pathogenicity of this T cell response was demonstrated by the adoptive transfer of S100β-specific T cells which induced an inflammatory response in the CNS and eye of naive syngeneic recipients. The distribution of lesions in this novel model of EAE resembles that seen in some patients with MS, suggesting that the initial autoimmune insult in MS may be directed against a non-myelin antigen co-expressed in the CNS and extra-neural tissues.

Introduction

Dysregulation of an autoreactive, myelin-specific T cell response is thought to be involved in the immunopathogenesis of multiple sclerosis (MS). This concept derives almost entirely from murine models of experimental autoimmune encephalomyelitis (EAE) in which an "MS-like" pathology is induced by the adoptive transfer of myelin antigen-specific T cell lines or clones (Martin et al., 1992). In the Lewis rat, at least three myelin proteins can induce EAE, myelin basic protein (MBP) (Ben-Nun et al., 1981), proteolipid protein (PLP) (Yamamura et al., 1986) and myelin oligodendrocyte glycoprotein (MOG) (Linington et al., 1993). However, the immunodominant encephalitogenic T cell response in the Lewis rat is directed against a single epitope of MBP located between amino acid residues 68 to 88. In contrast to PLP- or MOG-specific T cells which are only weakly encephalitogenic, the adoptive transfer of MBP-specific T cells initiates an intense, dose dependent inflammatory response in the CNS of naive Lewis rats and results in severe neurological dysfunction (Lassmann et al., 1988).

This model of EAE mediated by the adoptive transfer of MBP-specific T cells (tEAE) is generally accepted as a paradigm for the initial inflammatory response in MS. However, there is still no direct evidence that a similar myelin- or MBPspecific T cell response is involved in the immunopathogenesis of this human disease. On the contrary, several observations suggest that immune responses directed against non-myelin antigens may play a significant role in the etiology of MS. A significant population of patients with MS develop inflammatory lesions in the retina, which in man is not myelinated and should therefore not be involved in a myelin-specific disease (Berger and Leopold, 1968; Lucarelli et al., 1991). In established MS it is perhaps possible to discount this retinal inflammation as a secondary autoimmune response triggered by tissue damage in the CNS. However, abnormalities in the retinal endothelium are also a significant risk factor for MS in patients with acute isolated optic neuritis (Lightman et al., 1987). The identity of the (auto-) antigen responsible for these retinal abnormalities in man is unknown, but interestingly EAE can be induced by immunisation with isolated cerebral endothelial cell membranes (Tsukada et al., 1989) indicating that non-myelin antigens do indeed have an encephalitogenic potential. These observations prompted us to investigate the autoimmune potential of a defined non-myelin CNS protein, S100b.

S100β – a novel autoantigen

S100β is a calcium binding protein expressed in the cytosol of astrocytes (Cocchia, 1981) and Schwann cells (Spreca et al., 1989) in the nervous system. This protein is a major component of the CNS where it accounts for approximately 0.4% of the total soluble protein. However unlike MBP, S100β is expressed in many different cell types outside of the nervous system (Kuwano et al., 1987; Zimmer and van Eldik, 1987; Stefansson et al., 1982) inculding Müller cells in the retina (Kondo et al., 1984), adipocytes (Barbatelli et al., 1993) and a variaty of cells within immune the immune system inculding the thymus, spleen, lymph nodes and T cells (Cocchia et al., 1983; Higley and Rowden, 1984; Ushiki et al., 1984; Takahashi et al., 1987). Moreover, whereas MBP is tightly associated with the cytoplasmic face of the myelin membrane (Omlin et al., 1982), S100β is also secreted into the extracellular fluid (van Eldik and Zimmer, 1987) and can be detected in the cerebrospinal fluid (Shashoua et al., 1984).

This pattern of tissue expression lead us to believe that self-tolerance to S100β would be firmly established and difficult to break. Expression of S100β in the thymus should result in the removal of auto-reactive, S100β-specific T cells from the developing T cell repertoire by negative selection. Any S100β-specific T cells which escape this initial selection process would then be silenced in the periphery, by either "clonal deletion" or the induction of non-responsiveness, anergy (Arnold et al., 1993).

This proved not to be the case, immunisation with S100β elicts a strong proliferative T cell response in the Lewis rat, associated with an inflammatory response in the CNS, but no overt clinical signs of neurological dysfunction (Kojima et al., 1994). Self-reactive S100β-specific T cells are therefore a normal component of the T cell repertoire and are not effectively silenced by clonal deletion in either the thymus or periphery. Immunisation with S100β in Freunds complete adjuvant overwhelms the immunoregulatory mechanisms which must theretore act suppress the pathogenic potential of this response in vivo and expands the S100β-specific T cell population. How the S100β response is suppressed in the periphery is unknown but two possibilities are anergy and immunoregulatory T cell circuits similar to those implicated in the regulation of the autoimmune response to MBP by T cell vaccination studies (Ben-Nun et al., 1981; Lider et al., 1988; Sun et al., 1988) and the use of CD8 knockout mice (Koh et al., 1992).

Characteristics of S100β-specific T cell lines

S100β-specific T cell lines derived from the draining lymph nodes of Lewis rats immunised with bovine S100β exhibit the CD4+ CD8− TCR-αβ+ membrane phenotype and their antigen-specific response is found to be class II MHC restricted. These T cell lines also express gamma-interferon, TNF-α and IL-2+, but not IL-4 suggesting that they belong to the Th-1 subset of T cells. The phenotype of these S100β-specific T cell lines is therefore indistinguishable from that reported for many encephalitogenic, myelin-specific T cell lines (Mustafa et al., 1993; Linington et al., 1993; Ando et al., 1989). However, unlike encephalitogenic MBP-specific T cells, which despite their CD4+ phenotype, are cytotoxic, lysing syngeneic astrocytes, or other APCs, in vitro in an antigen dependent manner (Sun et al., 1986), S100β-specific T cells exhibited no cytotoxic activity (Kojima et al., submitted).

The most striking feature of the encephalitogenic MBP-specific T cell response in the Lewis rat is however the immunodominance of a single MBP epitope, amino acid residues 68 to 88, recognition of which is associated with a preferential usage of the Vβ8.2 T cell receptor (TCR) (Chluba et al., 1989; Burns et al., 1989). Interestingly, the same preferential usage of Vβ8.2 TCR genes occurs in encephalitogenic, MBP-specific T cell lines derived from the PL/J mouse, despite the fact that a different MBP T cell epitope is involved. This restricted usage of TCR Vβ gene products by encephalitogenic MBP-specific T cells allowed the development of therapeutic strategies for EAE based on the selective neutralisation Vβ8.2+ T cells (Acha-Orbea et al., 1988; Urban et al., 1988; Howell et al., 1989; Vandenbark et al., 1989; Zaller et al., 1990).

Epitope mapping revealed that the T cell response to S100β is also dominated by a single epitope, located within the carboxyl terminal amino acid sequence of S100β, residues 76–91. However, unlike the T cell response to MBP, the T cell response to S100β is not associated with the preferential usage of any particular Vβ TCR gene.

The histopathology of S100β-mediated EAE

As stated above immunisation with S100β failed to induce any clinical signs of disease, although inflammatory infiltrates were observed in the CNS. The

K. Kojima et al.

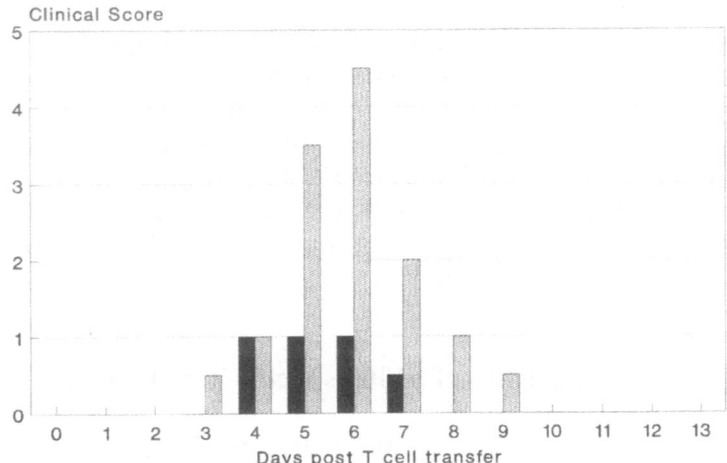

Fig. 1. Clinical course of disease induced by the adoptive transfer of MBP- or S100β-specific T line cells. Naive Lewis rats were injected in the tail vein with either 10^7 S100β-specific T line cells (n = 6; solid columns), or 5×10^6 MBP-specific T line cells (n = 5; hatched columns). Disease was scored on the following scale: grade 1 complete loss of tail tone; grade 2, hind limb weakness; grade 3, hind limb paralysis; grade 4, moribund; grade 5, dead

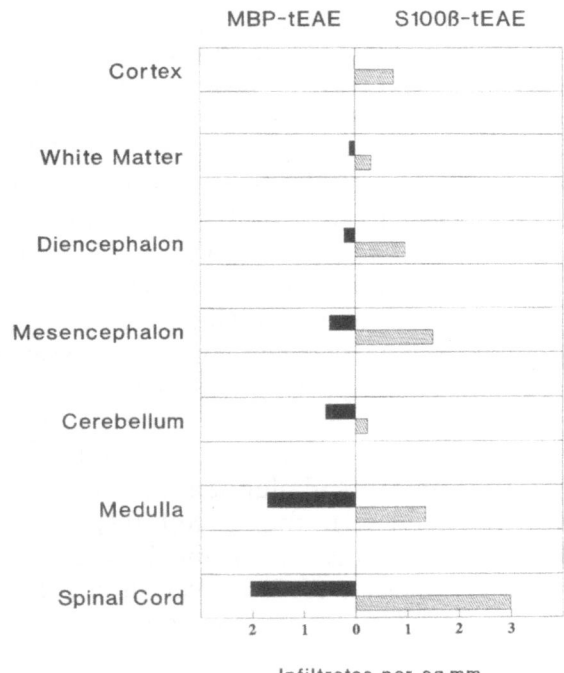

Fig. 2. Topographic distribution of inflammatory infiltrates in the CNS following the adoptive transfer of MBP- or S100β-specific T line cells. Lewis rats were injected intravenously with either MBP- (n = 7), or S100β-specific (n = 6) freshly activated T line cells and perfused six days later with 4% paraformaldehyde in phosphate buffered saline. The number of inflammatory infiltrates per mm² determined in multiple tissue sections stained with hemotoxylineosin

pathogenic potential of S100β- and MBP-specific T cell lines was therefore compared directly by the adoptive transfer of freshly activated T cell blasts into naive syngeneic recipients. As shown in Fig. 1, the adoptive transfer of 5×10^6 MBP-specific T cell blasts induces severe clinical EAE which is lethal in about 30% of the animals injected. In contrast the adoptive transfer of twice as many S100β-specific T line cells induced only a transient loss of tail tone. No signs of limb weakness or paralysis were observed.

Paradoxically, histopathological analysis revealed that the incidence of inflammatory lesions in these animals was infact higher than observed in rats with severe MBP-induced tEAE (Fig. 2). Moreover, whilst inflammatory lesions in MBP tEAE were most prominant in the lumbar spinal cord, the whole of the CNS was involved in the inflammatory response triggered by the adoptive tranfer of S100β-specific T line cells. This difference in the frequency of lesions between the two models was most marked in the the cortex and white matter of the brain (Fig. 2).

A possible explanation for the inability of the S100β-specific T cells to cause severe neurological dysfunction was revealed by a detailed immunohistochemical analysis of the lesions. In the classical model of MBP-mediated tEAE in the rat, the dominant infiltrating cell type is the ED1+ macrophage. These cells enter the CNS and then migrate away from the perivascular space into the parenchym. Macrophage depletion experiments have demonstrated that it is this cell type that is responsible for the clinical deficit in EAE

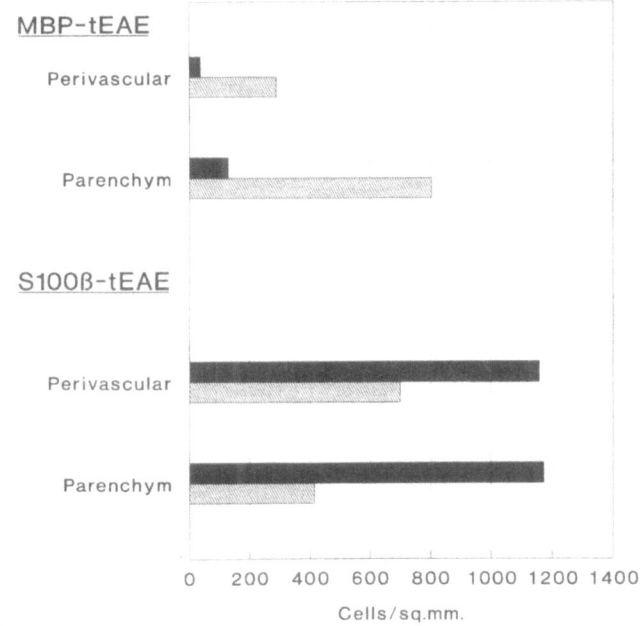

Fig. 3. Quantitation of macrophage and T cell infiltration into the CNS in the MBP-tEAE and S100β-tEAE. Multiple sections of spinal cord were stained immunohistochemically for T cells (solid bars) and macrophages (hatched bars) and the density of cells in the perivascular compartment and parenchym quantitated

(Brosnan et al., 1981; Huitinga et al., 1990). In the S100β-mediated model of tEAE the composition of the inflammatory infiltrates and their distribution between the perivascular compartment and the parenchym is very different to that seen in the classical model of MBP-mediated EAE. In the lesions induced by the adoptive transfer of S100β-specific T line cells, T cells are the predominating cell population and reach a density in the parenchym 10-fold greater than that seen in MBP-mediated tEAE (Fig. 3). In contrast the number of ED1+ macrophages is reduced and those which do enter the CNS appear to remain associated with the perivascular compartment (Fig. 3). Moreover, expression of ED1 is low suggesting those macrophages that do enter the CNS only become poorly activated.

These observations suggest that MBP- and S100β-specific T cells differ dramatically in their ability to trigger macrophage migration and activation in the CNS. We believe this reflects differences in the ability of the T cell lines to trigger the expression of chemotactic chemokines and/or cytokines the CNS. However these differences must be very selective or subtle. Both MBP- and S100β-specific T cell lines express the appropriate surface molecules to enable them to interact with the luminal surface of the CNS endothelium (Simmons et al., 1992, 1993; Baron et al., 1993) and hydrolytic enzymes (Naparstek et al., 1984) required to degrade extracellular matrix and enter the perivascular space. Moreover, in both models the permeability barrier provided by the BBB to serum proteins is disrupted and microglia within the lesions are activated to comparable levels (Kojima et al., 1994).

As anticipated by the tissue distribution of S100β, inflammatory infiltrates were not only observed in the CNS, but also in the eye and peripheral nerve. Approximately 80% of rats with S100β-induced EAE also developed uveitis which in some cases was associated with scleritis, iridocyclitis and periphlebitis retinae (Kojima et al., 1994). Pathological abnormalities were not observed in the eyes of animals with classical MBP-specific T cell mediated EAE.

Discussion

This new model of autoimmune mediated CNS inflammation requires that we reassess our conceptual view of the role autoimmunity may play in the immunopathogenesis of MS. The observation that lesions are found in both the CNS and eye (retina and uvea) in S100β-mediated tEAE is similar to that reported for a sub-set of MS patients (Berger and Leopold, 1968; Shaw et al., 1987; Lucarelli et al., 1991). This observation is difficult to reconcile with the concept that MS is a myelin-specific autoimmune disease, as the retina is not myelinated in man. Moreover, we demonstrate that the induction of an autoimmune disease of the CNS does not require that the target antigen is CNS-specific. In the case of S100β, an apparently tissue-specific disease is induced by S100β-specific T cells although this protein is expressed in many tissues, including thymus, spleen and lymph nodes. The initial immunological events responsible for triggering MS could therefore involve autoantigens co-expressed in the CNS and other extra-neural tissues.

Acknowledgements

This work was supported by the Deutsche Forschungsgemeinschaft ("Biology of Glia" research program) and the Austrian Science Foundation.

References

Acha-Orbea H, Mitchell DJ, Timmermann L, Wraith DC, Tausch GS, Waldor MK, Zamvil SS, McDevitt HO, Steinman L (1988) Limited heterogeneity of T cell receptors from T lymphocytes mediating autoimmune encephalomyelitis allows specific immune intervention. Cell 54: 263–273

Ando DG, Clayton J, Kono D, Urban JL, Sercarz EE (1989) Encephalitogenic T cells in the B10.PL model of experimental allergic encephalomyelitis (EAE) are of the Th1 lymphokine subtype. Cell Immunol 124: 132–143

Arnold B, Schönrich G, Hämmerling GJ (1993) Multiple levels of peripheral tolerance. Immunol Today 14: 12–14

Barbatelli G, Morroni M, Vinesi P, Cinti S, Michetti F (1993) S-100 protein in rat adipose tissue under different functional conditions: a morphological, immunocytochemical and immunochemical study. Exp Cell Res 208: 226–231

Baron JL, Madri JA, Ruddle NH, Hashim GA, Janeway CA (1993) Surface expression of α4 integrin by CD4 T cells is required for their entry into brain parenchyma. J Exp Med 177: 57–68

Ben-Nun A, Wekerle H, Cohen IR (1981) Vaccination against autoimmune encephalomyelitis using attenuated cells of a T lymphocyte line reactive against myelin basic protein. Nature 292: 60–61

Berger BC, Leopold IH (1968) The incidence of uveitis in multiple sclerosis. Am J Opthalmol 62: 540

Brosnan CF, Bornstein MB, Bloom BR (1981) The effects of macrophage depletion on the clinical and pathologic expression of experimental allergic encephalomyelitis. J Immunol 126: 614–620

Burns FR, Li X, Shen H, Offner H, Chou YK, Vandenbark AA, Heber-Katz E (1989) Both rat and mouse T cell receptors specific for the encephalitogenic determinant of myelin basic protein use similar Vα and Vβ chain genes even though the major histocompatibility complex and encephalitogenic determinants being recognized are different. J Exp Med 169: 27–40

Chluba J, Steeg C, Becker A, Wekerle H, Epplen JT (1989) T cell receptor β chain usage in myelin basic protein-specific rat T lymphocytes. Eur J Immunol 19: 279–284

Cocchia D (1981) Immunocytochemical localization of S-100 protein in the brain of adult rat. Cell Tissue Res 214: 529–534

Cocchia D, Tiberio G, Santarelli R, Michetti F (1983) S-100 protein in "follicular dendritic" cells of rat lymphoid organs. An immunohistochemical and immunocytochemical study. Cell Tissue Res 230: 95–103

Higley HR, Rowden G (1984) Thymic interdigitating reticulum cells demonstrated by immunocytochemistry. Thymus 6: 243–253

Howell MD, Winters ST, Olee T, Powell HC, Carlo DJ, Brostoff SW (1989) Vaccination against experimental allergic encephalomyelitis with T cell receptor peptides. Science 246: 668–671

Huitinga I, Van Rooijen N, De Groot CJA, Uitdehaag BMJ, Dijkstra CD (1990) Suppression of experimental allergic encephalomyelitis in Lewis rats after elimination of macrophages. J Exp Med 172: 1025–1033

Koh DR, Fung-Leung WP, Ho A, Gray D, Acha-Orbea H, Mak TW (1992) Less mortality but more relapses in experimental allergic encephalomyelitis in CD8-/- mice. Science 256: 1210–1213

Kojima K, Berger T, Lassmann H, Hinze-Selch D, Zhang Y, Gehrmann J, Reske K, Wekerle H, Linington C (1994) Experimental autoimmune panencephalo-myelitis and uveoretinitis transferred to the Lewis rat by T cells specific for the S100β molecule, a calcium binding protein of astroglia. J Exp Med 180

Kondo H, Takahashi H, Takahashi Y (1984) Immunohistochemical study of S-100 protein in the postnatal development of Müller cells and astrocytes in the rat retina. Cell Tissue Res 238: 503–508

Kuwano R, Usui H, Maeda T, Araki K, Yamakuni T, Kurihara T, Takahashi Y (1987) Tissue distribution of rat S-100α and β subunit mRNAs. Mol Brain Res 2: 79–82

Lassmann H, Brunner C, Bradl M, Linington C (1988) Experimental allergic encephalo-myelitis: the balance between encephalitogenic T lymphocytes and demyelinating antibodies determines size and structure of demyelinated lesions. Acta Neuropathol 75: 566–576

Lider O, Reshef T, Béraud E, Ben-Nun A, Cohen IR (1988) Anti-idiotypic network induced by T-cell vaccination against experimental autoimmune encephalomyelitis. Science 239: 181–183

Lightman S, McDonald WI, Bird AC, Francis DA, Hoskins A, Batchelor JR, Halliday AM (1987) Retinal nervous sheathing in optic neuritis. Its significance for the pathogenesis of multiple sclerosis. Brain 110: 405–414

Linington C, Berger T, Perry L, Weerth S, Hinze-Selch D, Zhang Y, Lu H, Lassmann H, Wekerle H (1993) T cells specific for the myelin-oligodendrocyte glycoprotein mediate an unusual autoimmune inflammatory response in the central nervous system. Eur J Immunol 23: 1364–1372

Lucarelli MJ, Pepose JS, Arnold AC, Foos RY (1991) Immunopathological features of retinal lesions in multiple sclerosis. Ophthalmology 98: 1652–1656

Martin R, McFarland HF, McFarlin DE (1992) Immunological aspects of demyelinating diseases. Annu Rev Immunol 10: 153–187

Mustafa M, Vingsbo C, Olsson T, Höjeberg B, Holmdahl R (1993) The major histocompatibility complex influences myelin basic protein 63-88-induced T cell cytokine profile and experimental autoimmune encephalomyelitis. Eur J Immunol 23: 3089–3095

Naparstek Y, Cohen IR, Fuks Z, Vlodavsky I (1984) Activated T lymphocytes produce a matrix-degrading heparan sulfate endoglycosidase. Nature 310: 241–244

Omlin FX, Webster H de F, Palkovits CG, Cohen SR (1982) Immunocytochemical localization of basic protein (MBP) in the major dense line regions of central and peripheral myelin. J Cell Biol 95: 242–248

Shashoua VE, Hesse GW, Moore BW (1984) Proteins of the brain extracellular fluid: evidence for release of S-100 protein. J Neurochem 42: 1536–1541

Shaw PJ, Smith NM, Inca PG, Bates D (1987) Chronic periphlebitis retinae in multiple sclerosis. A histological study. J Neurol Sci 77: 147–154

Simmons RD, Cattle BA (1993) Sialyl ligands facilitate lymphocyte accumulation during inflammation of the central nervous system. J Neuroimmunol 41: 123–130

Simmons RD, Cattle BA, Hugh AR, Willenborg DO (1992) Lymphocytes utilize sialylated surface molecules to accumulate in developing lesions of autoimmune encephalomyelitis. Autoimmun 14: 17–21

Spreca A, Rambotti MG, Rende M, Saccardi C, Aisa MC, Giambanca I, Donato R (1989) Immunocytochemical localization of S-100β protein in degenerating and regenerating rat sciatic nerves. J Histochem Cytochem 37: 441–446

Stefansson K, Wollmann RL, Moore BW (1982) Distribution of S-100 protein outside the central nervous system. Brain Res 234: 309–317

Sun D, Wekerle H (1986) Ia-restricted encephalitogenic T lymphocytes mediating EAE lyse autoantigen-presenting astrocytes. Nature 320: 70–72

Sun D, Ben-Nun A, Wekerle H (1988) Regulatory circuits in autoimmunity: recruitment of counterregulatory CD8+ T cells by encephalitogenic CD4+ T line cells. Eur J Immunol 18: 1993–2000

Takahashi K, Yoshida T, Hayashi K, Sonobe H, Ohtsuki Y (1987) S-100 beta positive human T lymphocytes: their characteristics and behavior under normal and pathologic conditions. Blood 70: 214–220

Tsukada N, Koh C-S, Yanagisawa N, Okano A, Taketomi T (1989) Autoimmune encephalomyelitis in rhesus monkeys induced by immunization with cerebral endothelial cell membranes. Acta Neuropathol 77: 39–46

Urban JL, Kumar V, Kono DH, Gomez C, Horvath SJ, Clayton J, Ando DG, Sercarz EE, Hood L (1988) Restricted use of T cell receptor V genes in murine autoimmune encephalomyelitis raises possibilities for antibody therapy. Cell 54: 577–592

Ushiki T, Iwanaga T, Masuda T, Takahashi Y, FujitaT (1984) Distribution and ultrastructure of S-100-immunoreactive cells in the human thymus. Cell Tissue Res 235: 509–514

Van Eldik LJ, Zimmer DB (1987) Secretion of S-100 from rat C6 glioma cells. Brain Res 436: 367–370

Vandenbark AA, Hashim G, Offner H (1989) Immunization with a synthetic T-cell receptor V-region peptide protects against experimental autoimmune encephalomyelitis. Nature 341: 541–544

Yamamura T, Namikawa T, Endoh M, Kunishita T, Tabira T (1986) Passive transfer of experimental allergic encephalomyelitis induced by proteolipid apoprotein. J Neurol Sci 76: 269–275

Zaller DM, Osman G, Kanagawa O, Hood L (1990) Preventio and treatment of murine experimental allergic encephalomyelitis with T cell receptor Vβ-specific antibodies. J Exp Med 171: 1943–1956

Zimmer DB, Van Eldik LJ (1987) Tissue distribution of rat S-100α and S-100β and S-100 binding proteins. Am J Physiol 252: 285–289

Authors' address: C. Linington, Department of Neuroimmunology, Max-Planck Institute for Psychiatry, Am Klopferspitz 18a, D-82152 Martinsried, Federal Republic of Germany.

Immunological aspects of experimental allergic encephalomyelitis and multiple sclerosis and their application for new therapeutic strategies

R. Martin

Department of Neurology, University of Tübingen Medical School, Tübingen, Federal Republic of Germany

Summary. The etiology of multiple sclerosis (MS), a demyelinating disorder of the central nervous system (CNS), is not yet known. Immunological, clinical and pathological studies suggest, however, that T lymphocytes directed against myelin antigens are involved in the pathogenesis of MS. The examination of an experimental animal model for MS, experimental allergic encephalomyelitis (EAE), demonstrated that myelin basic protein- (MBP) or proteolipidprotein- (PLP) specific T cells mediate the destruction of CNS myelin. In recent years, elegant studies in EAE showed that encephalitogenic T cells recognize short peptides of MBP or PLP in the context of MHC/HLA-class II molecules, express a restricted number of T cell receptor (TCR) molecules and secrete interferon-γ and tumor necrosis factor-α/β. Understanding the pathogenetic steps of demyelination at the molecular level led to highly specific immunotherapies of EAE targeting each individual molecule. MBP- and PLP-specific T cells with similar properties could also be isolated from MS patients and control individuals. Due to their heterogeneity in terms of specificity, function and TCR usage, it was difficult, however, to draw definite conclusions from these results, so far. The recent approval of interferon-β, a cytokine that antagonizes a number of the effects of interferon-γ, for the treatment of MS has raised great interest in examining novel strategies for immunotherapies in MS. The basic concepts as well as the current candidates for such new immunotherapies will be outlined in this brief article.

Introduction

Multiple sclerosis (MS) is the most frequent demyelinating disease of the central nervous system (CNS) in Northern Europeans and -Americans (McFarlin and McFarland, 1982a, b). Although its cause is still unknown an autoimmune pathogenesis is likely. This hypothesis is based on the composition of inflammatory CNS infiltrates consisting of lymphocytes and macrophages (Prineas, 1985) and also on striking parallels with a T cell-mediated animal model for demyelinating diseases, experimental allergic encephalomyelitis (EAE) (Raine, 1983), and on the beneficial effect of immunosuppressive and anti-inflammatory therapies (Mackin et al., 1992). Furthermore, similar to other autoimmune diseases, MS and EAE are associated with genes of the major histocompatibility complex (MHC in general or HLA in humans) which contribute to disease susceptibility (Martin et al., 1992a; Fritz et al., 1985). In recent years, enormous progress has been made in understanding which factors are involved in the demyelinating process in EAE (Fritz and McFarlin, 1989; Zamvil and Steinman, 1990). Based on this, novel therapeutic strategies have been introduced for EAE. Currently, there is great interest to employ similar approaches in MS. This report will briefly summarize the most important characteristics of MS, outline immunological findings in EAE and mention how T cell mediated demyelinating diseases may be therapeutically modulated.

Pathology and clinical characteristics of MS

Multiple sclerosis is a descriptive term characterizing the gross pathology of late manifestations of the disease. Numerous sclerotic lesions or plaques are scattered throughout the white matter of the brain and spinal cord (Prineas, 1985; Raine, 1983; Raine and Scheinberg, 1988). Predilection sites are the optic nerve and tract, the periventricular white matter, brain stem, cerebellum and spinal cord. Histopathological examination shows disseminated inflammatory lesions that are primarily located around small venules. The lesions are usually well demarcated from the surrounding tissue and characterized by loss of myelin and relative sparing of axons. During early stages of plaque development, infiltrates composed of various subsets of lymphocytes and macrophages and blood-brain-barrier (BBB) leakage predominate. Subsequently, the extent of inflammation decreases, but, at the same time, glial scar tissue as well as incomplete remyelination may be observed. The myelin sheath that physiologically insulates the axon and facilitates fast, saltatory nerve conduction is produced by oligodendrocytes in the CNS. Both, myelin sheath and oligodendrocytes may be the target of the immunological attack that leads to demyelination. As a consequence, an MS patient may clinically experience almost any CNS sign and symptom at some time during the disease including deficits in vision, coordination and sensation, weakness, or loss of sphincter control. The etiology of MS is unknown, and there is no single specific diagnostic parameter. The diagnosis is therefore still based on clinical findings and disease course (McFarlin and McFarland, 1982a, b). The presence of at least two clinical exacerbations affecting at least two different CNS regions or a progressive deterioration over more than six months are required. Several laboratory findings including elevated levels of cerebrospinal fluid immunoglobulin (Ig) secretion leading to oligoclonal Ig bands (Tourtellotte, 1985) and delayed evoked visual or somatosensory potentials support the diagnosis. In recent years, magnetic resonance imaging (MRI) of the brain and spinal cord has allowed the direct visualization of early inflammatory changes as well as the later stages of demyelinating lesions (McFarland et al., 1992).

According to its clinical course, MS may be separated into two major forms, i.e. relapsing-remitting (RR) and chronic progressive (CP) MS. The former usually begins in young adulthood (20–40 years) affecting women more often (female/male ratio = 2 : 1) than men. CP is characterized by an insidious onset later in adulthood, an almost even sex distribution and a poorer prognosis (McFarlin and McFarland, 1982a, b). Usually, the disease starts with acute exacerbations such as impairment or loss of vision or sensation followed by an often incomplete recovery within a few weeks. At any point, a relapsing-remitting course may evolve into a secondary chronic progressive one, but the disease may also run a more benign course with few and mild relapses resulting in only minimal disability.

Immunological aspects of experimental allergic encephalomyelitis (EAE)

The hypothesis that MS is an autoimmune disease finds support from studies of an animal model for demyelinating diseases called experimental allergic encephalomyelitis (EAE) (Martin et al., 1992a). Following the description of post-vaccinal encephalomyelitis (PVE) after rabies virus vaccination, Rivers was the first to demonstrate that a demyelinating disease can be reliably induced by injection of components of spinal cord homogenate (Rivers et al., 1933). Later, this model was refined, and it is now well known that EAE can be induced in various inbred animal strains by inoculation of whole myelin or defined myelin proteins such as myelin basic protein (MBP) and proteolipid protein (PLP) in complete Freund's adjuvant (CFA) (Fritz and McFarlin, 1989; Zamvil and Steinman, 1990; Tuohy et al., 1988). Attempts to transfer EAE by humoral or cellular components indicated that it is cell- and not antibody-mediated (Richert et al., 1979; Pettinelli and McFarlin, 1981; Ben Nun and Cohen, 1982). More specifically, it was shown that EAE can be transferred with lymph node cells from MBP-immunized animals that were restimulated in vitro into naive recipient animals (adoptive transfer or AT-EAE) (Pettinelli and McFarlin, 1981). During recent years, the pathogenetic factors involved in EAE have been studied in detail. It was not only shown that EAE can be induced by different myelin proteins, but also that EAE varies in susceptible animal strains with respect to course and pathology of the lesions (Fritz and McFarlin, 1989). An acute form of EAE reminiscent of PVE and characterized by inflammatory infiltrates in the lumbar spinal cord and brain stem with very little demyelination is observed in Lewis rats (Ben Nun and Cohen, 1982). In contrast, a chronic relapsing form with considerable demyelination can be produced in SJL/J mice (Pettinelli and McFarlin, 1981). Due to these differences, it is still controversial whether some forms of EAE are useful models for MS or only represent certain stages of the disease process (for example the early inflammatory stage in Lewis rat EAE). Demyelination is not induced by humoral components alone, myelin-specific antibodies are, however, able to enhance the extent of myelin damage (Linington et al., 1988).

Susceptibility for EAE is under the control of MHC genes, although their exact role is not yet completely understood (Fritz et al., 1985). In the last decade, the characteristics of encephalitogenic MBP- and PLP-specific T cells have been described in detail with respect to activation requirements, fine specificity, MHC restriction, T cell receptor (TCR) usage, phenotype and lymphokine profiles. These elegant studies documented that encephalitogenic T cells are CD4+, T helper 1 (TH1)-like cells [secrete tumor necrosis factor (TNF)α/β and interferon (IFN)-γ] (Ando et al., 1989; Broome Powell et al., 1990) and able to mediate delayed type hypersensitivity (DTH) responses. Depending on the animal strain used, encephalitogenic T cells are specific for a short encephalitogenic peptide of MBP or PLP that is recognized in the context of MHC-class II (IA, or rarely IE) antigens (Martin et al., 1992a; Zamvil and Steinman, 1990). Furthermore, they express a restricted number of TCR chains (Acha-Orbea et al., 1989). In the chronic relapsing EAE model in SJL/J mice, for example, encephalitogenic T cells primarily respond to MBP peptide 89–100 and are restricted by IA^S. Although encephalitogenic T cells are critical for the demyelinating process during EAE, their presence alone is not sufficient for disease. Transgenic mice that carry a TCR derived from encephalitogenic T cells from B10 mice do not develop spontaneous EAE when they are housed under germ free conditions although the majority of peripheral T cells expresses the transgenic TCR (Goverman et al., 1993). When these animals are housed under normal non-germ free conditions or are injected with unspecific stimulators of the immune system, such as pertussis toxin, EAE will occur in a considerable percentage of animals. This observation argues for exogenous factors that are probably important for non-specific activation of T cells and induction of adhesion molecules and upregulation of MHC expression on endothelial cells, macrophages and tissue-specific antigen-presenting cells (APC). A similar state of activation can be induced by injection of suboptimal doses of encephalitogenic T cells followed by bacterial superantigens (Brocke et al., 1993; Schiffenbauer et al., 1993). These are substances known to provide a strong, but non-specific stimulus to T cells. These observations have not only extended our knowledge about autoimmune diseases in general, but also helped to design a number of highly specific immunotherapies. Before specific therapeutic strategies will be described, the pathogenetic events that are involved in the demyelinating process shall be outlined.

Susceptibility for EAE or MS is associated with the immunogenetic background of an individual. It is clear from genetic studies in MS that multiple genes are likely to confer susceptibility and that one or several genes of the MHC/HLA complex are associated with disease both in MS (Tiwari and Terasaki, 1985) and EAE (Fritz et al., 1985). It can be concluded from concordance rates of identical twins (25–30%) (Sadovnick et al., 1993) that MS is not caused by a single gene and that exogenous factors such as viral infections are required for disease, even in individuals with the appropriate immunogenetic background. Experiments in EAE as well as in the transgenic mouse model that was referred to before have shown that the activation of autoantigen-specific T cells is critical and that this step occurs in the periphery rather than the target organ. During activation, an encephalitogenic T cell interacts with its TCR with autoantigenic peptide determinants embedded in the antigen-binding groove of disease-associated MHC/HLA-class II molecules. This complex of molecules consisting of a specific TCR, encephalitogenic peptide and MHC molecule is referred to as trimolecular complex of antigen recognition (Hohlfeld, 1989). In EAE, each component of this complex is well characterized at the molecular level. In addition, the induction and engagement of other surface receptors and their ligands including CD4, lymphocyte function-associated molecule (LFA-1) / intercellular adhesion molecule-1 (ICAM-1), and alpha-4 integrin (Yednock et al., 1992; Baron et al., 1993) are required for proper activation.

Following this step, encephalitogenic T cells are able to interact with cerebral vascular endothelial cells and migrate into the CNS parenchyma (Raine et al., 1990; Cross et al., 1990). BBB-leakage and damage of endothelial cells is probably caused by direct cytolysis mediated by cytotoxic, encephalitogenic T cells, but also by lymphokines secreted by these cells, in particular TNF-α/β (Broome Powell et al., 1990). The upregulation of adhesion receptors and MHC/HLA molecules on endothelial, but also glial cells is probably to a large extent mediated by IFN-γ that is locally produced by encephalito-

Table 1. Summary of experimental immunotherapies that have successfully been used in EAE and are currently being planned or have already been established for the treatment of MS

Target structure	Type of therapy/ Mode of action	EAE	MS
MHC/HLA molecule	Blocking with antibody	Successful	Not useful (general immunosuppression)
	Blocking peptide	Successful	Planned
	Copolymer-1	Successful	Multicenter Trial finished
	Peptide/MHC complexes	Successful	Planned
T Cell Receptor/ Autoantigen-specific T cell	TCR peptide vaccination	Successful	In progress
	T cell vaccination	Successful	In progress
	TCR antagonist peptides	In progress	Planned
Oral tolerization	Induction of anergy/ suppression	Successful	In progress
Adhesion molecules	Antibodies against ICAM-1 or VLA-4	Successful	Planned
CD4	Monoclonal antibodies/ design peptides	Successful	In progress
Modulation of lymphokine release (TNF-α/IFN-γ)	Antibodies against TNF-α	Successful	Planned
	IFN-β	Successful	Approved for MS
	TGF-β	Successful	In progress

genic T cells (McCarron et al., 1993). Such cells predominate in the target tissue during the early phase of inflammation (Cross et al., 1993; Offner et al., 1993). They may cause tissue damage via direct lysis of cells presenting myelin antigens, but also by enhancing the extent of inflammation. Secretion of IFN-γ and TNF-α/β may lead to upregulation of MHC-molecules, induction of heat shock protein expression, activation and recruitment of macrophages, microglia and other T cells and direct damage to myelin membranes and oligodendroglial cells. Factors produced by macrophages/microglia including oxygen radicals, nitric oxide and IL-1 may further increase the extent of tissue damage, before activated macrophages start to strip damaged myelin from axons and phagocytose the debris (Hartung, 1993; Raine and Scheinberg, 1988). Later cellularity decreases and both glial proliferation and incomplete remyelination may occur.

Immunotherapies of EAE

Each individual step of the above mentioned mechanisms has been modified by experimental therapies, and these strategies have greatly contributed to our understanding of the pathogenesis of EAE, but also of other autoimmune diseases. The different treatments will now be grouped into:

1. Blockade of antigen presentation and T cell activation
2. T cell- and T cell receptor peptide vaccination
3. Modulation of lymphokine release
4. Different routes to administer myelin antigens

The experimental data will be mentioned first, before the approaches are evaluated with respect to their potential use in MS.

1. Blockade of antigen presentation and T cell activation

Several molecules are involved in the activation of encephalitogenic T cells: the components of the trimolecular complex, MHC molecules with bound autoantigenic peptide and TCR, but also structures important for adhesion and costimulation such as CD4, alpha-4 integrin and LFA-1/ICAM-1 (see above).

At the level of the MHC molecule, modifications (removal of acetylation at the N-terminus or insertion of alanine substitutions) of the encephalitogenic, N-terminal MBP peptide (Ac1-9) resulted in peptides that still bind to the restriction element (IA^U), but are not recognized by encephalitogenic T cell clones (Wraith et al., 1989; Urban et al., 1989). Treatment of mice with these peptides binding with high-affinity to the MHC molecule were able to block AT-EAE. This treatment is probably based on competition between the modified and the encephalitogenic peptide for binding to the disease-associated MHC-class element. It was shown not only for modified encephalitogenic peptide, but also for unrelated peptides (Lamont et al., 1990). Both approaches are not specific for a single antigen, but have the potential for general immunosuppression since they block a restriction element important for T helper cells. A polypeptide consisting of four amino acids in random order called copolymer 1 (COP-1) has been developed for the same purpose, and it was demonstrated that it not only interferes with antigen presentation by blocking MHC-class II molecules (Racke et al., 1992), but also induces suppressor T cells (Teitelbaum et al., 1988). Recently, these therapies were tailored more specifically to interfere with recognition of a single myelin antigen either by using monoclonal antibodies specific for the complex of MHC-class II and encephalitogenic peptide (Aharoni et al., 1991), by immunizing animals with inactivated APC coupled with myelin proteins or whole myelin (Kennedy et al., 1990) or by administration of MHC-peptide complexes (Sharma et al., 1991). Adhesion molecules mediate accessory signals during antigen recognition by T cells, but their differential expression on endothelia is also important for migration to and homing in specific organs such as the CNS. Monoclonal antibodies to CD4 (Brostoff and Mason, 1984; Waldor et al., 1985), LFA-1 and/or ICAM-1 (Cannella et al., 1993; Archelos et al., 1993) as well as to the alpha-4 integrin (Yednock et al., 1992) were able to block or modulate EAE, but their effects were strongly dependent on the administered doses and timing.

2. T cell- and T cell receptor peptide vaccination

It is now well established that tolerance against autoantigens is generated by a multistep process that involves not only selective maturation and deletion of T cells in the thymus, but also mechanisms acting in peripheral organs. Experiments in animal models as well as studies of the human autoantigen-specific T cell response have shown that such cells are part of the mature T cell repertoire, but usually do not cause disease. The idea that autoimmune diseases might be treatable by vaccination with autoantigen-specific T cells is based on the concept that an idiotypic network of T cells regulates the response to self antigens. Such a network requires T cells specifically interfering with the function of autoantigen-specific T cells. To test this hypothesis, early experiments in EAE used encephalitogenic T cells which had been inactivated by pressure treatment or fixatives to immunize animals in analogy to vaccination against an infectious agent (Ben Nun et al., 1981). Animals treated by this T cell vaccination protocol subsequently became resistant to the induction of EAE.

Following the description that encephalitogenic T cells employ a restricted number of TCR in the recognition of the encephalitogenic peptide (Acha-Orbea et al., 1988; Burns et al., 1989; Chluba et al., 1989), vaccination with whole inactivated T cells was replaced by using peptides homologous to parts of the variable chain Vβ8.2 (Vandenbark et al., 1989) or the junctional region of the TCR (Howell et al., 1989). Both approaches were able to modulate disease when given before induction of EAE, and also to some extent when administered at later stages. In the model using the Vβ8.2 peptide, it was shown that the effects are probably mediated by anti-Vβ8.2 antibodies and CD8+ T cells recognizing the Vβ8.2 peptide in the context of MHC-class I molecules (Offner et al., 1991). It is currently not clear whether such idiotype-specific T cells act by direct lysis as shown by earlier experiments or by secretion of modulating lymphokines such as IL-4 and TGF-β.

Still, T cell- or TCR peptide vaccination are fascinating approaches because they demonstrate that a regulatory network exists. Some of the results could, however, not be reproduced in EAE and experimental allergic uveitis (Jung et al., 1993; Kawano et al., 1991) due to experimental difficulties. One possible reason for that may be the spreading of the immune response to various different myelin determinants resulting in a broader TCR usage with longer disease duration (Sercarz et al., 1993; Sun et al., 1992).

3. Modulation of lymphokine release

Encephalitogenic T cells belong to the subgroup of T helper-1 (Th1) cells (Ando et al., 1989), and tissue damage is mediated through a DTH response initiated by such cells. Based on the observation that encephalitogenic T cells secrete IFN-γ and TNFα/β (see above), attempts were made to antagonize these proinflammatory lymphokines that are important in the effector phase of the demyelinating process. Consequently, EAE was improved or abrogated by directly neutralizing secreted TNF with a monoclonal anti-TNF antibody (Ruddle et al., 1990) or by modulating the Th1 response with suppressive lymphokines such as IL-4 (M. Racke, personal communication) and particularly TGF-β (Racke et al., 1991; Kuruvilla et al., 1991). Blockade of another lymphokine, IL-1, that is important for antigen presentation and is expressed in large amounts in MS plaques (Wucherpfennig et al., 1992) can also modulate EAE (Jacobs et al., 1991). Compared to the highly specific immunotherapies mentioned before, the modulation of lymphokine release is attractive because it does not require the knowledge of fine specificity and TCR usage of encephalitogenic T cells.

4. Different routes to administer myelin antigens

EAE is induced by subcutaneous injection of myelin antigens in complete Freund's adjuvant. Different routes such as oral or intranasal administration will not result in EAE, but rather in a state of unresponsiveness to the induction of EAE (Whitacre et al., 1991; Higgins and Weiner, 1988; Metzler and Wraith, 1993). The exact mechanisms leading to this anergic state are not yet completely clear. It was, however, demonstrated that, depending on the doses administered, oral application of myelin or other autoantigens will result in anergy of autoantigen-specific T cells (high dose) or the induction of T cells that secrete immunoregulatory lymphokines, probably IL-4 and TGF-β (Gregerson et al., 1993). Tolerization with peptide causes unresponsiveness not only against the tolerizing peptide but also against other epitopes of the same autoantigen (Gregerson et al., 1993). Moreover, the resistance induced by oral tolerization can be transferred to naive animals by transferring T cells.

Different from oral tolerization, intravenous injection of high doses of MBP resulted in the induction of programmed cell death or apoptosis of encephalitogenic T cells and, as a consequence, only mild EAE (Critchfield et al., 1994). Apoptosis can, however, only be achieved if encephalitogenic T cells are preactivated and in the S-phase of the cell cycle. It was concluded from a study that documented DNA fragmentation in the CNS by in situ nick translation, that apoptotic cell death is involved in the physiological downregulation of immune responses in the tissue (Schmied et al., 1993).

Can experimental immunotherapies be transferred to MS?

The above mentioned immunotherapies had been developed on the basis of detailed knowledge of the pathogenetic mechanisms in EAE. It is clear that experiments in humans can only indirectly address these questions and that evidence for factors involved in the demyelinating process in MS must therefore be circumstantial. In summary, T cells specific for myelin antigens such as MBP, PLP or myelin oligodendroglia protein (MOG) have been documented in MS patients and controls, and it is controversial as to which myelin antigen is most important (Martin et al., 1990, 1992a; Pette et al., 1990a; Olsson et al., 1990; Ota et al., 1990; Richert et al., 1989; Kerlero de Rosbo et al., 1993; Meinl et al., 1993). Similar to EAE, the immune response to MBP has been characterized in greatest detail. The majority of MBP-specific T cells are CD4+ (Martin et al., 1990; Pette et al., 1990a; Ota et al., 1990), Th1-like cells that secrete large amounts of IFN-γ and TNF-α/β (Voskuhl et al., 1993), are cytotoxic (Richert et al., 1989; Weber et al., 1989; Martin et al., 1990) and restricted by HLA-class II molecules. HLA-DR molecules that are closely associated with MS in different ethnic groups, in particular HLA-DR2, serve as restriction element most often (Martin et al., 1990; Pette et al., 1990a; Chou et al., 1989; Ota et al., 1990). The analysis of large numbers of T cell lines (TCL) for fine specificity showed that immunodominant regions are located in the middle (MBP peptides 84–102, 87–106 or 81–99; Ota et al., 1990; Pette et al., 1990b; Martin et al., 1991), the C-terminus (MBP peptides 143–168 and 154–172) (Martin et al., 1990; Richert et al., 1989; Ota et al., 1990) and probably also the N-terminus of the molecule (Martin et al., 1990; Pette et al., 1990a; Olsson et al., 1992). The recent characterization of the peptide binding motifs for example for HLA-DR2 alleles allowed to explain high affinity binding at the molecular level (Wucherpfennig et al., 1994; Vogt et al., 1994). Studies of the TCR repertoire of MBP-specific TCL and of T cells in CNS infiltrates of MS patients have yielded conflicting results with some studies documenting restricted TCR usage (Wucherpfennig et al., 1990; Kotzin et al., 1991) and others finding a high degree of diversity (Martin et al., 1992b; Giegerich et al., 1992) in peptide specificity and TCR usage. Comparison of TCR sequences from brain lesions of patients with the MS-associated HLA-type HLA-DR2 DQw6 with TCR sequences derived from a DR2-restricted, cytotoxic TCL (Oksenberg et al., 1993) or with TCR from encephalitogenic TCL in rodents (Allegretta et al., 1994) have, however, shown striking similarities. In addition, the analysis of the TCR Vα chain repertoire in identical twins that were concordant or discordant for MS documented a skewed TCR repertoire in the discordant twin sets, underlining that TCR usage is likely to be involved in the pathogenesis of MS (Utz et al., 1993). Recent studies showed that certain T cell specificities and TCRs can be followed over longer periods of time in individual MS patients (Salvetti et al., 1993; Meinl et al., 1993) and that TCR usage tends to become more diverse the longer the disease runs (Utz et al., 1994).

It is clear from these studies that there are parallels between encephalito-genic T cells and myelin-specific T cells in humans. The encephalitogenic MBP epitopes in different animals strains (Ac1-9 in PUJ and B10.PL mice, 89–100 in SJL/J mice, 69–86 in Lewis rats, 153–165 in rhesus monkeys) for example are largely overlapping with the MBP epitopes that were found immunodom-inant in humans (Fritz and McFarlin, 1989; Martin et al., 1992a). Moreover, the cells are very similar in terms of phenotype and in vitro function. Looking for analogies is, however, complicated from a number of reasons. The human HLA complex is more diverse than the rodent MHC and several genes apparently contribute to susceptibility for MS. In contrast to the well-defined animal models, several clinical forms of MS exist. Their etiology is not known, and the importance of individual myelin antigens for the pathogenesis in individual patients is not yet clear. Finally, very few, if any studies have tried to correlate carefully the immune responses with the different clinical forms of MS and disease activity in individual patients. With the fast increase of the knowledge about the molecular interactions involved in antigen recognition as well as the better clinical characterization of MS patients, there is hope, however, that we will soon be able to address some of these questions more specifically in the near future.

Considering the knowledge about the myelin-specific T cell response in MS patients, can we transfer experimental immunotherapies from EAE to MS? At this point, the question can be answered as follows. Treatments that would block a disease-associated HLA-DR molecule by a monoclonal antibody or a high-affinity binding peptide are likely to act as general immunosuppressants. In addition, it will hardly be possible to block all HLA molecules unless high doses of peptides are frequently administered intravenously. Any monoclonal antibody whether directed against HLA molecules, CD4 or adhesion molecules has the potential to induce antibodies against it unless it would be humanized. The highly specific immunotherapies such as T cell vaccination or TCR peptide vaccination are complicated by a number of problems. Since we do not know which T cells and which TCR are most important in MS, it will be necessary to determine these components and specifically tailor the therapy for each indi-vidual patient, a procedure that seems hardly feasible for larger groups of pa-tients. Oral tolerization, another interesting strategy, has already been started in MS, but its efficacy cannot be assessed so far (Weiner et al., 1993).

Presently, the approaches that target the effector phases of the demyelinat-ing process, in particular the secretion of lymphokines such as IFN-γ and TNF-α/β, are the best candidates for the therapy of MS. For such therapies we do not need to know the antigen specificity and TCR usage of myelin-specific T cells as long as we assume that TH1-like cells are important for disease. In addition, they will be active even when the disease process is already established whereas some of the treatments mentioned above will only work when initiated before onset or very early in the course of the disease. IFN-β, a lymphokine with antiviral and immunomodulatory activities primarily acts by antagonizing the effects of IFN-γ. Its efficacy in RR-MS has recently been demonstrated by a controlled, double-blind multi-center trial that showed a 30% reduction in exacerbation rate and a dramatic decrease of inflammatory activity when MRI imaging was chosen as a readout (The IFN-β Study Group, 1993; Paty et al., 1993). These results led to the rapid approval of IFN-β by the Food and Drug Administration in the USA as a new therapy for RR-MS. Although a 30% reduction in exacerbation rate is still far from ideal, IFN-β is currently the most efficient drug for the treatment of RR-MS and, at the same time, showed few side effects. It is anticipated that drugs antagonizing TNF will be even more efficient. Since therapies that interfere with lymphokine secretion, however, potentially interfere with beneficial immune responses against infectious agents we will hopefully be able to eventually treat MS with more specific therapies or even prevent the development at very early stages.

Acknowledgements

R. Martin is a Heisenberg Fellow of the Deutsche Forschungsgemeinschaft (Ma 965/4-1).

References

Acha-Orbea H, Mitchell L, et al (1988) Limited heterogeneity of T cell receptors from lymphocytes mediating autoimmune encephalomyelitis allows specific immune intervention. Cell 54: 263–273

Acha-Orbea H, Steinman L, et al (1989) T cell receptors in murine autoimmune diseases. Annu Rev Immunol 7: 371–406

Aharoni R, Teitelbaum D, et al (1991) Immunomodulation of experimental allergic encephalomyelitis by antibodies to the antigen-1a complex. Nature 351: 147–150

Alegretta M, Albertini RJ, et al (1994) Homologies between T cell receptor junctional sequences unique to multiple sclerosis and T cells mediating experimental allergic encephalomyelitis. J Clin Invest 94: 105–109

Ando DG, Clayton J, et al (1989) Encephalitogenic T cells in the B10.PL model of experimental allergic encephalomyelitis (EAE) are of the Th-1 lymphokine subtype. Cell Immunol 124: 132–143

Archelos JJ, Jung S, et al (1993) Inhibition of experimental autoimmune encephalomyelitis by an antibody to the intercellular adhesion molecule ICAM-1. Ann Neurol 34: 145–154

Baron JL, Madri JA, et al (1993) Surface expression of α4 integrin by CD4 T cells is required for their entry into brain parenchyma. J Exp Med 177: 57–68

Ben Nun A, Cohen IR (1982) Experimental autoimmune encephalomyelitis (EAE) mediated by T cell line: process of selection of lines and characterization of the T cells. J Immunol 129: 303–308

Ben Nun A, Wekerle H, et al (1981) Vaccination against autoimmune encephalomyelitis with T-lymphocyte line cells reactive against myelin basic protein. Nature 293: 60–61

Brocke S, Gaur A, et al (1993) Induction of relapsing paralysis in experimental by bacterial superantigen. Nature 365: 642–644

Broome Powell M, Mitchell D, et al (1990) Lymphotoxin and tumor necrosis factor-alpha production by myelin basic protein-specific T cell clones correlates with encephalitogenicity. Int Immunol 2: 539–544

Brostoff SW, Mason DW (1984) Experimental allergic encephalomyelitis: successful treatment in vivo with a monoclonal antibody that recognizes T helper cells. J Immunol 133: 1938–1942

Burns FR, Li X, et al (1989) Both rat and mouse T cell receptors specific for the encephalitogenic determinant of myelin basic protein use similar Vα and Vβ chain genes even though the major histocompatibility complex and encephalitogenic determinants being recognized are different. J Exp Med 169: 27–39

Cannella B, Cross AH, et al (1993) Anti-adhesion molecule therapy in experimental autoimmune encephalomyelitis. J Neuroimmunol 46: 43–55

Chluba J, Steeg C, et al (1989) T cell receptor β chain usage in myelin basic protein-specific rat T lymphocytes. Eur J Immunol 19: 279–284

Chou YK, Vainiene M, et al (1989) Response of human T lymphocyte lines to myelin basic protein: association of dominant epitopes with HLA-class II restriction molecules. J Neurol Sci 23: 207–216

Critchfield JM, Racke MK, et al (1994) T cell deletion in high antigen dose therapy of autoimmune encephalomyelitis. Science 263: 1139–1143

Cross AH, Cannella B, et al (1990) Homing to central nervous system vasculature by antigen specific lymphocytes. I. Localization of 14C-labeled cells during acute, chronic and relapsing experimental allergic encephalomyelitis. Lab Invest 63: 162–170

Cross AH, O'Mara T, et al (1993) Chronologic localization of myelin-reactive cells in the lesions of relapsing EAE: implications for the study of multiple sclerosis. Neurology 43: 1028–1033

Fritz RB, McFarlin DE (1989) Encephalitogenic epitopes of myelin basic protein. Chem Immunol 46: 101–125

Fritz RB, Skeen MJ, et al (1985) Major histocompatibility complex-linked control of the murine immune response to myelin basic protein. J Immunol 134: 2328–2332

Giegerich G, Pette M, et al (1992) Diversity of T cell receptor alpha and beta chain genes expressed by human T cells specific for similar myelin basic protein peptide/major histocompatibility complexes. Eur J Immunol 22: 753–758

Goverman J, Woods A, et al (1993) Transgenic mice that express a myelin basic protein-specific T cell receptor develop spontaneous autoimmunity. Cell 72: 551–560

Gregerson DS, Obritsch WF, et al (1993) Oral tolerance in experimental autoimmune uveoretinitis. Distinct mechanisms of resistance are induced by low dose vs high dose feeding protocols. J Immunol 151: 5751–5761

Hartung H-P (1993) Immune-mediated demyelination. Ann Neurol 33: 563–567

Higgins PJ, Weiner HJ (1988) Suppression of experimental autoimmune encephalomyelitis by oral administration of myelin basic protein and its fragments. J Immunol 140: 440–445

Hohlfeld R (1989) Neurological autoimmune disease and the trimolecular complex of T lymphocytes. Ann Neurol 25: 531–538

Howell MD, Winters ST, et al (1989) Vaccination against experimental allergic autoimmune encephalomyelitis with T cell receptor peptides. Science 246: 668–670

Jacobs CA, Baker PE, et al (1991) Experimental autoimmune encephalomyelitis is exacerbated by IL-1α and suppressed by soluble IL-1 receptors. J Immunol 146: 2983–2989

Jung S, Schluesener HJ, et al (1993) Modulation of EAE by vaccination with T cell receptor peptides: Vβ8 T cell receptor peptide-specific CD4+ lymphocytes lack direct imunoregulatory activity. J Neuroimmunol 45: 15–22

Kawano Y-I, Sasamoto Y, et al (1991) Trials of vaccination against experimental autoimmune uveoretinitis with a T-Cell receptor peptide. Curr Eye Res 10: 789–795

Kennedy MK, Tan L-J, et al (1990) Inhibition of murine relapsing experimental autoimmune encephalomyelitis by immune tolerance to proteolipid protein and its encephalitogenic peptides. J Immunol 144: 909–915

Kerlero de Rosbo N, Milo R, et al (1993) Reactivity to myelin antigens in multiple sclerosis. Peripheral blood lymphocytes respond predominantly to myelin oligodendrocyte glycoprotein. J Clin Invest 92: 2602–2608

Kotzin BL, Karuturi S, et al (1991) Preferential T-cell receptor Vβ-chain variable gene use in myelin basic protein-reactive T-cell clones from patients with multiple sclerosis. Proc Natl Acad Sci USA 88: 9161–9165

Kuruvilla AP, Shah R, et al (1991) Protective effect of transforming growth factor β1 on experimental autoimmune diseases in mice. Proc Natl Acad Sci USA 88: 2918–2921

Lamont AG, Sette A, et al (1990) Inhibition of experimental autoimmune encephalomyelitis induction in SJL/J mice by using a peptide with high affinity for IA^S molecules. J Immunol 145: 1687–1693:

Linington C, Bradl M, et al (1988) Augmentation of demyelination in rat acute allergic encephalomyelitis by circulating mouse monoclonal antibodies directed against a myelin/oligodendrocyte glycoprotein. Am J Pathol 130: 443–454

Mackin GA, Dawson DM, et al (1992) Treatment of multiple sclerosis with cyclophosphamide. In: Rudick RA, Goodkin DE (eds) Treatment of multiple sclerosis. Springer, London, pp 199–216

Martin R, Howell MD, et al (1991) A myelin basic protein peptide is recognized by cytotoxic T cells in the context of four HLA-DR types associated with multiple sclerosis. J Exp Med 173: 19–24

Martin R, Jaraquemada D, et al (1990) Fine specificity and HLA restriction of myelin basic protein-specific cytotoxic T cell lines from multiple sclerosis patients and healthy individuals. J Immunol 145: 540–548

Martin R, McFarland HF, et al (1992a) Immunological aspects of demyelinating diseases. Annu Rev Immunol 10: 153–187

Martin R, Utz U, et al (1992b) Diversity in fine specificity and T cell receptor usage of the human CD4+ cytotoxic T cell response specific for the immunodominant myelin basic protein peptide 87–106. J Immunol 148: 1359–1366

McCarron RM, Wang L, et al (1993) Cytokine-regulated adhesion between encephalitogenic T lymphocytes and cerebrovascular endothelial cells. J Neuroimmunol 43: 23–30

McFarland HF, Frank JA, et al (1992) Using gadolinium-enhanced magnetic resonance imaging lesions to monitor disease activity in multiple sclerosis. Ann Neurol 32: 758–766

McFarlin DE, McFarland HF (1982a) Multiple sclerosis, part 1. N Engl J Med 307: 1183–1188

McFarlin DE, McFarland HF (1982b) Multiple sclerosis, part 2. N Engl J Med 307: 1246–1251

Meinl E, Weber F, et al (1993) Myelin basic protein-specific T lymphocyte repertoire in multiple sclerosis. Complexity of the response and dominance of nested epitopes due to recruitment of multiple T cell clones. J Clin Invest 92: 2633–2643

Metzler B, Wraith DC (1993) Inhibition of experimental autoimmune encephalomyelitis by inhalation but not oral administration of the encephalitogenic peptide: influence of MHC binding affinity. Int Immunol 5: 1159–1165

Offner H, Hashim GA, et al (1991) T cell receptor peptide therapy triggers autoregulation of experimental encephalomyelitis. Science 251: 430–432

Offner H, Buenafe AC, et al (1993) Where, when, and how to detect biased expression of disease-relevant Vβ genes in rats with experimental autoimmune encephalomyelitis. J Immunol 151: 506–517

Oksenberg JR, Panzara MA, et al (1993) Selection for T-cell receptor Vβ-Dβ-Jβ gene rearrangements with specificity for a myelin basic protein peptide in brain lesions of multiple sclerosis. Nature 362: 68–70

Olsson T, Wei Zhi W, et al (1990) Autoreactive T lymphocytes in multiple sclerosis determined by antigen-induced secretion of interferon-γ. J Clin Invest 86: 981–985

Olsson T, Sun J, et al (1992) Increased numbers of T cells recognizing multiple myelin basic protein epitopes in multiple sclerosis. Eur J Immunol 22: 1083–1087

Ota K, Matsui M, et al (1990) T-cell recognition of an immunodominant myelin basic protein epitope in multiple sclerosis. Nature 346: 183–187

Paty DW, Li DKB, et al (1993) Interferon beta-1b is effective in relapsing-remitting multiple sclerosis. II. MRI analysis results of a multicenter, randomized, double-blind, placebo-controlled trial. Neurology 43: 662–667

Pette M, Fujita K, et al (1990a) Myelin basic protein-specific T lymphocyte lines from MS patients and healthy individuals. Neurology 40: 1770–1776

Pette M, Fujita K, et al (1990b) Myelin autoreactivity in multiple sclerosis: recognition of myelin basic protein in the context of HLA-DR2 products by T lymphocytes of multiple sclerosis patients and healthy donors. Proc Natl Acad Sci USA 87: 7968–7972

Pettinelli CB, McFarlin DE (1981) Adoptive transfer of experimental allergic encephalomyelitis in SJL/J mice after in vivo activation of lymph node cells by myelin basic protein: requirement for Lyt-1+2- T lymphocytes. J Immunol 127: 1420–1423

Prineas JW (1985) The neuropathology of multiple sclerosis. In: Vinken PJ, Bruyn GW, Klawans HL, Koetsier JC (eds) Demyelinating diseases. Elsevier, Amsterdam New York, pp 213–257 (Handbook of Clinical Neurology 3, 47)

Racke MK, Dhib-Jalbut S, et al (1991) Prevention and treatment of chronic relapsing experimental allergic encephalomyelitis by transforming growth factor-β1. J Immunol 146: 3012–3017

Racke MK, Martin R, et al (1992) Copolymer-1-induced inhibition of antigen-specific T cell activation: interference with antigen presentation. J Neuroimmunol 37: 75–84

Raine CS (1983) Multiple sclerosis and chronic relapsing EAE: comparative ultrastructural neuropathology. In: Hallpike JF, Adams CW, Tourtellotte WW (eds) Multiple sclerosis. Williams & Wilkins, Baltimore, pp 413–478

Raine CS, Scheinberg LC (1988) On the immunopathology of plaque development and repair in multiple sclerosis. J Neuroimmunol 20: 189–201

Raine CS, Cannella B, et al (1990) Homing to central nervous system vasculature by antigen-specific lymphocytes. II. Lymphocyte/endothelial cell adhesion during the initial stages of autoimmune demyelination. Lab Invest 63: 476–489

Richert JR, Driscoll BG, et al (1979) Adoptive transfer of experimental allergic encephalomyelitis: incubation of rat spleen cells with specific antigen. J Immunol 122: 494–496

Richert JR, Robinson ED, et al (1989) Human cytotoxic T-cell recognition of a synthetic peptide of myelin basic protein. Ann Neurol 26: 342–346

Rivers TM, Sprunt DH, et al (1933) Observations on attempts to produce acute disseminated encephalomyelitis in monkeys. J Exp Med 58: 39–53

Ruddle NH, Bergman CM, et al (1990) An antibody to lymphotoxin and tumor necrosis factor prevents transfer of experimental allergic encephalomyelitis. J Exp Med 172: 1193–1200

Sadovnick AD, Armstrong H, et al (1993) A population-based study of multiple sclerosis in twins: update. Ann Neurol 33: 281–285

Salvetti M, Ristori G, et al (1993) Predominant and stable T cell responses to regions of myelin basic protein can be detected in individual patients with multiple sclerosis. Eur J Immuol 23: 1232–1239

Schiffenbauer J, Johnson HM, et al (1993) Staphylococcal enterotoxins can reactivate experimental allergic encephalomyelitis. Proc Natl Acad Sci USA 90: 8543–8546

Schmied M, Breitschopf H, et al (1993) Apoptosis of T lymphocytes in experimental autoimmune encephalomyelitis: evidence for programmed cell death as a mechanism to control inflammation in the brain. Am J Pathol 143: 446–452

Sercarz EE, Lehmann PV, et al (1993) Dominance and crypticity of T cell antigenic determinants. Annu Rev Immunol 11: 729–766

Sharma SD, Nag B (1993) Antigen-specific therapy of experimental allergic encephalomyelitis by soluble class II major histocompatibility complex-peptide complexes. Proc Natl Acad Sci USA 88: 11465–11469

Sun D, Gold DP, et al (1992) Characterization of rat encephalitogenic T cells bearing non-Vβ8 T cell receptors. Eur J Immunol 22: 591–594

Teitelbaum D, Aharoni R, et al (1988) Specific inhibition of the T-cell response to myelin basic protein by the synthetic copolymer Cop-1. Proc Natl Acad Sci USA 85: 9724–9728

The IFN-β Multiple Sclerosis Study Group (1993) Interferon beta-1b is effective in relapsing-remitting multiple sclerosis. I. Clinical results of a multicenter, randomized, double-blind, placebo-controlled trial. Neurology 43: 655–661

Tiwari JL, Terasaki PI (1985) HLA and disease associations. Springer, New York, pp 152–167

Tourtellotte WW (1985) The cerebrospinal fluid in multiple sclerosis. In: Vinken PJ, Bruyn GW, Klawans HL, Koetsier JC (eds) Demyelinating diseases. Elsevier, Amsterdam New York, pp 79–130 (Handbook of Clinical Neurology 3, 47)

Tuohy VK, Sobel RA, et al (1988) Myelin proteolipid protein-induced experimental allergic encephalomyelitis. Variations of disease expression in different strains of mice. J Immunol 140: 1868–1873

Urban JL, Horvath SJ, et al (1989) Autoimmune T cells: immune recognition of normal and variant peptide epitopes and peptide-based therapy. Cell 59: 257–271

Utz U, Biddison WE, et al (1993) Skewed T cell receptor repertoire in genetically identical twins with multiple sclerosis correlates with disease. Nature 364: 243–247

Utz U, Brooks JA, et al (1994) Heterogeneity of T-cell receptor α-chain complementarity-determining region 3 in myelin basic protein-specific T cells increases with severity of multiple sclerosis. Proc Natl Acad Sci USA 91: 5567–5571

Vandenbark AA, Hashim G, et al (1989) Immunization with a synthetic T-cell receptor V-region peptide against experimental autoimmune encephalomyelitis. Nature 341: 541–544

Vogt AB, Kropshofer H, et al (1994) Ligand motifs of HLA-DRB5*0101 and DRB1*1501 molecules delineated from self-peptides. J Immunol 153: 1665–1673

Voskuhl RR, Martin R, et al (1993) T Helper 1 (TH1) functional phenotype of human myelin basic protein-specific T lymphocytes. Autoimmunity 1 5: 137–143

Waldor MK, Sriram S, et al (1985) Reversal of experimental allergic encephalomyelitis with a monoclonal antibody to a T cell subset marker (L3T4). Science 227: 415–417

Weber WEJ, Vandermeeren MMPP, et al (1989) Human myelin basic protein-specific cytolytic T lymphocyte clones are functionally restricted by HLA class II gene products. Cell Immunol 120: 145–153

Weiner HL, Mackin GA, et al (1993) Double-blind pilot trial of oral tolerization with myelin antigens in multiple sclerosis. Science 259: 1321–1324

Whitacre CC, Gienapp IE, et al (1991) Oral tolerance in experimental autoimmune encephalomyelitis. III. Evidence for clonal anergy. J Immunol 147: 2155–2163

Wraith DC, Smilek DE, et al (1989) Antigen recognition in autoimmune encephalomyelitis and the potential for peptide-mediated immunotherapy. Cell 59: 247–255

Wucherpfennig KW, Ota K, et al (1990) Shared human T cell receptor V beta usage to immunodominant regions of myelin basic protein. Science 248: 1016–1019

Wucherpfennig KW, Newcombe J, et al (1992) T cell receptor Vα-Vβ repertoire and cytokine gene expression in active multiple sclerosis lesions. J Exp Med 175: 993–1002

Wucherpfennig KW, Sette A, et al (1994) Structural requirements for binding of an immunodominant myelin basic protein peptide to DR2 isotypes and for its recognition by human T cell clones. J Exp Med 179: 279–290

Yednock TA, Cannon C, et al (1992) Prevention of experimental autoimmune encephalomyelitis by antibodies against α4β1 integrin. Nature 356: 63–66

Zamvil SS, Steinman L (1990) The T lymphocyte in experimental allergic encephalomyelitis. Annu Rev Immunol 8: 579–621

Author's address: Dr. R. Martin, Department of Neurology, University of Tübingen Medical School, Hoppe-Seyler Strasse 3, D-72076 Tübingen, Federal Republic of Germany.

Nigrostriatal neuronal death in Parkinson's disease – a passive or an active genetically-controlled process?

I. Ziv[1], **A. Barzilai**[2], **D. Offen**[1], **N. Nardi**[2], and **E. Melamed**[1]

[1]Department of Neurology and Felsenstein Research Institute,
Beilinson Medical Center, Petah-Tiqva, and
[2]Department of Biochemistry, The George Wise Faculty of Life Sciences,
Neuroscience, Tel-Aviv University, Tel-Aviv, Israel

Summary. The cause for the rather selective degeneration of the nigrostriatal dopaminergic (DA) neurons in Parkinson's disease (PD) is still enigmatic. The major current hypothesis suggests that nigral neuronal death in PD is due to excessive oxidant stress generated by auto- and enzymatic oxidation of DA, formation of neuromelanin and presence of high concentrations of iron. Such cell death is generally regarded as a passive, necrotic process, mainly resulting from membrane lipid peroxidation, leading to its dysfunction and rupture and then to neuronal disintegration.

We suggest a novel approach, that views neuronal degeneration in PD as an active process that occurs mainly the nuclear level. Our concept is based on the following observations: (1) Nigral histopathology in PD is characterized by a slow, protracted degeneration of individual neurons. We propose that it may be due to apoptosis [programmed cell-death (PCD), an active, genetically-controlled, intrinsic program of cell "suicide"] rather than to necrotic cell death. (2) DA exerts antitumor effect on melanoma and neuroblastoma cells. (3) Many anticancer drugs, trigger PCD by causing DNA damage. (4) DA has been shown to be genotoxic. (5) We recently first showed that DA, the endogenous neurotransmitter in the nigra, can trigger apoptosis in cultured, postmitotic sympathetic neurons. (6) We have also shown that PC-12 cells, transfected with the bcl-2 gene (a proto-oncogene that inhibits PCD) are relatively resistant to DA-apoptotic effect.

Degeneration of nigrostriatal neurons in PD may therefore be linked to dysregulation of the control mechanisms that normally restrain the PCD-triggering-potential of their own neurotransmitter.

Parkinson's disease (PD) is a progressive neurological disorder that results from a selective degeneration of the pigmented, dopaminergic (DA) neurons in the pars compacta of the substantia nigra (SNPC) (Hirsch, 1992). To date, the etiology of this specific neuronal death is still unknown. It is currently suggested that nigrostriatal neuronal degeneration is due to excessive oxidant stress, associated with dopamine metabolism (Olanow, 1993). Lipid peroxidation of cellular membranes, membrane dysfunction and rupture has been hypothesized as the main pathway leading to nigral cell death (Fahn and Cohen, 1992; Jenner et al., 1992). However, histopathology of SNPC in PD does not support such mode of cell death, i.e., by necrosis. Such process is relatively rapid, with cell swelling, uncontrolled spillage of cellular contents to extracellular space and a significant inflammatory reaction (Boobis et al., 1989). In contrast, nigral neuronal loss in PD is a slow, "silent" process, with a seemingly protracted dysfunction and loss of individual neurons.

Apoptosis (programmed-cell-death, PCD) is a recently-discovered unique mode of cell death. It is an active, genetically-controlled process of cell "suicide" or "self-sacrifice". When activated, apoptosis involves a characteristic chain of events: cells shrink and lose contacts with neighbouring cells, their chromatin condenses and fragments, and there is an extensive blebbing of cell membranes. Cells are ultimately degraded to membrane bound organelles (apoptotic bodies), containing nuclear particles, cellular organelles and residual cytoplasm. Apoptotic bodies are subsequently rapidly phagocytized by macrophages, thus avoiding inflammatory response. The hallmark of apoptosis is the early nuclear events in the death process, with early DNA fragmentation at internucleosomal sites due to endonuclease activation (Wyllie, 1980; Bursch et al., 1992).

This mode of cell death, first described by Wyllie in thymocytes, has also been shown to occur in neurons, including sympathetic neuronal cells, following deprivation of nerve growth factor (NGF) (Martin et al., 1988). It is considered of major importance in the development of the nervous system, serving as a major tool through which strict negative selection of "unfit" neuronal cells is performed, thus ensuring the high order of organization of neuronal networks (Barinaga, 1993). However, it is theoretically possible, that inappropriate, or inadvertent activation of this inherent death program, may be involved in the pathogenesis of degenerative neurological disorders of older age.

The endogenous neurotransmitter dopamine as an anticancer agent

Dopamine (DA), the endogenous neurotransmitter in the nigrostriatal neurons, in recent years has been the focus of interest in an entirely different field, i.e. melanoma research. Wick (1978, 1980), in his effort to develop novel antimelanoma therapies, found that DA, l-dopa and their analogues can exert a substantial anticancer effect, and are toxic to melanoma cells both in vitro and in vivo. These agents were found to be synergistic in their toxic effects to irradiation therapy. The antitumor effect of DA has been attributed to its genotoxicity (Graham et al., 1978; Moldeus et al., 1983). DA, probably mainly through products of its oxidative metabolism, has been shown to cause substantial DNA damage manifested mainly by strand breaks and base modifications. In addition, Wick (1989) also found that DA genotoxicity is further augmented by its inhibitory activity on several enzymatic pathways operative in the DNA repair process, such as DNA polymerase, ribunucleotide reduct-ase and thymidylate synthase. Thus, DA can both cause DNA damage and interfere with its repair. The observations on DA toxicity were extended to other cellular systems such as neuroblastoma (Graham et al., 1978), and also to postmitotic cultured cortical and mesencephalic neurons (Michel and Hefti, 1990).

DNA damage, trigger of apoptosis, and anticancer activity

In recent years, a number of anti-cancer drugs (e.g. cysplatin, methotrexate) have been shown to induce their anti-tumor effect through a DNA-damaging mechanism, ultimately leading to triggering of PCD. Eastman (1990) has shown this phenomenon with more than ten anti-cancer drugs and concluded that this sequence of events might be the common denominator of most anti-cancer compounds. Still unclear, however, is the signalling mechanism by which DNA damage induces apoptosis. It is suggested that the DNA repair process, with its enormous nucleotide consumption, initiates this process (Bursch et al., 1992). Increases in free calcium levels (Orrenius et al., 1992), and activation of poly (ADP-ribose) polymerase (Schraufstatter et al., 1986) also seem to play a vital role in the induction of the apoptotic process. The peculiar anti-cancer activity of the neurotransmitter DA, may therefore result from its capability to initiate PCD, in similarity to other antitumor chemother-apeutic agents. Moreover, it is possible that this endogenous neurotransmitter, if not properly "restrained", may induce inappropriate, or "inadvertent" activation of PCD in the nigrostriatal neurons, leading to degeneration of this neuronal population and evolvement of PD.

Dopamine induces apoptosis in cultured chick sympathetic neurons

We recently were the first to show that DA is capable of inducing apoptosis in cultured, postmitotic sympathetic neurons (Ziv et al., 1994). Paravertebtral sympathetic ganglia of chick embryos at embryonic day 9 (E9) were dissected out and cells were dissociated by trypsinization. Cells were then grown for four days in serum free medium with nerve growth factor (NGF), which is essential for survival of this neuronal system. Fluorodeoxyuridine (FDU) and uridine were used to kill nonneuronal dividing cells. On the fourth day in culture, cells were subjected to 24 hour exposure to one of the following treatments: (1) Control: No further treatment. (2) DA, in a concentration range of 0.1–1 mM. This range was chosen because it is the estimated level range of this neurotransmitter within cell bodies of nigrostriatal neurons (Michel and Hefti, 1990). (3) DA + dithiothreitol (DTT); to test the effect of antioxidant therapy. (4) DA + actinomycin-D or cycloheximide, to test the role of de-novo protein synthesis in the dopamine effect. (5) NGF-depriva-tion, by addition of anti-NGF antibody, was used for comparison as a model of apoptosis in sympathetic neurons.

Study parameters included: (1) Assessment of cell viability by the MTT [3-(4,5-dimethylthiazol-2-yl)-2,5-diphenyl-tetrazolium bromide] test, measuring extent of mitochondrial MTT conversion to formazan. (2) Morphological studies, by Nomarski light- and scanning-electron microscopy. (3) Nuclear alterations were evaluated by flow-cytometric (FACS) analysis of purified, propidium-iodide-stained cell nuclei. Flow-cytometric analysis was focused on the two main features of apoptosis: One is an increase in nuclear granularity, reflected by an increase in the side-scatter value. The other is an orderly nuclear fragmentation with creation of a distinct population of nuclear particles with reduced DNA content, measurable as a distinct subdiploid, apoptotic peak on the FL_2 scale.

The MTT test revealed a marked, dose-dependent toxicity of DA, characterized by an LD^{50} (lethal dose for 50% of cells, exposure for 24 hours) of 0.1 mM, i.e., in vitro concentrations that are well within the estimated physiological range of DA within nigrostriatal neuronal cell bodies. Based on this curve, exposure to 0.3 mM of DA for 24 hours was chosen for further characterization of the death process.

The features of DA-induced death process were highly characteristic of apoptosis, in all three different methods employed. Moreover, this process was similar to the effect of NGF-deprivation that served as a model of neuronal PCD.

Light microscopy morphological studies

Cells exposed to DA manifested severe condensation and shrinkage, along with extensive thinning and multifocal disruption of the neuritic network (Fig. 1).

Scanning-electron-microscopic (SEM) studies

SEM studies revealed, in addition to the above alterations extensive formation of membranal blebs, with cells practically turning to clusters of membrane-bound blebs or apoptotic bodies, seemingly ready to be engulfed by macrophages (if these were present in the culture dish).

These morphological alterations are considered as the hallmarks of apoptosis, and were identical to the changes observed following NGF-deprivation. Morphological evidence for DA-induced apoptosis were further substantiated by the FACS analysis of purified, neuronal nuclei.

FACS studies

Following exposure to DA, a marked, well-defined hypodiploid apoptotic peak was observed on the FL_2 scale, reflecting a large (80% of events) population of apoptotic nuclei, with reduced DNA content. A significant increase was also observed in the side-scatter (SC) value, reflecting increased

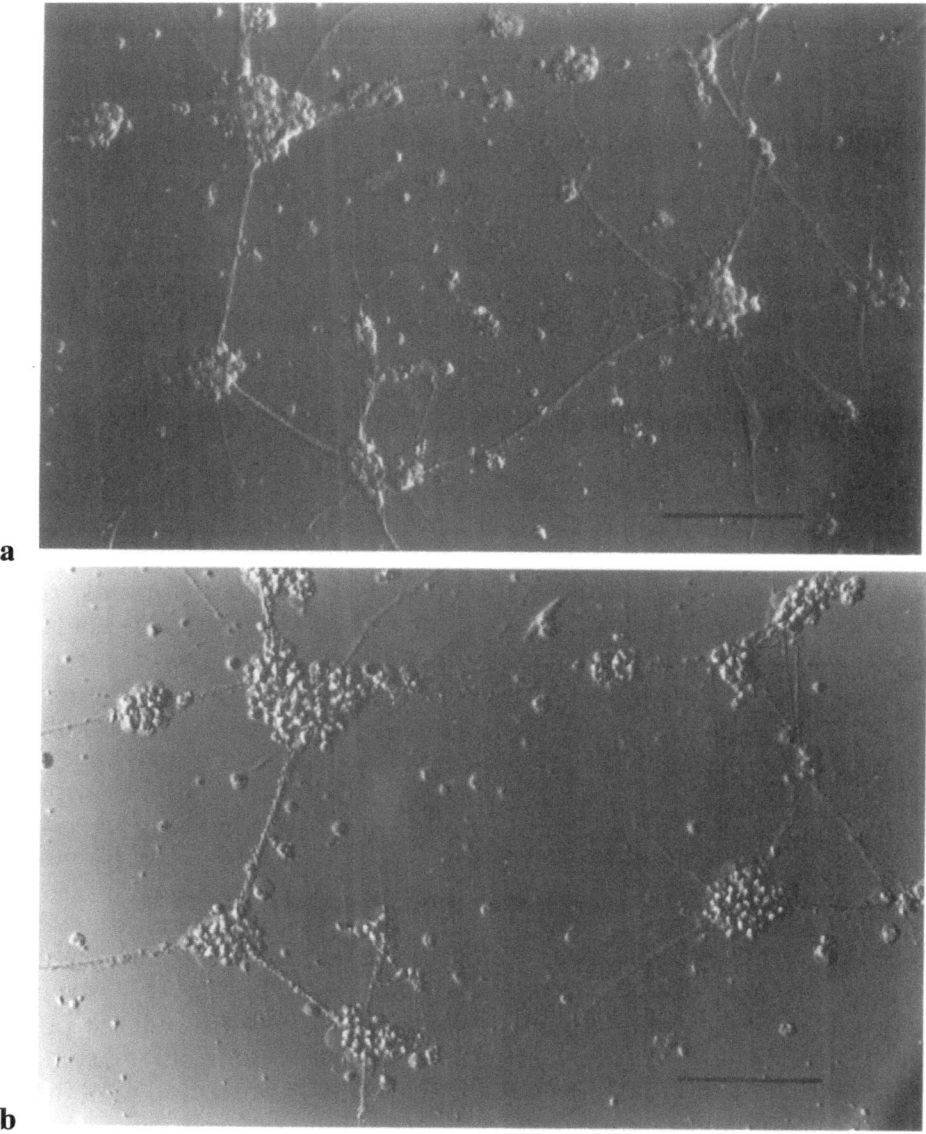

Fig. 1. Dopamine (DA) effect on neuronal morphology: Nomarski optic morphology of cultured chick sympathetic neurons. **a** Before exposure: well developped neuronal network, bright cell bodies. **b** Same cells following exposure to DA (0.3 mM, 24 hours): extensive neurite thinning and disruption, severe shrinkage and condensation of cell bodies (× 300, bar = 100 nm)

nuclear granularity. When SSC was plotted against FL_2, DA effect could be observed in two main nuclear populations: One was the development of a large population of nuclei with increased granularity and reduced DNA content, apoptotic nuclei. The second was a small population with increased side-scatter but normal FL_2, probably representing nuclei in the beginning of the apoptotic process, already with increased granularity, but not yet succumbing to fragmentation. As previously described for neuronal morphology, nuclear effects of DA were also identical to the effect of NGF deprivation including both the apoptotic peak and the characteristic alterations in nuclear populations.

The DA-induced death process could be prevented by concomitant treatment with the antioxidant DTT, whereas inhibition of protein synthesis did not have any effect. DA can therefore induce apoptosis in cultured, postmitotic sympathetic neurons, an effect that is probably mediated by products of its oxidative metabolism, and [similar to other models (Ueda and Shah, 1992; Cohen, 1993)] does not require de novo protein synthesis for its execution. Based on the above considerations, it is possible that DA-induced DNA damage might have a role in the activation of the death program by this endogenous neurotransmitter.

Possible implications for the pathogenesis of Parkinson's disease

We may assume that potent in vivo protective mechanisms normally restrain the apoptotic potential of DA, thus enabling humans to utilize this neurotransmitter without developing PD. Some of these mechanisms may act to prevent DA genotoxicity, e.g., DA-vesiculation and presence of intra- and extra-nuclear antioxidant natural protective mechanisms. Currently less explored is the competence of DNA-repair apparatus in PD. It is noteworthy that there is some evidence associating impairment of DNA-repair with PD, as implied by enhanced susceptibility of fibroblasts from these patients to the DNA-damaging effects of x-ray irradiation and N-methyl-n nitrosoguanosine (Scudiero et al., 1982; Robbins et al., 1985). Growth factors (Tooyama et al., 1993; Lin et al., 1993) and several proto-oncogenes such as bcl-2 (Garcia et al., 1992) are other recently-discovered control systems of apoptosis. It is conceivable that normally, the interaction between these systems versus the PCD-triggering effect of DA and/or oxidative metabolites acts to inhibit the death program and promotes nigrostriatal neuronal survival. We recently first exemplified such interaction, by demonstrating that PC-12 cells, transfected with the bcl-2 gene, manifest enhanced resistance to the toxic effect of DA, as compared to controls (Ziv et al., submitted for publication). However, it is also possible, that in patients with PD there is dysregulation, either inherited or acquired, in any of these restraints of the apoptotic process in nigral neurons, culminating in inappropriate activation of programmed-cell-death in these cells and neuronal degeneration.

This line of thought raises two intriguing questions. The first is the long-debated question of the effect of levodopa therapy on survival of the remaining nigrostriatal neurons in PD. If indeed there is a defect in PD patients in the control of apoptosis, than levodopa therapy, beyond providing symptomatic relief for the motor disability, might actually hasten the neuronal degenerative process.

The other question that we are currently trying to address is whether and to what relative extent can other monoaminergic neurotransmitters [noradrenaline (NA), serotonin (5-HT)] induce apoptosis. Our preliminary results

(Ziv et al., in preparation) show that such potential does exist, with the relative potency of DA > NA > 5-HT. This grading correlates well with the relative involvement of dopaminergic/noradrenergic/serotoninergic pathways in PD, and thus may further support the concept of a common, multisystem underlying abnormality in apoptosis-control mechanisms in these patients.

Our studies therefore suggest, that the nigrostriatal neuronal cell nucleus and associated control systems of PCD may be the major scene of events in the pathogenesis of Parkinson's disease, rather than the cellular membranes. Research in this field may yield new insights on the pathogenesis of PD and development of novel and more specific therapeutic approaches.

Acknowledgements

Supported, in part, by the National Parkinson Foundation, Miami, Florida, U.S.A., Teva Pharmaceutical Industries, Ltd. and the National Institute of Psychobiology, Israel.

References

Barinaga M (1993) Death gives birth to the nervous system. But how? Science 259: 762–763

Boobis AR, Fawthrop DJ, Davies DS (1989) Mechanisms of cell death. Trends Pharmacol Sci 10: 275–280

Bursch W, Oberhammer F, Sculte-Hermann R (1992) Cell death by apoptosis and its protective role against disease. Trends Pharmacol Sci 13: 245–251

Cohen JJ (1993) Apoptosis. Immunol Today 14: 126–130

Eastman A (1990) Activation of programmed cell death by anticancer agents: cisplatin as a model system. Cancer Cells 2: 275–280

Fahn S, Cohen G (1992) The oxidant stress hypothesis in Parkinson's disease: evidence supporting it. Ann Neurol 32: 804–812

Garcia I, Martinou I, Tsushimoto Y, Martinou JC (1992) Prevention of programmed cell death of sympathetic neurons by the bcl-2 proto-oncogene. Science 258: 302–304

Graham DJ, Tiffany SM, Bell WR, Gutknecht WF (1978) Autoxidation versus covalent binding of quinones as the mechanism of toxicity of dopamine, 6-hydroxydopamine, and related compounds toward c1300 neuroblastoma cells in vitro. Mol Pharmacol 14: 644–653

Hirsch EC (1992) Why are nigral catecholaminergic neurons more vulnerable than other cells in Parkinson's disease? Ann Neurol 32: S88–S93

Jenner P, Dexter DT, Sian J (1992) Oxidative stress as a cause of nigral cell death in Parkinson's disease and incidental Lewy body disease. Ann Neurol 32: S82–S87

Lin LFH, Doherty DH, Lile JD (1993) GDNF: a glial cell line-derived neurotrophic factor for midbrain dopaminergic neurons. Science 260: 1131–1132

Martin DP, Schmidt RE, Distefano PS, Johnson EN (1988) Inhibitors of protein synthesis and RNA synthesis prevent neuronal death caused by NGF deprivation. J Cell Biol 106: 829–843

Michel PP, Hefti F (1990) Toxicity of 6-hydroxydopamine and dopamine for dopaminergic neurons in culture. J Neurosci Res 26: 428–435

Moldeus P, Nordenskjold M, Bolcsfoldi G (1983) Genetic toxicity of dopamine. Mut Res 124: 9–23

Olanow CW (1993) A radical hypothesis for neurodegeneration. Trends Neurosci 16: 439–444

Orrenius S, Burkitt M, Kass GEN (1992) Calcium ions and oxidative cell injury. Ann Neurol 32: S33–S42

Robbins Jh, Otsuka F, Nee LE (1985) Parkinson's disease and Alzheimer's disease: hypersensitivity to x-rays in cultured cell lines. J Neurol Neurosurg Psychiatry 48: 916–923

Schraufstatter IU, Hyslop PA, Hinshaw DB (1986) Hydrogen peroxide-induced injury of cells and its prevention by inhibitors of poly (ADP-ribose) polymerase. Proc Natl Acad Sci 83: 4908–4912

Scudiero DA, Tarone RE, Robbins JH (1982) Parkinson's disease and Alzheimer's disease fibroblasts are hypersensitive to killing by MNNG. Clin Res 30: 857(A)

Tooyama I, Kawamata T, Walker D, Mcgeer PL (1993) Loss of basic fibroblast growth factor in substantia nigra neurons in Parkinson's disease. Neurology 43: 372–376

Ueda N, Shah SV (1992) Endonuclease-induced DNA damage and cell death in oxidant injury to renal tubular epithelial cells. J Clin Invest 90: 2593–2597

Wick MM (1978) Dopamine: a novel antitumor agent active against B-16 melanoma in vivo. J Invest Dermatol 71: 163–164

Wick MM (1980) Levodopa and dopamine analogs as DNA polymerase inhibitors and antitumor agents in humsn melanoma. Cancer Res 40: 1414–1418

Wick MM (1989) Levodopa/dopamine analogs as inhibitors of DNA synthesis in human melanoma cells. J Invest Dermatol 92: 329S–331S

Wyllie AH (1980) Glucocorticocoid-induced thymocyte apoptosis is associated with endogenous endonuclease activation. Nature 284: 555–557

Ziv I, Melamed E, Nardi N, Luria D, Achiron A, Offen D, Barzilai A (1994) Dopamine induces apoptosis-like cell death in cultured sympathetic neurons – a possible novel pathogenetic mechanism in Parkinson's disease. Neurosci Lett 170: 136–140

Authors' address: E. Melamed, M.D., Department of Neurology, Beilinson Medical Center, 49100 Petah-Tiqva, Israel.

Animal model and in vitro studies of anti neurofilament antibodies mediated neurodegeneration in Alzheimer's disease

L. Oron[1], **V. Dubovik**[1], **L. Novitsky**[1], **D. Eilam**[2], and **D. M. Michaelson**[1]

Departments of [1]Neurobiochemistry and [2]Zoology,
Tel Aviv University, Ramat Aviv, Israel

Summary. Alzheimer's disease (AD) is associated with serum antibodies directed specifically against phosphorylated epitopes highly enriched in the heavy neurofilament protein NF-H of cholinergic neurons. Prolonged immunization of rats with these molecules but not with other NF-H isoforms results in cognitive impairments. This animal model, termed experimental autoimmune dementia (EAD), supports a role for such antibodies in neurodegeneration in AD.

In the present study we investigated the cellular and immunological mechanisms underlying the cognitive defects in EAD. Immunohistochemical studies revealed that IgG accumulate in the septum, hippocampus and in the entorhinal cortex of the EAD rats. This is accompanied by a marked reduction in the density of septal cholinergic neurons. An inverse correlation was observed between the level of IgG in the septum of individual EAD rats and the density of their septal cholinergic neurons. Time course studies revealed that the decrease in the density of cholinergic neurons in the septum of EAD rats and the accumulation of IgG in this brain area have the same time course and are both significant by three to four months following the initiation of immunization with cholinergic NF-H. The cognitive deficits of the EAD rats evolve more slowly and are pronounced only after six months following the initation of immunization. In vitro studies revealed that anti NF-H IgG bind to the outer surface of neurons in tissue cultures of rat forebrain and can affect neuronal viability. These AD and in vitro findings provide model systems for studying the mechanisms underlying the neuropathological effects of specific anti NF-H antibodies.

Introduction

Previous studies suggest that different neurons contain distinctly phosphorylated isoforms of the heavy neurofilament protein NF-H and that neurodegeneration in Alzheimer's disease (AD) and in Down's syndrome is associated with serum antibodies (IgG) directed specifically against phosphorylated epitopes highly enriched in the NF-H protein of mammalian and *Torpedo* cholinergic neurons (Chapman et al., 1989; Hassin-Baer et al., 1992; Tcherna-kov et al., 1992; Soussan et al., 1994).

Although anti NF-H antibodies (Abs) may be a secondary phenomenon resulting from an initial non-immunological insult on cholinergic neurons, they could play a role in furthering the degeneration of these neurons. In order to investigate this possibility, we recently developed an animal model. This model, termed Experimental Autoimmune Dementia (EAD), revealed that prolonged immunization of rats with cholinergic NF-H results in specific memory dysfunctions (Chapman et al., 1991). These observations support a role for such Abs in neurodegeneration in AD. In the present study we investigate, by means of in vivo and in vitro experiments, the cellular and immunological mechanisms underlying the cognitive deficits in EAD.

I. IgG mediated loss of forebrain cholinergic neurons in EAD

Forebrain cholinergic neurons were visualized immunohistochemically utiliz-ing both a mAb to the low affinity nerve growth factor receptor (Korsching and Thoenen, 1983) and an anti choline acetyltransferase mAb. Incubation of forebrain sections of control rat brain with the anti low affinity nerve growth factor receptor mAb resulted in specific staining of the medial septal and the diagonal band (Fig. 1A). Examination of the level of staining in a similar brain section of an EAD rat revealed that it was markedly lower than that of the sham injected control and that the decrease was most pronounced in the medial septum and vertical limb of the diagonal band (MS+DBv) areas (Fig. 1B). In contrast, brain sections of rats which were immunized with chemically heterogeneous NF-H (NFHC rats) stained as intensely as those of control rats (Fig. 1A, C). Examination of the sections at higher magnification revealed that staining in the three rat brains was localized to perikarya and neurites whose density was lower in the EAD brain than in those of the two controls. Quantitative morphometric analysis revealed that whereas the den-sity of MS+DBv cholinergic neurons of the control and NFHC groups were similar (respectively 34.0 ± 1.2 and 33.4 ± 2.3 neurons/mm^2; mean $\pm$ SEM; $n = 4$) it was over 20% lower in the EAD rats (26.3 ± 2.4 neurons/mm^2) ($p < 0.05$). The density of cholinergic neurons of the horizontal limb of the diagonal band was also lower than those of the other groups except that in this brain area the effect was less pronounced. Similar results were obtained when the forebrain cholinergic neurons were visualized with a mAb directed against choline acetyltransferase. This cholinergic marker revealed a density of re-spectively 40.7 ± 4.3 and 38.1 ± 0.9 cholinergic neurons/mm^2 in the MS+DBv of the sham injected and NFHC rats (mean $\pm$ SEM; $n = 4$) and about a 20% lower density in the MS+DBv of the EAD rats (33.2 ± 1.7 neurons/mm^2) ($p < 0.05$). Time course studies revealed that the specific decrease in the density of MS+DBv cholinergic neurons of the EAD rats was detectable by three to four months following the initiation of immunization, and was more pronounced by six to eight months.

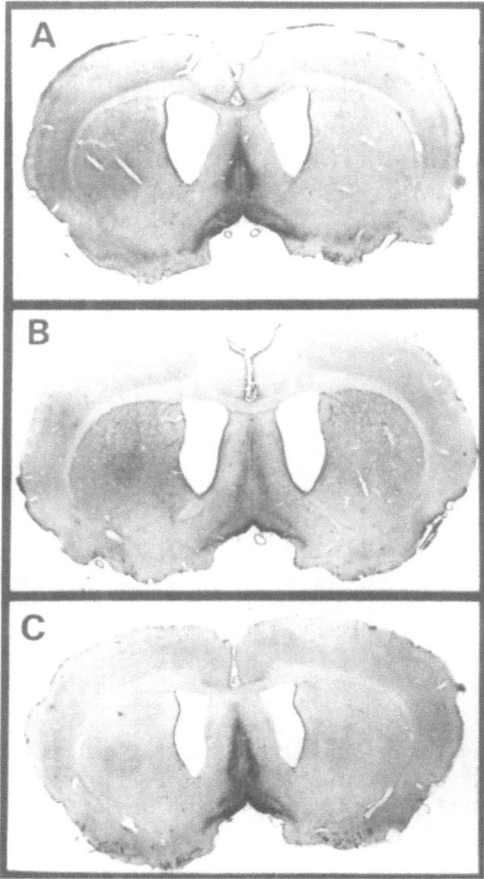

Fig. 1. Staining of forebrain cholinergic neurons by anti low affinity nerve growth factor receptor mAb. Coronal sections (40 µM) of a sham injected control rat (**A**), an EAD rat (**B**) and an NFHC rat (**C**) which were immunized for six months with respectively cholinergic and chemically heterogeneous NF-H were stained in parallel as previously described (Dubovik et al., 1993). Reprinted by permission from Neuroscience

Immunization of EAD rats with cholinergic NF-H for six months resulted in the appearance of IgG containing neurons in the MS+DBv (Fig. 2). IgG containing neurons were also observed in neurons in the horizontal limb of the diagonal band except that in this area their density was lower than in the MS+DBv (not shown). Quantitation of this effect and examination of its specificity revealed that the density of IgG containing neurons in the MS+DBv of EAD rats (43.8 ± 6.1; mean ± SEM; n = 4) was twofold higher than that of the NFHC rats (21.35 ± 5.4; n = 4) and that no such neurons were detectable in the sham injected controls. Similar results were observed when the sections were stained with biotinylated anti rat Fc and anti rat Fab Abs, suggesting that the immunohistochemical staining is indeed due to the presence of IgG and not to cross-reactivity with other molecules. Time course studies revealed that IgG containing neurons are detectable in the MS+DBv of EAD rats by three months following the initiation of immunization.

The possibility that the decreased density of cholinergic neurons in the MS+DBv of EAD rats is related to the accumulation of IgG in this brain area was investigated. Figure 3 depicts the relationship obtained for four EAD and four NFHC rats which were immunized for six months. As can be seen, a clear inverse relationship exists between levels of IgG containing neurons in the MS+DBv and the densities of cholinergic neurons in this brain area. The data is best fit by a curve which extrapolates at zero IgG containing neurons, to the

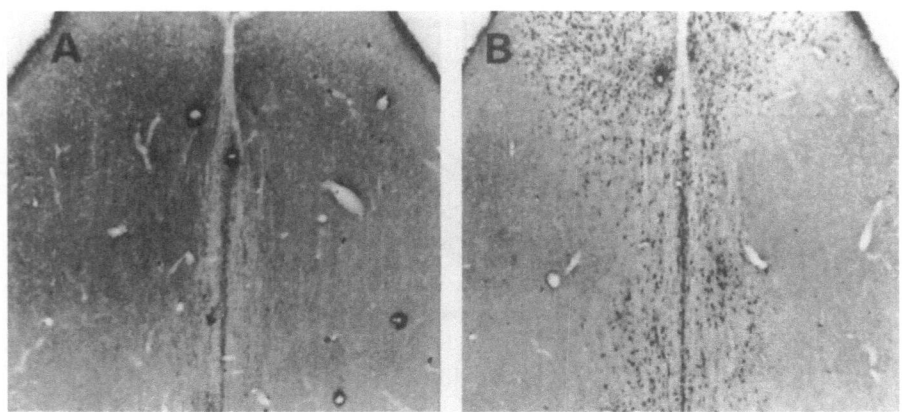

Fig. 2. IgG in neurons in the medial septum and the vertical limb of the diagonal band of an EAD rat immunized with cholinergic NF-H for six months (**B**) and of a sham injected control rat (**A**). Sections were stained with biotinylated anti rat IgG. Original light microscopy magnification × 40

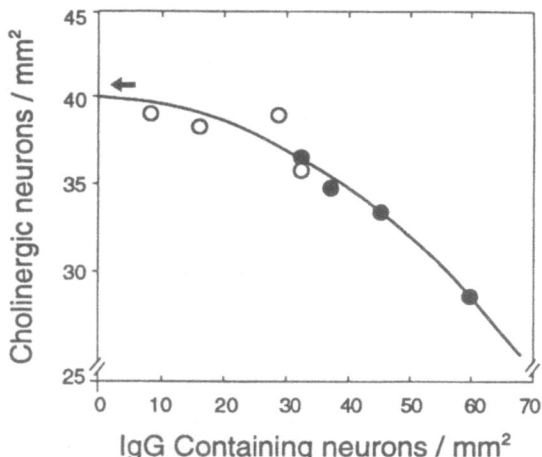

Fig. 3. Comparison of the densities of cholinergic neurons and of IgG containing neurons in the medial septum and vertical limb of the diagonal band of four EAD rats (●) and four NFHC control rats (○). The arrow indicates the average level of cholinergic neurons of sham injected controls. Alternate secitons were stained immunohistochemically with either an anti choline acetyltransferase mAb or with biotinylated anti rat IgG utilizing a Vecstatin ABC kit (Dubovik et al., 1993). Results presented are the average of ten fields obtained from at least three sections of each brain

average density of MS+DBv cholinergic neurons of the controls (see arrow in Fig. 3). This suggests that MS+DBv cholinergic neurons are affected by high levels of IgG but that low levels of IgG, such as those found in the NFHC rats, have no detectable effect. It should be noted that the density of IgG containing neurons in the MS+DBv is somewhat larger than that of the cholinergic neurons (Fig. 3), and that IgG may have thus also accumulated in non cholinergic neurons.

Two possible mechanisms may be considered regarding the entry of IgG into the brain. They are leakage from the blood into the cerebrospinal fluid and intraneuronal space, or endocytosis by neurons Projecting outside the blood brain barrier. Previous studies indicate that the cerebrospinal fluid of EAD rats contains anti NF-H IgG (Michaelson et al., 1991). However, examination of posterior brain areas revealed that IgG also accumulate specifically in the hippocampus and in the entorhinal cortex of EAD rats (not shown). These brain areas and the septum are connected by major pathways. Furthermore trajectories exist from the hypothalamus, where the blood brain barrier is leaky, to the entorhinal cortex (Paxinos, 1985). Additional experiments, including time course studies, are necessary for elucidating whether IgG accumulates in the brain by leakage through the blood brain barrier or via specific neuronal pathways.

II. Cognitive dysfunction in EAD rats

Previous studies utilizing T-maze alternation paradigms revealed that the short term memory of EAD rats is impaired (Chapman et al., 1991) and that this impairment is reversed by the acetylcholinesterase inhibitor physostigmine (Michaelson et al., 1991). Dementia is a multidimensional process which in addition to memory loss involves severe deterioration in the spatio-temporal organization of behavior. In order to examine whether such changes occur in the EAD model, we analyzed the organization of their behavior in an open field. The experiment was performed six months after the initation of immunization as previously described (Eilam et al., 1990). Analysis of the open field behavior of freely moving EAD rats revealed that the spatio-temporal organization of their behavior differed dramatically from those of both the sham injected and the NFHC controls rats (Fig. 4). While rats of the two control groups explored about 80% of the open field, the EAD rats explored only 25% of the field. Measurements of the frequency and duration of stops revealed that each rat of the three groups had a location where it spent most of its stopping time and where it performed the highest frequency of the grooming and rearing behaviors characteristic of a "home base". The EAD rats however displayed a profound change in another behavioral measure of the home base – the number of visits. Whereas rats of the two control groups performed numerous round trips which started at the home base and covered most of the open field, the EAD rats performed short sequences of stops of varying duration and rarely returned to the home base (Eilam et al., 1993). These behavioral derangements and the short term memory deficits of the

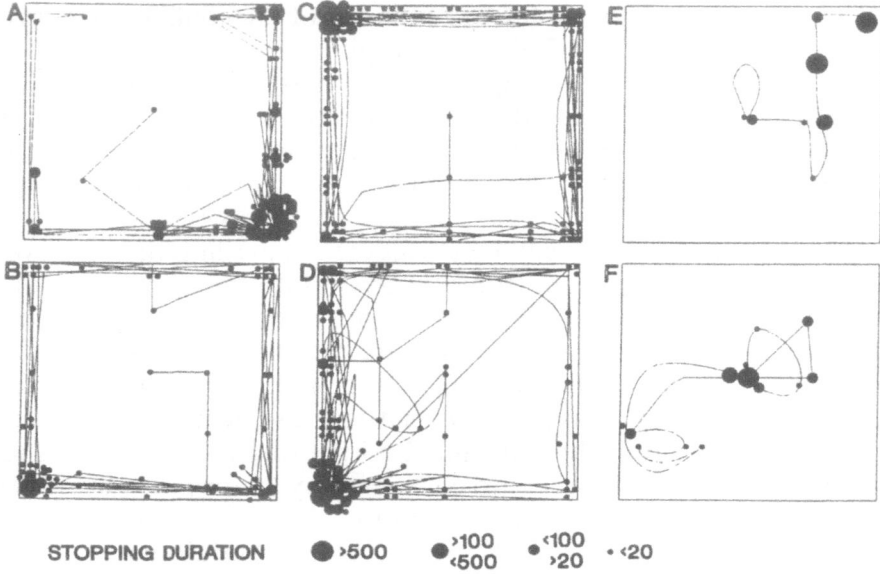

Fig. 4. Representative trajectories of locomotion in the open field of two control rats (**A**, **B**), two NFHC rats (**C, D**) and two EAD rats (**E, F**). The square enclosures correspond to the open field. Each circle corresponds to a single stop and its diameter represents the duration of the stop (as indicated in the figure). Lines represent the path of locomotion between successive steps. Reprinted by permission from Neurosciences

EAD rats evolve more slowly than the immunohistochemical changes in their brains and are significant only after more than six months of immunization with cholinergic NF-H (Chapman et al., 1989; Dubovik et al., 1993). Taken together, the morphological and behavioral data suggest that the derangement of EAD brain cholinergic neurons is due to the accumulation of anti NF-H IgG in the forebrain, and that it plays a role in the observed cognitive deficits.

III. Tissue culture studies of the anti neuronal effect of anti NF-H antibodies

In this section we describe results of in vitro studies aimed at examining whether the pathological effects observed with EAD rats can be reproduced and investigated at the cellular level. This was explored by studying the effects of anti NF-H antisera on rat forebrain embryonic cultures enriched in cholinergic neurons. Incubation of the cultures at 4°C with rabbit anti sera prepared against the NF-H protein of bovine ventral root nerves resulted in binding of IgG to the cultures. The extent of this binding was markedly higher than that of the preimmune serum and it was half maximal at a dilution of 1:1000. Conditions which disrupt antigen-antibody interactions but do not permeabilize the cells (i.e. pH 2.5 for 5 min at 4°C) (Pelchen-Matthews et al., 1989) virtually liberated all the bound anti NF-H IgG (> 90%). Further experiments revealed that this treatment also released all the bound anti NF-H IgG from

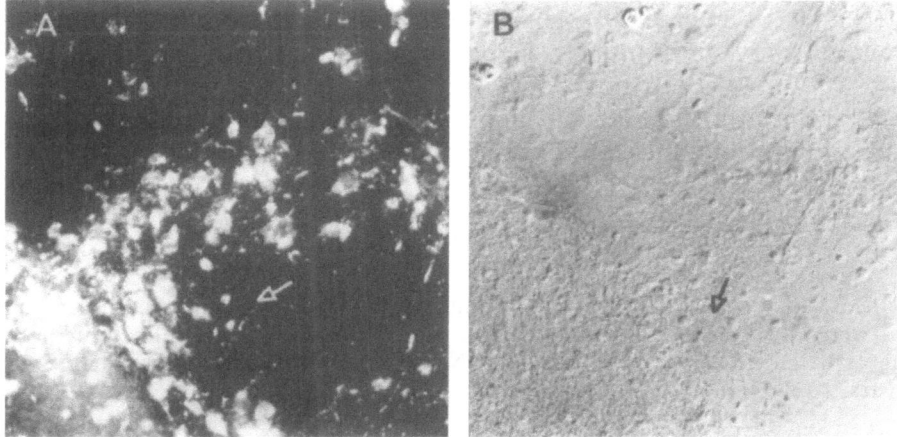

Fig. 5. Immunofluorescence (**A**) and Nomarsky optics (**B**) views of anti NF-H IgG bound to a primary culture of rat forebrain neurons. Binding of anti NF-H anti sera was performed as described in the text and the bound IgG was visualized utilizing a second antibody tagged with the fluorescent dye YC-3

cultures which were incubated at 37°C for up to 30 min. These results show that the anti NF-H IgG were not internalized even though the cultures were incubated at 37°C.

Immunofluorescence microscopy was employed to confirm the above binding data and to examine whether anti NF-H IgG bind similarly to all the neurons. This was performed by incubating the cultures at 4°C with either preimmune or anti NF-H antisera and by visualizing the bound Abs with anti rabbit IgG second Abs tagged with the fluorescent dye CY-3. As can be seen in Fig. 5A, anti NF-H IgG lit up both perikarya and neurites. Comparison of fields of fluorescently labelled neurons to those of their contours, as seen by Nomarsky optics (Fig. 5A and B), revealed that the labeling was selective and that only a subpopulation of the neurons were labeled. An example of such neurons is indicated by the arrows in Fig. 5. Further support for the assertion that anti NF-H IgG bind to the neuronal outer surface is provided by a pilot study which suggests that incubation of the cultures with anti NF-H antisera in the presence of complement results in marked deterioration of the neurons. Previous studies utilizing a neuroblastoma cell line have shown that an anti NF-H mAb cross reacts with a 62–68 kD neuronal surface protein (Sadiq et al., 1991). It is tempting to suggest that rat primary cultures contain analogous surface proteins which mediate the interactions of the anti NF-H Abs with the neuronal exterior.

Conclusions

The present EAD animal model studies show that anti NF-H IgG similar to those of AD patients induce specific neuronal and behavioral derangements. These deficits may replicate pathogenic processes in AD and support a role for such Abs in neuronal degeneration in the disease. The tissue culture experiments show that anti NF-H IgG can interact with the neuronal outer surface. They provide a novel in vitro system for studying the interactions between anti neurofilament Abs and intact neurons and for investigating the pathological consequences and specificity of these interactions.

Acknowledgements

This work was supported in part by grants to DMM from the Joseph K. and Inez Eichenbaum Foundation Limited and from the Simon Revah-Kabelli Fund, by grants from the Israeli Ministry of Absorption to LN and from the Branco Weiss Fund to VD. We thank Mrs. A. Cohen for her editorial assistance.

References

Chapman J, Bachar O, Korczyn AD, Wertman E, Michaelson DM (1989) Alzheimer's disease antibodies bind speclfically to a neurofilament protein in Torpedo cholinergic neurons. J Neurosci 9: 2710–2717

Chapman J, Alroy G, Weiss Z, Faigon M, Feldon J, Michaelson DM (1991) Anti neuronal antibodies similar to those found in Alzheimer's disease induce memory dysfunction in rats. Neurosci 40: 297–305

Dubovik V, Faigon M, Feldon J, Michaelson DM (1993) Decreased density of forebrain cholinergic neurons in experimental autoimmune dementia. Neuroscience 56: 75–82

Eilam D, Golani I (1990) Home base behavior of tame wild rats (Rattus norvegicus) injected with amphetamine. Behav Brain Res 36: 161–170

Eilam D, Szechtman H, Faigon M, Dubovik V, Feldon J, Michaelson DM (1993) Disintegration of the spatial organization of behavior in experimental autoimmune dementia. Neuroscience 56: 83–91

Hassin-Baer S, Wertman E, Raphael M, Stark V, Chapman J, Michaelson DM (1992) Antibodies from Down syndrome patients bind to the same cholinergic neurofilament protein recognized by Alzheimer's disease antibodies. Neurology 42: 551–555

Korsching S, Thoenen H (1983) Nerve growth factor in sympathetic ganglia and corresponding target organs of the rat: correlation with density of sympathetic innervation. Proc Natl Acad Sci USA 80: 3513–3516

Michaelson DM, Alroy G, Soussan L, Chapman J, Feldon J (1991) Experimental autoimmune dementia (EAD): an immunological model of memory dysfunction and Alzheimer's disease. In: Giacobini E, Becker RE (eds) Pharmacological basis of cholinergic therapy in Alzheimer's disease. Birkhäuser, Boston, pp 126–133

Paxinos G (1985) The rat nervous system. Academic Press, Sydney, p 535

Pelchen-Matthews A, Armes JE, Marsh M (1989) Internalization and recycling of CD4 transfected into Hela and NIH3T3 cells. EMBO J 8: 3641–3649

Sadiq SA, Van-den Berg LH, Thomas FP, Kilidivias K, Hays AP, Latov N (1991) Human monoclonal antineurofilament antibody cross-reacts with a neuronal surface protein. J Neurosci Res 29: 319–325

Soussan L, Barzilai A, Michaelson DM (1994) Distinctly phosphorylated neurofilaments in different classes of neurons. J Neurochem 62: 770–776

Tchernakov K, Soussan L, Hassin-Baer S, Wertman E, Michaelson DM (1992) Alzheimer's disease and Down's syndrome antibodies bind to the heavy neurofilament protein of cholinergic neurons. Res Immunol 6: 583–588

Authors' address: Dr. D. M. Michaelson, Department of Neurobiochemistry, Tel Aviv University, Ramat Aviv 69978, Israel.

Cop 1 as a candidate drug for multiple sclerosis

D. Teitelbaum, R. Arnon, and **M. Sela**

Department of Immunology, Weizmann Institute of Sciene, Rehovot, Israel

Summary. Copolymer 1 (Cop 1), a synthetic copolymer of amino acids, is very effective in suppression of experimental autoimmune encephalomyelitis (EAE), the animal model for multiple sclerosis (MS). Cop 1 was found incapable of inducing EAE, yet it suppressed EAE in a variety of animal species, including primates. The immunological cross-reaction between the myelin basic protein (MBP) and Cop 1 serves as the basis for the suppressive activity of Cop 1 in EAE, by the induction of antigen-specific suppressor cells and competition with MBP for binding to major histocompatibility complex (MHC) molecules. Clinical trials with Cop 1, both Phase II and Phase III, were performed in relapsing-remitting (E-R) patients. The latter, a two-year multi-center double blind trial with 251 participating patients was conducted at 11 leading medical centers in the USA. It demonstrated a significant beneficial effect of Cop 1 in both diminishing the rate of exacerbations and improving the clinical status. The side effects of Cop 1 were only minimal. The cumulative results indicate that Cop 1 is a promising candidate drug for multiple sclerosis.

Introduction

Multiple sclerosis (MS) is a chronic inflammatory demyelinating disease of the central nervous system, in which infiltrating lymphocytes, predominantly T cells and macrophages, lead to damage of the myelin sheath. The aetiology of MS is still unknown, but according to the leading hypothesis, it is an autoimmune disease probably associated with an early viral infection. Experimental allergic encephalomyelitis (EAE) serves as the experimental animal model for the autoimmune processes in MS (Waksman, 1985). EAE is an acute neurological autoimmune disease mediated by CD4+ autoreactive T cells which recognize the encephalitogenic antigen(s) in the CNS (Zamvil and Steinman, 1991). Three myelin antigens were demonstrated to be encephalitogenic, myelin basic protein (MBP), proteolipid protein (PLP) and myelin oligodendrocyte glycoprotein (MOG) and all three were also implicated as putative autoantigens in MS (Bernard and Kerlero de Rosbo, 1992).

In view of the immunological nature of EAE, attempts have been made in several laboratories to suppress the disease in animals challenged with MBP, by desensitization procedures using the specific antigens relevant to the system. Thus, it has been shown that MBP, if given in high doses in incomplete Freund's adjuvant (IFA), is highly effective in preventing EAE in guinea pigs when administered before sensitization or in suppressing EAE if given after sensitization. It should be mentioned that not only the encephalitogenic protein was effective in such treatment, but other non-encephalitogenic related substances, such as myelin non-encephalitogenic basic proteins, altered MBP, and non-encephalitogenic degradation products or synthetic fragments of MBP had a similar protective effect. On the other hand, several peptides with encephalitogenic activity were incapable of suppressing EAE (Arnon, 1981). This represents corroborating evidence for the assumption that the site responsible for the immunological inhibition of EAE is not necessarily identical to the site responsible for induction of the disease.

Suppression of EAE by synthetic polymers

In view of the findings that EAE can be suppressed or inhibited not only by MBP but by non-encephalitogenic materials as well, we synthesized in our laboratory several random basic copolymers of amino acid composition, approaching to a certain extent that of the natural encephalitogen, and have tested their activity in either inducing or suppressing EAE. Whereas, none of these synthetic materials possessed any encephalitogenic activity, some of them showed high efficacy in suppressing the disease (Teitelbaum et al., 1971, 1973; Webb et al., 1976). Most of our work has been carried out with a copolymer, denoted Cop 1, composed of L-alanine, L-glutamic acid, L-lysine and L-tyrosine, in a residue molar ratio of 6.0/1.9/4.7/1.0. This copolymer did not exert encephalitogenic activity when injected into guinea pigs in doses of 10 µg to 5 mg. On the other hand, it had a marked suppressive effect on EAE when injected either in IFA or even in aqueous saline solution after initial challenge with a disease- inducing dose of BP (Teitelbaum et al., 1973).

Cop 1 is not the only suppressive material – several other synthetic copolymers related to it, in which either the glutamic acid was replaced by aspartic acid, or the tyrosine was replaced by tryptophan or omitted, were also effective in suppressing EAE. None of these copolymers is encephalitogenic, and they do not exhibit any general immunosuppressive activity.

In contrast to the known diversity in the response among various susceptible species to different encephalitogenic determinants, and in their susceptibility to MBP of various origins, the suppressive effect of Cop 1 is a general phenomenon. Thus, Cop 1 is equally efficient in the suppression of EAE induced in guinea pigs by encephalitogen of either human or bovine origin (Teitelbaum et al., 1973). Furthermore, in a detailed study we have demonstrated that suppression of EAE by Cop 1 could be achieved in several other species, such as rabbit, mouse, and in two species of primate – rhesus monkey and baboon. The results clearly indicate that the degree of suppression of

EAE by Cop 1 is remarkable in all species studied. It is thus apparent that Cop 1 does not manifest species specificity, either for the source of the encephalitogen or for the test animal (Sela et al., 1990).

The experiments in primates may prove relevant to demyelinating diseases in man, not only because of the closer phylogenetic relationship, but also due to the closer resemblance between the manifestation of EAE in primates and the human disease. In both the rhesus monkey and baboon, Cop 1 can suppress EAE even when administered to the animals after the onset of clinical symptoms. Monkeys treated with daily injections of Cop 1 in IFA over a period of 15–20 days, showed a reversal of the disease state with full recovery, whereas the control group (treated with IFA alone or untreated) exhibited rapid deterioration ending in death within 3–9 days of the onset of the symptoms. A significant observation was that all monkeys had histological damage typical of EAE, except in those fully recovered from the disease, where either no histopathological lesions were detected or only small foci were observed (Teitelbaum et al., 1974, 1977).

Cop 1 was also effective in both suppressing and preventing the chronic relapsing EAE (CR-EAE) in guinea pigs induced by whole spinal cord homogenate (Keith et al., 1979). This disease was shown to simulate in many ways the clinical and pathological features of human MS (Wisniewski and Keith, 1977). CR-EAE can be induced also in mice by the purified PLP and MOG proteins, or synthetic peptides based on their sequences. Cop 1 blocked CR-EAE induced by the encephalitogenic peptides PLP 139–151 and PLP 178–191 (Teitelbaum et al., 1996) and the disease induced by a MOG encephalitogenic peptide, MOG 35–55 (Ben-Nun et al., 1996). Thus, the suppressive effect of Cop 1 in EAE is a general phenomenon and is not restricted to a certain species, disease type or the encephalitogen used for EAE induction.

Immunological mechanisms involved in suppression of EAE

In an effort to elucidate the mechanism of suppression of EAE by the synthetic polymers, and bearing in mind the autoimmune nature of EAE and the involvement of cell-mediated immunity in its induction, the cross-reactivity between the synthetic copolymers and MBP has been evaluated. At the humoral level, a slight cross-reaction between rabbit anti-Cop 1 antibodies and bovine MBP could be detected, when sensitive assays such as passive cutaneous anaphylaxis were used. Recently, using monoclonal antibodies raised against MBP, we could demonstrate clearly that several monoclonal anti-MBP antibodies reacted with Cop 1 and vice versa. Some of the antibodies had a heteroclitic response and reacted better with the cross-reactive antigen than with the immunizing antigen (Teitelbaum et al., 1991).

At a cellular level, a marked cross-reaction was observed both in vivo in the delayed hypersensitivity skin test, and in vitro by measuring lymphocyte transformation (Webb et al., 1973). Of interest is the very good correlation between the extent of immunological cross-reactivity and the suppressive effect on EAE of the various materials. Thus, a polymer resembling Cop 1 in all parameters except that built of D-amino acids rather than L-amino acids, does not cross-react with MBP and has no suppressing activity whatsoever (Webb et al., 1976). On the basis of these findings, it seems that immunological cross-reactivity of MBP with several basic copolymers may serve as a basis for explaining their suppressive effect on EAE, and that this relationship might contribute to the further understanding of the mechanism of EAE as such.

Studies in mice suggested two possible mechanisms for Cop 1 activity in EAE:

A. Induction of antigen-specific suppressor cells

It was demonstrated that mice pretreated with Cop 1 in ICFA became resistant to further EAE induction. This state of unresponsiveness could be adoptively transferred to normal recipients by spleen cells from Cop 1-treated donors, and the cells responsible for the suppressive activity were identified as T-lymphocytes (Lando et al., 1979). We have further demonstrated that the suppressor cells which mediate unresponsiveness to EAE regulate the cellular immune response to MBP in a specific manner (Lando et al., 1981a). Moreover, a soluble suppressor factor could be extracted from these cells that had the same biological activities as the suppressor cells it originated from, and was capable of interfering with the induction of EAE (Lando et al., 1981b).

Furthermore, we have demonstrated the generation of suppressor T hybridomas and cell lines from spleen cells of mice rendered unresponsive to EAE by Cop 1. Both cell types lead to in vitro inhibition of effector lines and in vivo inhibition of clinical EAE (Aharoni et al., 1993). Recent results revealed that these T suppressor cells secrete Th2 cytokines after exposure to either Cop 1 or MBP (Aharoni et al., 1996). These cytokines may mediate the therapeutic effect of Cop 1 in disease induced not only with MBP but also with PLP and MOG by the mechanism of "bystander suppression". These results establish the stimulation of specific suppressor cells as a mechanism underlying the therapeutic activity of Cop 1 in EAE.

B. Competition with MBP on the binding to MHC

It was demonstrated that Cop 1 can competitively inhibit the response to MBP of diverse MBP-specific murine T-cell lines and clones, which responded to different epitopes of the MBP. These results suggest that Cop 1 or peptides derived from it can bind to the relevant MHC molecules and competitively inhibit the binding of MBP (Teitelbaum et al., 1988). Indeed, co-injection of Cop 1 with the encephalitogenic emulsion prevented clinical development of EAE, indicating that this mechanism may operate also in vivo. These findings were recently extended also to the human MHC, demonstrating that Cop 1 competitively inhibited the proliferative response of various human MBP-specific T-cell clones, while having no effect on PPD-specific T-cell clones (Teitelbaum et al., 1992).

Recently we demonstrated the direct binding of Cop 1, using its biotinylated derivative, to MHC molecules on living antigen presenting cells (APC).

Binding of biotinylated MBP and peptide p 84–102 (an immunodominant epitope of MBP), was also demonstrated. Cop 1 and MBP bound in a promiscuous manner to different types of APC of various H-2 and HLA haplotypes. The specificity of the binding was confirmed by its inhibition with either the relevant anti-MHC class II antibodies or unlabeled analogs. Cop 1 exhibited the most extensive and fast binding to the APC. In addition, Cop 1 inhibited the binding of biotinylated derivatives of MBP and of p 84–102 to the MHC class II molecules, and even displaced these antigens when already bound (Fridkis-Hareli et al., 1994). Taken together, these results suggest that Cop 1 indeed competes with MBP for MHC binding, and thereby inhibits T cell responses to MBP. The binding of Cop 1 to different DR alleles, probably because of its multiple MHC binding motifs, may indicate its potential as a broad spectrum drug for MS.

The primary event of Cop 1 binding to the class II MHC antigens may lead to two different pathways of suppression. On the one hand, it blocks MBP binding and thereby induces inhibition of the T cell response to MBP, when the TCR of a specific clone does not cross react with Cop 1. On the other hand, Cop 1 binding to the relevant MHC may lead to activation of Ts cells which are stimulated by the suppressive epitopes present in both BP and Cop 1 and which are distinct from the encephalitogenic ones. Thus, Cop 1 manifests in its activity a dual combination of antigen-based therapy approaches, namely MHC blockade, as well as antigen induced suppression.

Clinical studies with Cop 1 in MS patients

In view of the putative resemblance between EAE and MS and assuming that MBP might be involved in the pathogenesis of MS, preliminary clinical studies using Cop 1 were conducted (Abramsky et al., 1977; Bornstein et al., 1982) in MS patients. These were begun after toxicity studies in experimental animals showed that Cop 1 was non-toxic in both acute and subchronic administration in mice, rats, rabbits and beagle dogs, and showed no significant uptake by any of the animal's organs. The first clinical application of Cop 1 involved 4 patients in the "terminal" stages of MS, two of whom showed some improvement in vision and speech, but more importantly, no side effects were noted.

In a subsequent open study with Cop 1, we examined its ability to alter the course of MS in 12 chronic-progressive (C-P) and 4 exacerbating-remitting (E-R) patients. Again, no undesirable side reactions were noted during their more than 2-year use of Cop 1, and in 3 of the C-P and 2 of the E-R patients, the disease was arrested and the patients improved. These results, which may have represented either a placebo effect or a response to therapy, or a combination of both, prompted the initiation of a blind controlled trial involving 50 E-R MS patients (Bornstein et al., 1987).

These patients were selected and pair-matched for age (20–35), disability (Kurtzke DSS 0–6) and exacerbation frequency (1 and preferably 2 in each of 2 years before entry). The results show that six of 23 placebo and 14 of 25 Cop 1-treated patients were exacerbation-free. Average exacerbations over 2 years were 2.7 for placebo and 0.6 for Cop 1 patients, and disability status changes also favoured COP-1. All these differences were more pronounced in the less involved patients of disability score 0–2, than in the more severe patients.

A phase III multi-center double-blind trial with Cop 1, on 251 patients, which took place in 11 centers in the USA was recently finished. This trial demonstrated a 32% decrease in relapse rate (p = 0.002) after 2.5 years and improvement in EDSS score (p = 0.001) in Cop 1 treated patients vs. placebo group (Johnson et al., 1995, 1996). In parallel an open phase III clinical trial involving 238 patients was conducted in 4 centers in Israel. Follow-up included clinical as well as humoral and cellular immunological parameters related to Cop 1 and MBP. The results of this study are summarized in the following abstract.

In all the trials the undesirable side effects were minimal and consisted mainly of local irritation at the injection sites. Thus, Cop 1 is rather promising as a potential low-risk drug for the treatment of MS, and indeed it has been approved in December 1996 by Food and Drug Administration of USA.

References

Abramsky O, Teitelbaum D, Arnon R (1977) Effect of a synthetic polypeptide (Cop 1) on patients with multiple sclerosis and with acute disseminated encephalomyelitis. J Neurol Sci 31: 433–438

Aharoni R, Teitelbaum D, Arnon R (1993) T-suppressor hybridomas and IL-2 dependent lines induced by copolymer 1 or by spinal cord homogenate down regulate experimental allergic encephalomyelitis. Eur J Immunol 23: 17–25

Aharoni R, Teitelbaum D, Sela M, Arnon R (1996) Copolymer 1 induces suppressor T cells of the Th2 type. J Neurol 243 [Suppl 2]: S17

Arnon R (1981) Experimental allergic encephalomyelitis-susceptibility and suppression. Immunol Rev 55: 5–30

Bernard CCA, Kerlero de Rosbo N (1992) Multiple sclerosis an autoimmune disease of multifunctional etiology. Curr Opin Immunol 2: 760–765

Ben-Nun A, Mendel I, Bakimeer R, Fridkis-Hareli M, Teitelbaum D, Arnon R, Sela M, Kerlero de Rosbo N (1996) The autoimmune reactivity to myelin oligodendrocyte glycoprotein (MOG) in multiple sclerosis is potentially pathogenic: effect of Copolymer 1 on MOG-induced disease. J Neurol 43 [Suppl 1]: S14–S22

Bornstein MB, Miller AJ, Teitelbaum D, Arnon R, Sela M (1982) Multiple sclerosis: trial of a synthetic polypeptide. Ann Neurol 11: 317–319

Bornstein MB, Miller A, Slagle S, Weitzman M, Crystal H, Drexler E, Keilson M, Merriam A, Wassertheil-Smoller S, Spada V, Weiss W, Arnon R, Jacobsohn I, Teitelbaum D, Sela M (1987) A pilot trial of Cop 1 in exacerbating-remitting multiple sclerosis. N Engl J Med 317: 408–414

Fridkis-Hareli, M, Teitelbaum D, Gurevich E, Pecht I, Brautbar C, Oh Joong K, Brenner T, Arnon R, Sela M (1994) Direct binding of myelin basic protein and synthetic copolymer 1 to class II major histocompatibility complex molecules on living antigen presenting cells-specificity and promiscuity. Proc Natl Acad Sci US 91: 4872–4876

Johnson KP, Brooks, BR, Cohen JA, Ford CC, Goldstein J, Lisak RP, Myers LW, Pannitch HS, Rose JW, Schiffer RB, Vollmer T, Weiner LP, Wolinsky JC (1995) Copolymer 1 reduces relapse rate and improves disability in relapsing – remitting multiple sclerosis. Results of a phase III multicenter double blind, placebo-controlled trial. Neurology 1: 65–70

Johnson KP, The Copolymer 1 Multiple Sclerosis Study Group (1996) Extended report of the positive multicenter phase III trial of copolymer 1 for the treatment of relapsing remitting multiple sclerosis. Neurology 46: A406

Keith AB, Arnon R, Teitelbaum D, Caspary EA, Wisniewski HM (1979) The effect of Cop 1, a synthetic polypeptide, on chronic relapsing experimental allergic encephalomyelitis in guinea pigs. J Neurol Sci 42: 267–274

Lando Z, Teitelbaum D, Arnon R (1979) Effect of cyclophosphamide on suppressor cell activity in mice unresponsive to EAE. J Immunol 123: 2156–2160

Lando Z, Teitelbaum D, Arnon R (1981) The immunological response in mice unresponsive to experimental allergic encephalomyelitis. J Immunol 126: 1526–1528

Lando Z, Dori Y, Teitelbaum D, Arnon R (1981) Unresponsiveness to experimental allergic encephalomyelitis in mice-replacement of suppressor cells by a soluble factor. J Immunol 127: 1915–1919

Sela M, Arnon R, Teitelbaum D (1990) Suppressive activity of Cop 1 in EAE and its relevance to multiple sclerosis. Bull Inst Pasteur 88: 303–314

Teitelbaum D, Meshorer A, Hirshfeld T, Arnon R, Sela M (1971) Suppression of experimental allergic encephalomyelitis by a synthetic basic copolymer. Eur J Immunol 1: 242–248

Teitelbaum D, Webb C, Meshorer A, Arnon R, Sela M (1973) Suppression by several synthetic polypeptides of experimental allergic encephalomyelitis induced in guinea pigs and rabbits with bovine and human basic encephalitogen. Eur J Immunol 3: 273–279

Teitelbaum D, Webb C, Bree M, Meshorer A, Arnon R, Sela M (1974) Suppression of experimental allergic encephalomyelitis in rhesus monkeys by a synthetic basic copolymer. Clin Immunol Immunopathol 3: 256–262

Teitelbaum D, Meshorer A, Arnon R (1977) Suppression of experimental allergic encephalomyelitis in baboons by Cop 1. Israel J Med Sci 13: 1038

Teitelbaum D, Aharoni R, Arnon R, Sela M (1988) Specific inhibition of the T-cell response to myelin basic protein by the synthetic copolymer Cop 1. Proc Natl Acad Sci USA 85: 9724–9728

Teitelbaum D, Aharoni R, Sela M, Arnon R (1991) Cross reactions and specificities of monoclonal antibodies against myelin basic protein and against the synthetic copolymer 1. Proc Natl Acad Sci USA 88: 9528–9532

Teitelbaum D, Milo R, Arnon R, Sela M (1992) Synthetic copolymer 1 inhibits human T-cell lines specific for myelin basic protein. Proc Natl Acad Sci USA 89: 137–141

Teitelbaum D, Fridkis-Hareli M, Arnon R, Sela M (1996) Copolymer 1 inhibits the onset of chronic relapsing experimental autoimmune encephalomyelitis and interferes with T cell responses to encephalitogenic peptides of myelin proteolipid protein. J Neuroimmunol 64: 209–217

Waksman BH (1985) Mechanisms in multiple sclerosis. Nature 318: 104–105

Webb C, Teitelbaum D, Arnon R, Sela M (1973) In vivo and in vitro immunological cross-reactions between basic encephalitogen and synthetic basic polypeptides capable of suppressing experimental allergic encephalomyelitis. Eur J Immunol 3: 279–286

Webb C, Teitelbaum D, Herz A, Arnon R, Sela M (1976) Molecular requirements involved in suppression of EAE by synthetic basic copolymers of amino acids. Immunochemistry 13: 333–337

Wisniewski HM, Keith AB (1977) Chronic relapsing experimental allergic encephalomyelitis: an experimental model of multiple sclerosis. Ann Neurol 1: 144–148

Zamvil SS, Steinman L (1990) The T lymphocyte in experimental allergic encephalomyelitis. Ann Rev Immunol 8: 579–621

Authors' address: Dr. D. Teitelbaum, Department of Immunology, Weizmann Institute of Science, Rehovot, Israel 761000.

Possible role of the cholinergic system and disease models

M. Weinstock

Department of Pharmacology, School of Pharmacy, Hebrew University Hadassah
Medical Centre, Jerusalem, Israel

Summary. Memory impairment associated with the loss of cortical cholinergic neurons in AD has stimulated the development of animal models based on blockade or destruction of these systems. Strategies include mechanical lesions, local injection of excitotoxic amino acids or ethylcholine aziridinium (AF 64A), which disrupt reference and working memory in rats, but lack specificity for cholinergic systems. Other models involving, reduction in cerebral blood flow and interference with oxidative metabolism of glucose, mimic those found in AD, and also interfere with working and long-term memory in the rat. Memory impairments can be reversed by acetylcholinesterase inhibitors and cholinergic agonists but beneficial effects of these agents in AD patients are small and inconsistent. This may be partly due to unfavorable pharmacokinetics and dose-limiting side effects of existing drugs. Newer, brain specific acetylcholinesterase inhibitors and M1 muscarinic agonists with a lower incidence of unwanted effects are currently being evaluated.

Alzheimer's dementia (AD) is a neurodegenerative disease of unknown etiology that is characterized by a progressive, irreversible decline in a wide range of cognitive abilities, including short- and long term memory, abstract thinking, attention, language and mental flexibility. AD sufferers also show a variety of non-cognitive symptomatology, such as personality changes, depression, delusions, hallucinations, agitation and sleep disturbances. This leads to a marked disruption in their social activity and ability to work and take care of themselves.

Forebrain cholinergic systems and cognitive impairment in AD

Postmortem analyses in brains of AD subjects reveal a characteristic pathology of widespread neuritic plaques and tangles in the hippocampus, nucleus basalis of Meynert (NBM) and neocortex (Khachaturian, 1985). There is also a loss of cortical pyramidal neurons and depletion of several cholinergic markers, such as choline acetyl transferase (CAT), high affinity choline uptake (Sims et al., 1983a) and acetylcholinesterase (AChE) (Perry, 1986).

Other neurotransmitter systems, such as glutamate (Chalmers et al., 1990) and serotonin (Quirion et al., 1986) are also disrupted in AD. However, biopsies from the neocortex at earlier stages of the disease show that the severity of symptoms on a dementia rating scale are strongly correlated to synaptic loss and a reduction in acetylcholine (ACh) synthesis, but not to CAT activity or other neurotransmitter systems (Francis et al., 1993). The failure of ACh synthesis in synaptic regions in the hippocampus and cortex could be due to a loss of acetyl-CoA, which is formed during glucose metabolism. This could occur through an impairment of the cerebral oxidative metabolism of glucose, which has been found at an early stage of AD (Hoyer, 1992). There is also a disruption of glucose-related amino acid metabolism, a fall in ATP and an increase in intracellular levels of Ca^{++}. Biopsies from AD brains also indicate that there is a partial uncoupling of oxidative phosphorylation (Sims et al., 1983b). These metabolic abnormalities are assumed to occur primarily in cholinergic and glutaminergic nerve terminals and may lead to abnormal cleavage of amyloid precursor protein and amyloid formation (Hoyer, 1992).

Animal models of spatial memory

The ability of an animal to keep track of its location in space by remembering where it has been involves a form of short term memory, called working memory that stores information as it is being processed (Olton, 1977). When rewards or punishments await them in a particular place, animals quickly learn to obtain the rewards and avoid the punishments. The majority of recent studies on the role of cholinergic and other systems on spatial memory has employed mazes of the T-, radial arm or Morris water type. In these, the animal has to learn to enter the arms in a certain sequence in order to obtain a reward, or to find a platform submerged below the surface of a bath of opaque fluid to avoid the necessity of swimming. These tests can be designed to assess both learning and cognitive performance, as well as motoric and motivational deficits and to separate working from reference memory (McNamara and Skelton, 1993).

Lesions of forebrain cholinergic systems. The correlation between the degree of impairment of cholinergic transmission and cognitive function in AD has stimulated the search for animal models of dementia based on pharmacological cholinergic blockade or lesions of the forebrain cholinergic systems. These models have provided the basis for the development of cholinergic replacement therapy, including anticholinesterase (AChE) agents, cholinergic agonists (Miyamoto et al., 1989), or ACh releasers (Lavretsky and Jarvik, 1992). In the rat, the nucleus basalis magnocellularis (nbm) is analogous to the NBM in humans and provides a major source of extrinsic cholinergic neurones entering the neocortex (Divac, 1975). The cholinergic innervation of the hippocampus is derived from nerve cells in the medial septal nucleus (MS) and diagonal band of Broca (McKinney et al., 1983). Different strategies employed for destroying these nuclei include electrolytic or radiofrequency lesions (Dubois et al., 1985), local injection of excitotoxic amino acids (EAA; glutamate receptor agonists), e.g. ibotenic acid (IBO), kainic acid

(KA), quisqualic acid (QA) (Dunett et al., 1991), or ethylcholine aziridinium (AF 64A) (Sandberg et al., 1984). Mechanical lesions of the nbm in rats impair a variety of learning tasks, but also destroy non-cholinergic systems, e.g. noradrenergic (NA) and dopaminergic (DA), and can damage the globus pallidus and increase the mortality rate (Nabeshima, 1993). In general, chemical lesions with neurotoxins are more selective since they only destroy neuronal perikarya at the injection site but not fibers passing through the area if the dose given is carefully controlled. Although AF 64A was claimed to be taken up selectively into cholinergic neurones and destroy the nerve terminals (Kasa et al., 1986), it was shown to cause gross non-specific tissue damage (Jarrard et al., 1984).

Destruction of the cortical cholinergic projection by lesion of the nbm in rats causes a greater impairment of long-term memory, while an MS lesion results in a more selective interference with short-term memory (Miyamoto, 1987). Nbm lesions reduce CAT activity and other cholinergic parameters in the cortex but do not affect the levels of other neurotransmitters (Dekker et al., 1991). However, more recent studies have questioned the specificity of the EAA-induced lesions for cholinergic neurons. Although four different EAAs produced similar reductions in CAT activity in the forebrain, only IBO acid disrupts spatial navigation in the Morris water maze test (Dunnett et al., 1991). Moreover, greater losses in CAT activity (> 70%) have been induced by QA without impairment of the aquisition or retention of memory. This suggests either that the functional deficits induced by IBO acid lesions in the forebrain are not due to a primary destruction of cholinergic neurones in the nbm, or that IBO and QA affect different subpopulations of cholinergic neurones by activating different glutamate receptors (Dekker et al., 1991).

Since uncoupling of oxidative phosphorylation and impairment of oxidative glucose metabolism occurs at an early stage of AD, a more valid animal model could be one which mimics these deficits, even though it is not known how they occur. Such a model may enable one to develop therapeutic strategies which can prevent the nerve destruction and loss of acetylCoA, consequent to a failure of energy mechanisms. For example, inhibition of brain cytochrome oxidase by sodium azide administered chronically to rats via osmotic minipumps, impairs hippocampal plasticity and learning in the radial arm and Morris water mazes, without affecting sensory or motor function (Bennett et al., 1992). It remains to be determined whether this treatment also reduces cholinergic transmission in cortical and hippocampal neurones, and whether the learning deficits can be removed by an AChE inhibitor.

Cholinesterase inhibitors and memory

Although there is not a good correlation between reductions in CAT activity and disruption of spatial memory induced by different EAAs (Dunnett et al., 1991), many of the behavioral deficits they induce can be significantly reversed by AChE inhibitors such as physostigmine, tetrahydroaminoacridine (Tacrine), heptyl-physostigmine (Eptastigmine), RA7 (SDZ ENA713) and muscarinic agonists (Dekker et al., 1991). These agents improve performance in spatial alternation in a T maze and restore reference and working memory in the Morris water and radial arm mazes. ACh levels in the cortex and hippocampus are also decreased in response to cerebral ischemia after carotid artery occlusion (Kakihana et al., 1984). This treatment impairs working memory in rats in a 3-panel runway task, which is almost completely restored by physostigmine and other antiAChEs (Yamamoto et al., 1993) (Table 1). This further supports the relationship between decrements in cholinergic transmission in selected sites and disruption of spatial memory. Part of the beneficial effect of AChE inhibitors may result from their ability to increase cerebral blood flow without a concomitant increase in glucose utilisation (Scremin et al., 1988; Linville et al., 1992).

Clinical studies with cholinesterase inhibitors

In spite of the fact that physostigmine and Tacrine produce clear improvements in memory and learning in a variety of animal models, the clinical experience with these agents is inconsistent and unimpressive (Summers et al., 1986; Sano et al., 1993). There are several possible reasons for this. Since it is so difficult to make a conclusive diagnosis of AD the studies have inevitably included patients who do not have this form of dementia. Alternatively, some of the patients may have so much neural degeneration that there is insufficient ACh release to prolong by inhibiting its hydrolysis. Another major factor is the method of assessment of the patient, and whether the evaluation is based on clinical impression, or also takes into account that of caregivers. Tacrine does not appear to produce a significant improvement if global assessments scales are used, since it only influences the cognitive component of this disease but not the other behavioral or functional abnormalities (Davis et al., 1992). However, physostigmine has been claimed to improve the delusions and hallucinations in an AD patient (Cumming et al., 1993). Not all studies take account of the rapid deterioration in some patients, or allow adequate time for washout if a double blind crossover procedure is used. Apart from the problems with the disease and its assessment, there are also many shortcomings in the two drugs, physostigmine and Tacrine, that have been used to treat

Table 1. Increases in errors and latency induced by ischemia three panel runway task

Treatment	Drug/dose	No. of errors	Latency (sec)
Control	none	2.3 ± 0.6	27.9 ± 1.9
Ischemia (5 min)	none	16.7 ± 1.3	76.7 ± 5.6
Ischemia	Physostigmine (0.1 mg)	$5.8 \pm 1.4*$	$41.2 \pm 4.0*$
Ischemia	Tacrine (3.2 mg)	$6.3 \pm 0.8*$	57.2 ± 7.0

* significant reduction cf no drug, $p < 0.01$ (from Yamamoto et al., 1993)

AD patients. The major disadvantages of physostigmine are its narrow therapeutic window, unpredictable oral bioavailability and short duration of action (Christie et al., 1981). The high incidence of side effects resulting from cholinergic hyperactivity, including nausea, vomiting, diaphoresis, muscle fasciculations or cramps, bradycardia and arrythmias are sufficient to prevent the attainment of an optimal dose in a large proportion of subjects (Stern et al., 1987). Although Tacrine has a longer duration of action, it still possesses the other disadvantages of physostigmine and also causes hepatotoxicity in up to 40% of subjects (Gautier and Gautier, 1993). Several new drugs are currently undergoing clinical trial. One of these, Eptastigmine has a long duration of action, is well absorbed orally, with an improved safety margin but it has been withdrawn because of hematological toxicity (Iversen, 1993). It is not known whether this is related to its AChE inhibitory activity or independent of it. Although both acetyl- and butyrylcholinesterase are present in developing blood cells (Paoletti et al., 1992), blood dyscrasias have not been reported in patients receiving either physostigmine or Tacrine. This could be because they may have a lower affinity than Eptastigmine for the form of enzyme present in developing blood cells.

SDZ ENA 713

In a search for safer AChE inhibitors with a more selective effect in the CNS, we synthesized a series of derivatives of N-N-dimethylamino-ethyl-phenyl carbamate. Several of these are well absorbed orally and show a long duration of AChE inhibition (Weinstock et al., 1986). The l-isomer of the methyl-ethyl

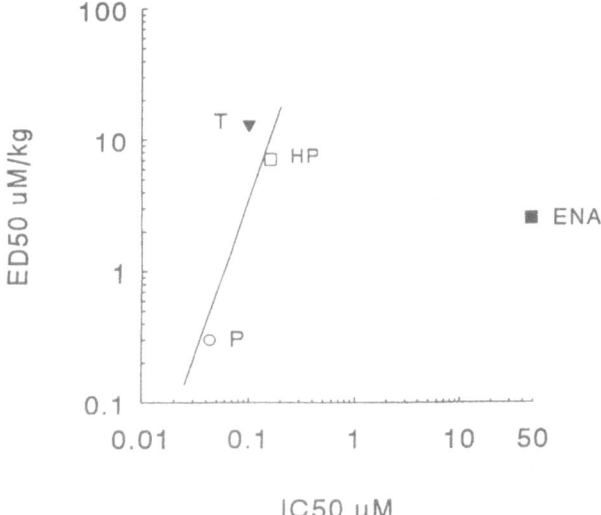

Fig. 1. Relationship between in vitro inhibition of human erythrocyte AChE and ex vivo inhibition of mouse brain AChE. *P* physostigmine; *T* tacrine; *HP* heptylphysosostigmine; *ENA* SDZ ENA713

derivative SDZ ENA713 (formerly known as RA7) has distinctive properties
when compared to the other derivatives, physostigmine and Tacrine. Although it has only 1/500th of the potency of physostigmine as an inhibitor of
red cell cholinesterase in vitro it is almost equipotent in blocking AChE in the
brain and producing central cholinergic activation after oral administration to
mice, rats and human subjects. The relationship between in vitro AChE
inhibition on the red cell enzyme and ex vivo, on brain enzyme after sc.
administration in mice is compared to other AChE drugs in Fig. 1. The
therapeutic index, LD_{50}/ED_{50}, where LD_{50} is the dose that is lethal in 50% of
the mice, and ED_{50}, the dose required to inhibit by 50% the AChE in mouse
brain, is considerably higher for SDZ ENA713 than for physostigmine or
Tacrine (Fig. 2). So also is the degree of protection afforded by pretreatment
with atropine, showing that SDZ ENA713 is potentially a much safer drug
than the other AChE inhibitors (Weinstock et al., 1994). After oral administration to rats, physostigmine inhibits AChE in different brain regions, heart
and skeletal muscle to a similar extent, but SDZ ENA713 has a much more
selective action in the cortex and hippocampus (Table 2). This explains its
lower relative toxicity and lack of effect on nicotinic cholinergic transmission
in skeletal muscle. In doses which show clear signs of central cholinergic
stimulation, SDZ ENA713 has no significant effects on the cardiovascular
system in rats or monkeys (Enz et al., 1993).

Muscarinic receptor stimulation in the hippocampus results in synchronization of theta waves in the EEG (Bevan, 1984). A single oral dose of SDZ
ENA713 (7.5 mg/kg) in the rat induces this characteristic hippocampal EEG

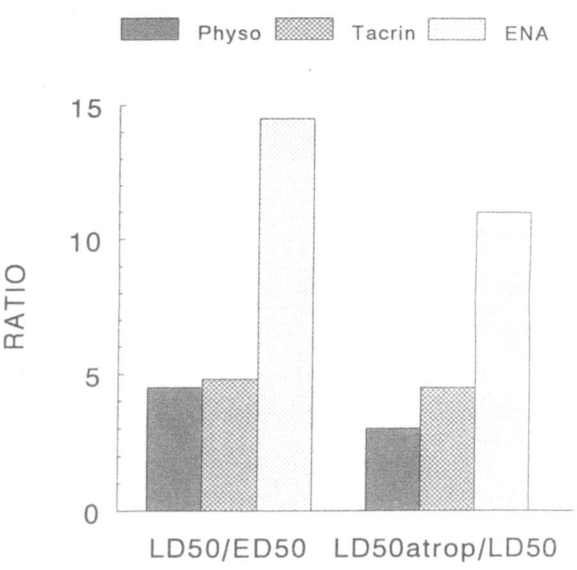

Fig. 2. Therapeutic ratios of AChE inhibitors with and without atropine pretreatment.
ED50 dose required to inhibit AChE in mouse brain by 50% after sc injection of the drug.
LD50 dose that is lethal in 50% of mice after sc injection of the drug. *LD50 atrop* dose that
is lethal in 50% of mice after pretreatment with atropine sulphate, 5 mg/kg, 15 min before
AChE inhibitor. *Physo* physostigmine; *tacrin* tacrine; *ENA* SDZ ENA713

Table 2. AChE inhibition ex vivo rat after oral administration of physostigmine and SDZ ENA713

Brain region or organ	Physostigmine IC$_{50}$ (μmoles/kg)	SDZ ENA713 IC$_{50}$ (μmoles/kg)
Cortex	1.20	2.69
Hippocampus	1.31	3.89
Striatum	1.25	7.12
Pons/medulla	1.44	7.21
Skeletal muscle	1.26	32.7
Heart	1.23	38.4

(from Weinstock et al., 1994)

pattern which lasts for over 6 hours without any signs of peripheral cholinergic hyperactivity. An equivalent degree of activation is seen after Tacrine (10 mg/kg), but is of much shorter duration (Enz et al., 1993). The selective central effect of SDZ ENA713 is also seen in normal human subjects. After an oral dose of 1.3 mg, the drug increases REM sleep density by about 50% without causing any significant effect on plasma ChE (Table 3). This is in contrast to findings of Lines et al. (1993) with Eptastigmine, which failed to reverse scopolamine-induced cognitive deficits in healthy human subjects at a dose that inhibited red cell AChE by > 37% and the plasma enzyme by > 25%, two hours after administration. This degree of plasma ChE inhibition by physostigmine is associated with a high incidence of adverse cholinergic effects. Smaller doses of this drug do not produce significant central cholinergic activation in AD patients (Sherman et al., 1987).

The distinctive brain selective property and low toxicity of SDZ ENA713 make it a more suitable candidate to test the hypothesis that cognitive impairments in the early stages of AD are associated with a loss of cholinergic transmission in the hippocampus and neocortex. Its selective inhibitory effect

Table 3. Effects of SDZ ENA713 in human subjects

Dose (mg) O/R	% inhibition of plasma ChE	Duration (hr)	% inc. in REM sleep density	Duration (hr)
1.0	< 5	0.5	51.1	6–7
2.0	5–8	1.0	50	6–8
3.0	17.5	10.0	–	–
4.6	30.0	22.5	–	–

(from Enz et al., 1991; Holsboer-Trachsler et al., 1993)

on the G1 form of AChE (Enz et al., 1993), which unlike the G4 form, is not lost in the brains of AD patients (Siek et al., 1990), could contribute to its beneficial effect in conditions of cholinergic hypofunction.

References

Bennett MC, Diamond DM, Parker WD, Stryker SL, Rose GM (1992) Inhibition of cytochrome oxidase impairs learning and hippocampal plasticity: a novel animal model of Alzheimer's disease. In: Meyer EM, et al (eds) Treatment of dementias. Plenum Press, New York

Bevan P (1984) Effect of muscarinic ligands on the electrical activity recorded from the hippocampus: a quantitative approach. Br J Pharmacol 82: 431–440

Chalmers DT, Dewar D, Graham DI, Brooks DN, McCulloch J (1990) Differential alterations of cortical glutamatergic binding sites in senile dementia of the Alzheimer type. Proc Natl Acad Sci USA 87: 1352–1356

Christie JE, Shering A, Fergusen J, Glen AIM (1981) Physostigmine and arecoline: efffects of intravenous infusions in Alzheimer presenile dementia. Br J Psychiatry 138: 46–50

Cummings JL, Gorman DG, Shapira J (1993) Physostigmine ameliorates the delusions of Alzheimer's disease. Biol Psychiatry 33: 536–541

Davis KL, Thal L, Gamzu E, Davis CS, Woolson RF, Gracon SI, et al (1992) A double-blind, placebo-controlled multicenter study of Tacrine for Alzheimer's disease. N Engl J Med 327: 1253–1259

Dekker AJAM, Connor DJ, Thal LJ (1991) The role of cholinergic projections from the nucleus basalis in memory. Neurosci Biobehav Rev 15: 299–317

Divac I (1975) Magnocellular nuclei of the basal forebrain project to neocortex, brain stem, and olfactory bulb. Review of some functional correlates. Brain Res 93: 385–398

Dubois B, Mayo W, Agid Y, Le Moal M, Simon H (1985) Profound disturbances of spontaneous and learned behaviors following lesions of the nucleus basalis magnocellularis in the rat. Brain Res 338: 249–258

Dunnett SB, Everitt BJ, Robbins TW (1991) The basal forebrain-cortical cholinergic system: interpreting the functional consequences of excitotoxic lesions. TINS 14: 494–501

Enz A, Boddeke H, Gray J, Spiegel R (1991) Pharmacologic and clinicopharmacologic properties of SDZ ENA 713, a centrally-selective acetylcholinesterase inhibitor. Ann NY Acad Sci 640: 272–275

Enz A, Amstutz R, Boddeke H, Gmelin G, Malanowski J (1993) Brain selective inhibition of acetylcholinesterase: a novel approach to therapy for Alzheimer's disease. Prog Brain Res 98: 431–438

Francis PT, Sims NR, Procter AW, Bowen DM (1993) Cortical pyramidal neurone loss may cause glutamatergic hypoactivity and cognitive impairment in Alzheimer's disease: investigative and therapeutic perspectives. J Neurochem 60: 1589–1604

Gauthier S, Gauthier L (1993) What we have learned from the THA trials to facilitate testing of new AChE inhibitors. Prog Brain Res 98: 427–429

Holsboer-Trachsler E, Hatsinger M, Syohler R, Hemmeter U, Gray J, Muller J, et al (1993) Effects of the novel acetylcholinesterase inhibitor SDZ ENA 713 on sleep in man. Neuropsychopharmacology 8: 87–92

Hoyer S (1992) Intermediary metabolism disturbance in AD/SDAT and its relation to molecular events. Prog Neuropsychopharmacol Biol Psychiatry 17: 199–228

Iversen LL (1993) Approaches to cholinergic therapy in Alzheimer's disease. Prog Brain Res 98: 423–426

Jerrard LE, Kant GJ, Meyerhoff JL, Levy A (1984) Behavioral and neurochemical effects of intraventricular AF64A administration in rats. Pharmacol Biochem Behav 21: 273–280

Kakihana M, Yamazaki N, Nagaoka A (1984) Effects of idebenone (CV-2619) on the concentrations of acetylcholine and choline in various brain regions of rats with cerebral ischemia. Jpn J Pharmacol 36: 357–363

Kasa P, Szerdahelyi P, Fisher A, Hanin I (1986) Histochemical and electronmicroscopic study of the brain of the AF64A-treated rat. In: Fisher A, Hanin I, Lachman C (eds) Alzheimer's and Parkinson's diseases: strategies for research and development. Plenum Press, New York

Khachaturian ZS (1985) Diagnosis of Alzheimer's disease. Arch Neurol 42: 1097–1105

Lavretsky EP, Jarvik LF (1992) A group of potassium-channel blockers – acetylcholine releasers: new potentials for Alzheimer's disease? J Clin Psychopharmacol 12: 110–118

Lines CR, Ambrose JH, Heald A, Traub M (1993) A double-blind, placebo-controlled study of the effects of eptastigmine on scopolamine-induced cognitive deficits in healthy male subjects. Hum Psychopharmacol 8: 271–278

Linville DG, Giacobini E, Arneric SP (1992) Heptyl-physostigmine enhances basal forebrain control of cortical cerebral blood flow. J Neurosci Res 31: 573–577

McKinnley M, Coyle JT, Hedreen JC (1983) Topographic analysis of the innervation of the rat neocortex and hippocampus by the basal forebrain cholinergic system. J Comp Neurol 217: 103–121

McNamara RK, Skelton RW (1993) The neuropharmacological and neurochemical basis of place learning in the Morris water maze. Brain Res Rev 18: 33–49

Miyamoto M, Kato J, Narumi S, Nagaoka A (1987) Characteristics of memory impairment following lesioning of the basal forebrain and medial septal nucleus in rats. Brain Res 419: 19–31

Miyamoto M, Narumi S, Nagaoka A, Coyle JT (1989) Effects of continous infusion of cholinergic drugs on memory impairment in rats with basal forebrain lesions. J Pharmacol Exp Ther 248: 825–835

Nabeshima T (1993) Behavioral aspects of cholinergic transmission: role of basal forebrain cholinergic system in learning and memory. Prog Brain Res 98: 405–411

Olton DS (1977) Spatial memory. Sci Am 236: 82–98

Paoletti F, Mocali A, Vanucchi A (1992) Acetylcholinesterase in murine erythroleukemia (Friend) cells: evidence for megakaryocyte-like expression and potential growth-regulatory role of enzyme activity. Blood 79: 2873–2879

Perry EK (1986) The cholinergic hypothesis – ten years on. Br Med Bull 42: 63–69

Quirion R, Marbel JC, Robitaille Y, Etienne P, Wood P, Nair NVP, Gauthier S (1986) Neurotransmitter and receptor deficits in senile dementia of the Alzheimer type. Can J Neural Sci 13: 503–510

Sandberg K, Sandberg PR, Hanin I, Fisher A, Coyle JT (1984) Cholinergic lesion of the striatum impairs acquisition and retention of a passive avoidance response. Behav Neurosci 98: 162–165

Sano M, Bell K, Marder K, Stricks L, Stem Y, Mayeux R (1993) Safety and efficacy of oral physostigmine in the treatment of Alzheimer's disease. Clin Neuropharmacol 16: 61–69

Scremin OU, Allen K, Torres C, Scremin E (1988) Physostigmine enhances blood flow – metabolism ratio in neocortex. Neuropsychopharmacol 1: 297–303

Sherman KA, Kumar V, Ashford JW, Murphy JM, Elble RJ, Smith R, Giacobini E (1987) Effect of oral physostigmine in senile dementia patients: utility of blood cholinesterase inhibition and neuroendocrine response to define pharmacokinetics and pharmacodynamics. In: Strong R, Woods WG, Burke WJ (eds) Central nervous system disorders of aging: strategies for intervention and research (aging). Raven Press, New York

Siek GC, Kats LS, Fishman EB, Korosi TS, Maquis JK (1990) Molecular forms of acetylcholinesterase in subcortical areas of normal and Alzheimer disease brain. Biol Psychiatry 27: 573–580

Sims NR, Bowen DM, Allen SJ, Smith CCT, Neary D, Thomas DJ, Davison AN (1983a) Presynaptic cholinergic dysfunction in patients with dementia. J Neurochem 40: 503–509

Sims NR, Bowen DM, Neary D, Davison AN (1983b) Metabolic processes in Alzheimer's disease: adenine nucleotide content and production of $^{14}CO_2$ from [U-^{14}C] glucose in vitro in human neocortex. J Neurochem 41: 1329–1334

Sano M, Bell K, Marder K, Stricks L, Stern Y, Mayeux R (1993) Safety and efficacy of oral physostigmine in the treatment of Alzheimer's disease. Clin Neuropharmacol 16: 61–69

Stern Y, Sano M, Mayeux R (1987) Effects of oral physostigmine in Alzheimer's disease. Ann Neurol 22: 306–310

Summers WK, Majovski LV, Marsh GM, Tachiki K, Kling A (1986) Oral tetrahydro-aminoacridine in long-term treatment of senile dementia, Alzheimer type. N Engl J Med 315: 1241–1245

Weinstock M, Razin M, Chorev M, Tashma Z (1986) Pharmacological activity of novel anticholinesterase agents of potential use in the treatment of Alzheimer's disease. In: Fisher A, Hanin I, Lachman C (eds) Advances in behavioral biology. Plenum Press, New York

Weinstock M, Razin M, Chorev M, Enz A (1994) Pharmacological evaluation of phenyl-carbamates as CNS-selective acetylcholinesterase inhibitors. J Neural Transm [Suppl] 43: 307–309

Yamamoto T, Ohno M, Kitajima I, Yatsugi SI, Ueki S (1993) Ameliorative effects of the centrally active cholinesterase inhibitor, NIK-247, on impairment of working memory in rats. Physiol Behav 53: 5–10

Author's address: Prof. M. Weinstock, Department of Pharmacology, School of Pharmacy, Hebrew University Hadassah Medical Centre, Ein Kerem, Jerusalem, Israel.

Loss of dopaminergic neurons in parkinsonism: possible role of reactive dopamine metabolites

T. G. Hastings and **M. J. Zigmond**

Departments of Neurology and Neuroscience, University of Pittsburgh,
Pittsburgh, PA, U.S.A.

Summary. Parkinson's disease affects one out of every 100 people above the age of 55. Its cause is unknown and although the symptoms can be treated, there is no cure. The disease is associated with the selective loss of neurons that contain biogenic amines, and among these it is the dopamine (DA) neurons of the nigrostraital projection that are the most consistently and severely affected (Bernheimer et al., 1973). In this review we discuss the possibility that DA may act as an endogenous neurotoxin, causing the degeneration of the very neurons that release it. We further suggest that although treatments which increase the synthesis and release of DA reduce the symptoms, they also may serve to exacerbate the neurodegenerative process. We propose that the treatments which increase the antioxidant capacity of brain may be protective.

Studies with 6-hydroxydopamine-treated rats may provide insights into the etiology of Parkinson's disease

A considerable amount of information regarding Parkinson's disease has come from animal studies using the selective neurotoxin 6-hydroxydopamine (6-OHDA). This compound is selectively accumulated in catecholamine neurons where it autoxidizes to form such toxic species as superoxide anions ($O_2^{\cdot}$), hydrogen peroxide (H_2O_2), hydroxyl radicals (OH·), and 6-OHDA quinones. Several parallels exist between 6-OHDA-treated rats and patients with Parkinson's disease, including (1) an apparent causal relation between the loss of nigrostriatal DA neurons and neurological symptoms, (2) a preclinical phase that persists until the loss of DA is almost complete, (3) temporary reversal of symptoms with stress and L-DOPA, and (4) an apparent increase in DA turnover in residual DA neurons (for review see Zigmond and Stricker, 1989; Zigmond et al., 1992). We generally have administered 6-OHDA via the lateral cerebroventricles (ivt.), and the behavioral sparing does not appear to be a result of regeneration, sprouting, or receptor supersensitivity. Moreover, since behavioral impairments can be elicited by drugs that interfere with dopaminergic transmission, the DA terminals that remain after 6-OHDA-induced lesions appear to play an important functional role.

The absence of gross neurological dysfunction despite extensive loss of DA neurons is associated with an increase in the capacity of the remaining DA neurons to deliver transmitter to denervated sites. Our group has examined this phenomenon in a variety of ways. Studies using neostriatal slices prepared from control and 6-OHDA-lesioned rats demonstrated that partial destruction of neostriatal DA terminals increased the fractional overflow of DA. The increase in overflow was related to lesion size, reaching 7-fold with losses of DA greater than 90%. Further studies have suggested that this apparent increase in DA efflux per terminal is due to a reduction in DA reuptake as well as an increase in DA release from residual terminals (Stachowiak et al., 1987; Snyder et al., 1990). This was shown to be accompanied by an increase in DA synthesis (Liang and Zigmond, 1993). The impact of partial lesions on the concentration of DA in the extracellular fluid of neostriatum was examined using in vivo microdialysis, and it was observed that in animals with less than an 80% loss of DA, the extracellular concentration of DA was not significantly different from normal (Abercrombie et al., 1990; see also Zhang et al., 1988; Robinson and Wishaw, 1988).

The extracellular concentration of DA in neostriatum also was examined by microdialysis after systemic administration of L-DOPA. Both intact rats and rats treated with 6-OHDA were studied. Basal levels of extracellular DA were reduced by 6-OHDA, although the reduction was much less than that of DA in the striatal tissue as determined post-mortem. In the presence of an inhibitor of peripheral decarboxylation, L-DOPA (100 mg/kg) caused a 2-fold increase in extracellular DA. However, in animals with large bilateral neostriatal DA depletions (–82%), L-DOPA increased extracellular DA by 14-fold, to a level that was several times higher than that seen in control animals (Abercrombie et al., 1990). Further insight into the mechanism of action of L-DOPA was derived from complementary studies using neostriatal slices. L-DOPA (10 mM) increased tissue levels of DA, elevated fractional DA overflow, and greatly elevated DOPAC levels, both in tissue and in the superfusate (Snyder and Zigmond, 1990). Collectively, these studies suggest that after substantial damage to DA neurons in the nigrostriatal bundle, aromatic L-amino acid decarboxylase located external to dopaminergic terminals can play a major role in the release of DA formed from exogenous L-DOPA.

DA readily forms reactive oxygen species and can act as a neurotoxin

The hypothesis that DA contributes to the etiology of Parkinson's disease was first suggested because DA, like 6-OHDA, is a reactive molecule that will oxidize to form free radicals and quinones (Graham, 1978). The formation of these reactive metabolites can occur by one of three routes. First, DA can undergo oxidation by 1 or 2 electron transfers, forming various reactive compounds. Second, brain can oxidize DA enzymatically (Grisham et al., 1987), including via the peroxidase activity contained in prostaglandin synthase (Hastings, 1995). And third, H_2O_2 is formed as a normal product of DA metabolism by monoamine oxidase; without adequate reduction by glutathione (GSH) peroxidase, the H_2O_2 can be broken down through interaction with metal ions into free radicals species such as $O_2^{\cdot}$ or $OH\cdot$ (Cohen, 1983). In short, conditions that increase the concentration and/or turnover of DA should increase the potential for the formation of reactive metabolites. Uncontrolled accumulation of reactive species formed from DA can be expected to cause cellular damage, including lipid peroxidation, DNA fragmentation, and modification of proteins (Halliwell and Gutteridge, 1985).

There is now a large body of evidence indicating that DA can, indeed, participate in neurotoxic events. For example, pharmacological or lesion-induced depletion of DA can attenuate the neurotoxic effects of a variety of challenges, including systemic administration of methamphetamine (Schmidt et al., 1985; Johnson et al., 1987), intracerebral administration of excitatory amino acids (Chapman et al., 1989), and ischemia (Weinberger et al., 1985). Moreover, the exposure of cells to DA has been shown to have neurotoxic consequence (Rosenberg, 1988; Michel and Hefti, 1990). However, these studies provide little or no evidence for *selective* effects of DA on dopaminergic neurons, and thus fall short of providing an explanation for the neurodegenerative profile of Parkinson's disease.

The cytotoxicity of DA and related catechol analogues has been attributed in part to the reactive quinone form of the oxidized molecule, at least under in vitro conditions (Graham et al., 1978). The electron-deficient DA quinone appears to react with nucleophilic sulfhydryl groups on proteins, inactivating enzymes and receptors. The DA quinone may also form conjugates with endogenous reducing agents, GSH or cysteine, by reacting with thiol groups of these molecules. Indices of in vivo DA oxidation have been obtained by measuring the amount of free cysteinyl-catechol derivatives formed in dopaminergic brain regions of several species, including humans (Fornstedt et al., 1986). Cysteinyl-catechol conjugates also increase with age (Fornstedt et al., 1990) and ascorbic acid deficiency (Fornstedt and Carlsson, 1991).

The rate of formation of reactive products of DA is determined by the relative concentration of DA and antioxidants

As described above, DA will oxidize to form free radicals and reactive DA quinones. The electrophilic quinones will react with nucleophilic sulfhydryl groups, a major source of which is provided by the cysteinyl residues of proteins. Thus, the binding of DA to cysteinyl residues may serve as an index of DA oxidation; it may also be a cytotoxic event. To examine conditions under which such binding might occur we incubated neostriatal slices with ^{3}H-DA under various buffer conditions and then determined the amount of radioactivity bound to the acid-precipitated protein. Results showed that the amount of tritium bound to protein was greatly influenced by the concentration of reducing agent, ascorbate or GSH, that is present in the incubation buffer. Following a 60 min incubation with 60 nM ^{3}H-DA, there was a 3.4-fold increase in the amount of ^{3}H-DA bound to protein in a Krebs-bicarbonate buffer without ascorbate when compared with the same buffer containing 0.85 mM ascorbate (Hastings and Zigmond, 1994). Binding to protein could be completely inhibited by 10 µM GSH, suggesting that the thiol group on GSH was competing with protein cysteinyl residues for the oxidized DA quinone. Following acid hydrolysis and alumina extraction of the labeled protein, we were able to detect the presence of both cysteinyl-^{3}H-DA and cysteinyl-^{3}H-DOPAC (Hastings and Zigmond, 1994).

We are currently seeking to determine where the DA binding to protein occurs. Our preliminary studies suggest that the great bulk of the binding is external to dopaminergic neurons. For example, inhibitors of high affinity DA uptake did not reduce ^{3}H-DA binding; moreover, binding also was high when cerebellar slices are employed, although cerebellum has a relatively low density of noradrenergic terminals and are virtually devoid of DA terminals (Hastings and Zigmond, 1993). This, then, may explain the lack of specificity of the toxicity induced by *exogenous* catecholamines in cell culture. On the other hand, endogenous DA would be expected to be localized primarily within and around DA neurons. Moreover, these neurons and/or their immediate environment may be uniquely deficient in antioxidant capacity, at least in Parkinson's disease (Agid et al., 1993). As a consequence, dopaminergic terminals could be exposed to high concentrations of reactive catechol metabolites, both in their cytoplasm and in the local extracellular compartment.

Selective neurotoxicity was associated with DA oxidation in vivo

We sought to determine whether the oxidation of DA and its binding to protein would also occur in vivo. High concentrations of DA (0.05–1.0 µmol in µl) were injected into neostriatum, and tissue surrounding the injection site was assayed 24 h later. Protein-bound cysteinyl-DA and cysteinyl-DOPAC rose in proportion to the concentration of DA injected, increasing 60–100-fold above baseline at the highest dose. In addition, cellular damage and gliosis were observed at the injection site. This was surrounded by a region of specific loss of tyrosine hydroxylase immunoreactivity with no detectable loss of synaptophysin immunoreactivity or alterations in cellular architecture. As in the case of the formation of cysteinyl-catechols, the extent of damage was dependent upon the DA concentration. Moreover, co-administration of DA with an equimolar concentration of ascorbate or GSH greatly reduced both protein cysteinyl-catechol formation and the loss of tyrosine hydroxylase immunoreactivity (Hastings et al., 1996). These findings suggest that DA can cause specific toxicity to dopaminergic neurons (also shown by Filloux and Townsend, 1993), that binding of DA to protein is correlated with this toxicity, and that both result from the oxidation of DA.

Acute administration of L-DOPA may increase the extracellular concentration of free radicals

Another approach to monitoring the formation of reactive metabolites is to focus on the free radicals formed when DA oxidizes. The major toxin in this group is OH·; however, this species is so reactive that it cannot be measured directly. Thus, we adapted the method of Floyd and colleagues (1984) in which OH· is trapped in vivo by exogenously administered salicylate, resulting in the stable and quantifiable products 2,3-dihydroxybenzoic acid (DHBA) and 2,5-DHBA (Giovanni et al., 1995). We utilized this approach to determine whether the large elevation of DA caused by the systemic administration of L-DOPA is associated with free radicals formation. 6-OHDA (4 µg) was placed along the nigrostriatal projection to produce a 90–95% loss of neostriatal DA terminals. One week later, a concentric microdialysis probe was implanted into the neostriatum, and 18 hr following the implantation salicylate (2 mM in the probe) was infused through the probe. Three hours after the onset of salicylate administration, a peripheral decarboxylase inhibitor was administered systemically, followed 30 min later by L-DOPA (100 mg/kg). 2,3-DHBA and DA were then measured in the dialysate (Liang et al., 1994).

L-DOPA caused a 10- to 20-fold increase in extracellular DA in the neostriatum of rats lesioned with 6-OHDA, but little or no change in the neostriatum of intact rats. Furthermore, there was a 69% increase in the extracellular levels of 2,3-DHBA in lesioned rats, but no significant change in control animals. Intact animals given an inhibitor of high affinity DA uptake showed intermediate increases in DA and 2,3-DHBA in response to L-DOPA. These results suggest that in partially lesioned animals, L-DOPA treatment may significantly increase the extracellular concentration of reactive DA metabolites, including OH· (Liang et al., 1994).

Hypothesis: DA-induced toxicity plays a role in Parkinson's disease

We hypothesize that there are at least two conditions under which toxic products of DA oxidation might lead to degeneration of DA neurons: an increase in the rate of formation of the DA oxidation products, and a decrease in the capacity of brain to inactivate these toxic compounds. There is reason to believe that these events might take place in the Parkinson's disease patient. Increased DA turnover, and thus the potential for increased oxidation, is indicated by the increased ratio of DA metabolites to DA that has been detected in post-mortem brain tissue (Bernheimer et al., 1973). The accumulation of neuromelanin, a polymerization product of oxidized DA, within DA neurons is an indication of oxidizing conditions (Graham et al., 1978). It is the neuromelanin-containing neurons of the substantia nigra that are most susceptible to degradation in Parkinson's disease (Hirsch et al., 1988), and the vulnerability of DA neurons appears to be related in part to their neuromelanin content (Kastner et al., 1992). In addition, a decreased capacity for free radical inactivation in Parkinson's disease is suggested by several observations, including: a decrease in the availability of such free radical defenses as GSH and GSH peroxidase (Kish et al., 1985; Perry et al., 1982); an increase in the apparent availability of free iron (Sofic et al., 1991); and an increase in malondialdehyde (MDA), an index of lipid peroxidation (Dexter et al., 1989) (for review see Agid et al., 1993).

Conclusions

The neurotoxic actions of 6-OHDA are mediated via the formation of free radicals and reactive quinones. Other conditions that promote the formation of free radicals within and around DA neurons also have the potential to cause neuronal damage. Since DA itself can form such compounds, conditions leading to high local concentrations of DA, increased DA turnover, or decreased capacity to inactivate these reactive metabolites, may cause neurodegeneration.

There are at least three points when a high local concentration of DA or its neurotoxic products may occur during the course of Parkinson's disease: (1) initiation of the disease, when conditions which cause sudden increases in extracellular DA, or a dysfunction in one or more of the buffering systems for removing free radicals formed during DA metabolism, may serve as an initiating factor in certain forms of the disease, (2) during the progression of the disease as a result of increased DA turnover and high local concentrations of the transmitter, and (3) during treatment of the disease due to increases in the availability of DA within and external to DA neurons that accompany L-DOPA therapy and, more recently, monoamine oxidase inhibitors. Further studies will be needed to determine whether reactive metabolites of endogenous DA do indeed contribute to the pathology of Parkinson's disease and, if so, whether conditions that increase the antioxidant capacity of brain will have therapeutic effects.

Acknowledgements

Many colleagues have contributed to the research and concepts summarized in this review, including: E. D. Abercrombie, A. Giovanni, D. A. Lewis, L. Ping Liang, A. Rabinovic, G. L. Snyder, and M. K. Stachowiak. The authors also thank S. D. Giegel and D. Malanosky for help in preparing this manuscript. This work was supported in part by U.S. Public Health Service Grants NS19608, MH43947, MH29670, MH45156, MH00058, and NS09076.

References

Abercrombie ED, Bonatz AE, Zigmond MJ (1990) Effects of L-DOPA on extracellular dopamine in striatum of normal and 6-hydroxydopamine-treated rats. Brain Res 525: 36–44

Agid Y, Ruberg M, Javoy-Agid F, et al (1993) Are dopaminergic neurons selectively vulnerable to Parkinson's disease? Adv Neurol 60: 148–163

Bernheimer H, Birkmayer W, Hornykiewicz O, Jellinger K, Seitelberger F (1973) Brain dopamine and the syndromes of Parkinson and Huntington: clinical morphological and neurochemical correlations. J Neurol Sci 20: 415–455

Chapman AG, Durmuller N, Lees GJ, Meldrum BS (1989) Excitoxicity of NMDA and kainic acid is modulated by nigrostriatal dopaminergic fibres. Neurosci Lett 107: 256–260

Cohen G (1983) The pathobiology of Parkinson's disease: biochemical aspects of dopamine neuron senescence. J Neural Transm [Suppl] 19: 89–103

Dexter DT, Carter CJ, Wells FR, Javoy-Agid F, Agid Y, Lees A, Jenner P, Marsden CD (1989) Basal lipid peroxidation in substantia nigra is increased in Parkinson's disease. J Neurochem 52: 381–389

Filloux F, Townsend JJ (1993) Pre- and post-synaptic neurotoxic effects of dopamin demonstrated by intrastriatal injection. Exp Neurol 119: 79–88

Fornstedt B, Carlsson A (1991) Vitamin C deficiency facilitates 5-S-cysteinyldopamine formation in guinea pig striatum. J Neurochem 56: 407–414

Fornstedt B, Rosengren E, Carlsson A (1986) Occurrence and distribution of 5-S-cysteinyl derivatives of dopamine dopa and dopac in the brains of eight mammalian species. Neuropharmacol 25: 451–454

Fornstedt B, Pileblad E, Carlsson A (1990) In vivo autoxidation of dopamine in guinea pig striatum increases with age. J Neurochem 55: 655–659

Giovanni A, Liang LP, Hastings TG, Zigmond MJ (1995) The estimation of hydroxyl radical content in rat brain using salicylate: impact of methamphetamine. J Neurochem 64: 1819–1825

Graham DG (1978) Oxidative pathways for catecholamines in the genesis of neuromelanin and cytotoxic quinones. Mol Pharmacol 14: 633–643

Graham DG, Tiffany SM, Bell WR jr, Gutknecht WF (1978) Autoxidation versus covalent binding of quinones as the mechanism of toxicity of dopamine 6-hydroxydopamine and related compounds toward C1300 neuroblastoma cells in vitro. Mol Pharmacol 14: 644–653

Grisham MB, Perez VJ, Everse J (1987) Neuromelanogenic and cytotoxic properties of canine brainstem peroxidase. J Neurochem 48: 876–882

Halliwell B, Gutteridge JMC (1985) Oxygen radicals and the nervous system. Trends Neurosci 8: 22–26

Hastings TG (1995) Enzymatic oxidation of dopamine: the role of prostaglandin H synthase. J Neurochem 64: 919–924

Hastings TG, Zigmond MJ (1994) Identification of catechol-protein conjugates in neostriatal slices incubated with ^{3}H-dopamine: impact of ascorbic acid and gluthathione. J Neurochem 63: 1126–1132

Hastings TG, Lewis DA, Zigmond MJ (1996) Role of oxidation in the neurotoxic effects of intrastriatal dopamine injections. Proc Natl Sci USA 93: 1956–1961

Hirsch E, Graybiel AM, Agid YA (1988) Melanized dopaminergic neurons are differentially susceptible to degeneration in Parkinson's disease. Nature 334: 345–348

Johnson M, Stone DM, Hanson GR, Gibb JW (1987) Role of dopaminergic nigrostriatal pathway in metamphetamine-induced depression of the neostriatal serotonergic system. Eur J Pharmacol 135: 231–234

Kastner A, Hirsch EC, Lejeune O, Javoy-Agid F, Agid Y (1992) Is the vulnerability of neurons in substantia nigra of patients with Parkinson's disease related to their neuromelanin content? J Neurochem 59: 1080–1089

Kish SJ, Morito C, Hornykiewicz O (1985) Glutathione peroxidase activity in Parkinson's disease brain. Neurosci Lett 58: 343–346

Liang LP, Zigmond MJ (1993) Dopamine synthesis in neostriatal slices after intraventricular 6-hydroxydopamine. Soc Neurosci Abstr 19: 401

Liang LP, Hastings TG, Zigmond MJ (1994) Hydroxyl radical formation is increased in rat striatum after L-DOPA treatment. Soc Neurosci Abstr 20: 413

Michel PP, Hefti F (1990) Toxicity of 6-hydroxydopamine and dopamine for dopaminergic neurons in culture. J Neurosci Res 26: 428–435

Perry TL, Godin DV, Hansen S (1982) Parkinson's disease: a disorder due to nigral glutathione deficiency? Neurosci Lett 33: 305–310

Robinson TE, Whishaw IQ (1988) Normalization of extracellular dopamine in striatum following recovery from a partial unilateral 6-OHDA lesion of the substantia nigra: a microdialysis study in freely moving rats. Brain Res 450: 209–224

Rosenberg PA (1988) Catecholamine toxicity in cerebral cortex in dissociated cell culture. J Neurosci 8: 2887–2894

Schmidt CJ, Ritter JK, Sonsalla PK, Hanson GR, Gibb JW (1985) Role of dopamine in the neurotoxic effects of methamphetamine. J Pharmacol Exp Ther 233: 539–544

Snyder GL, Zigmond MJ (1990) The effects of L-DOPA on in vitro dopamine release from striatum. Brain Res 508: 181–187

Snyder GL, Keller RW, Zigmond MJ (1990) Dopamine efflux from striatal slices after intracerebral 6-hydroxydopamine: evidence for compensatory hyperactivity of residual terminals. J Pharmacol Exp Ther 253: 867–876

Sofic E, Paulus W, Jellinger K, Riederer P, Youdim MBH (1991) Selective increase of iron in substantia nigra zona compacta of Parkinsonian brains. J Neurochem 56: 978–982

Stachowiak MK, Keller RW Jr, Stricker EM, Zigmond MJ (1987) Increased dopamine efflux from striatal slices during development and after nigrostriatal bundle damage. J Neurosci 7: 1648–1654

Weinberger J, Nieves-Rosa J, Cohen G (1985) Nerve terminal damage in cerebral ischemia: protective effect of alpha-methyl-para-tyrosine. Stroke 16: 864–870

Zhang WQ, Tilson HA, Nanry KP, Hudson PM, Hong JS, Stachowiak MK (1988) Increased dopamine release from striata of rats after unilateral nigrostriatal bundle damage. Brain Res 461: 335–342

Zigmond MJ, Stricker EM (1989) Animal models of parkinsonism using selective neurotoxins: clinical and basic implications. Int Rev Neurobiol 31: 1–79

Zigmond MJ, Hastings TG, Abercrombie ED (1992) Neurochemical responses to 6-hydroxydopamine and L-DOPA therapy: implications for Parkinson's disease. Ann NY Acad Sci 648: 71–86

Authors' address: Dr. T. G. Hastings, Department of Neurology, University of Pittsburgh School of Medicine, Pittsburgh, PA 15213, U.S.A.

Role of interferons in demyelinating disease

K. P. Johnson

School of Medicine, University of Maryland, Baltimore, MD, USA

Summary. Interferon beta 1b (Betaseron®) was licensed by the U.S. Federal Food and Drug Administration in July 1993 as the first treatment to alter the natural history of multiple sclerosis (MS). The drug, injected subcutaneously every other day, reduced the frequency of relapses and the expansion of central white matter pathology as measured by MRI. Twelve previous interferon trials in MS, employing a variety of interferon preparations, doses and routes of administration, preceded this trial and provided the scientific foundation for its success. Beta interferon therapy probably inhibits gamma interferon to achieve its therapeutic effect. Future MS therapy may require combination treatment with multiple agents with complimentary immunologic effects.

Introduction

The preponderance of evidence strongly suggests that multiple sclerosis (MS) is a neuroimmunologic disease centered on a T-cell mediated attack on central nervous system (CNS) myelin (Dhib-Jalbut and McFarlin, 1990). The concept of the trimolecular complex: antigen-presenting cell (APC), antigen and T-cell receptor (TCR), is critical in understanding the immunologic attack and, more particularly, the role played by interferons both in stimulating MS disease activity and in treating the disease process. This concept is expanded later.

The interaction between the immune system and the CNS in MS is based on several well established features. Cerebral spinal fluid (CSF) commonly shows a modest increase in cells, primarily lymphocytes and an elevation of the gamma globulin portion of CSF proteins. Electrophoresis studies show that the CSF immunoglobulins are primarily oligoclonal in nature and present in a pattern similar to that seen in chronic viral infections. Unfortunately, intense study of the CSF oligoclonal proteins have failed to show reactivity to a single antigen and has not contributed to the understanding of the etiology of MS. Tissue from MS plaques displays an immunopathologic reaction, especially in the early and more active MS plaques (Table 1).

This evidence has been combined to create the concept of an autoimmune model of MS described in Fig. 1. MS occurs more frequently in Caucasians of northern European descent and is over represented in some families. Identical

Table 1. Immunopathologic changes in the multiple sclerosis lesion

1. Perivascular inflammation
2. Activated T-lymphocytes
3. Activated macrophages
4. Complement bound to macrophages
5. Myelin fragmentation and phagocytosis
6. Expression of HLA-Class II molecules on astrocytes and endothelial cells
7. Plasma cells
8. Ig in extracellular fluid

twins are much more likely to be concordant for the disease than fraternal twins or other siblings. Certain major histocompatibility antigens are over-expressed in MS populations, indicating a clear genetic susceptibility to the disease. Epidemiologic studies strongly suggest that an event occurring in childhood initiates the immunopathic process. While this early event has not been clearly identified, one possible component is a childhood virus infection affecting CNS myelin which leads to recognition of nervous tissue (white matter) by the immune system. The active disease process we know as MS appears in young adult life, possibly years or even decades after the initial immunologic sensitization. Two common life events have been clearly identi-fied as risk factors in the initiation of renewed activity. These include birth of a child, and the post-partum period when MS attacks accelerate and, more importantly, common viral respiratory infections which frequently stimulate new MS attacks (Panitch and Bever, 1993).

The final pathway of these various genetic and immunologic events is the MS lesion which starts as a discreet acute inflammatory process progressing over time to contain gliosis, demyelination and loss of oligodendrogliocytes.

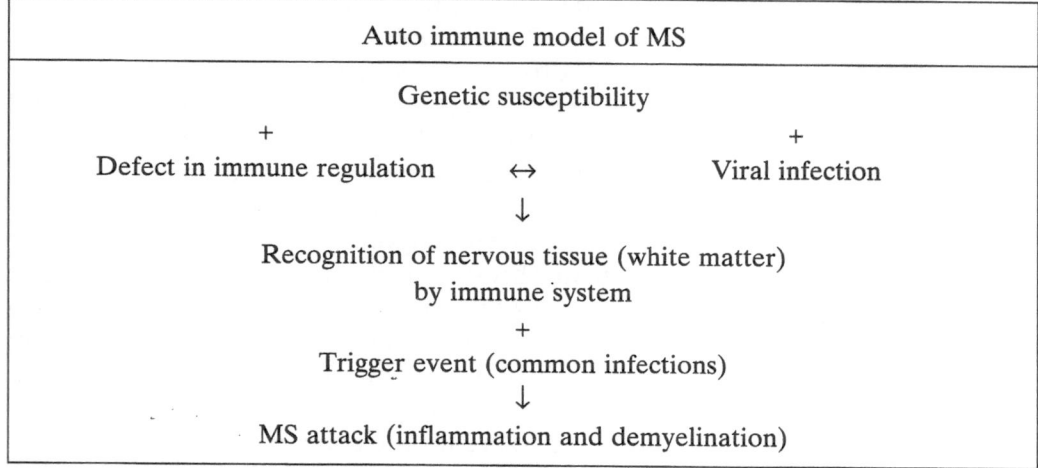

Fig. 1. A theoretical scheme relating genetic susceptibility with life events active in MS

History of MS interferon trials

An extensive clinical research effort, carried out over the past 15 years, has included 13 controlled human therapeutic trials employing all three species of human interferon (IFN), alpha and beta IFN known as type 1 and gamma IFN known as type 2. These trials have employed both natural and recombinant IFN preparations administered by several routes (subcutaneous, intramuscular, intravenous and intrathecal) at a variety of doses. This extensive therapeutic effort was recently reviewed in detail by Panitch (Panitch, 1992). Early studies employed natural interferons, alpha IFN given by subcutaneous injection (Knobler et al., 1984) and beta IFN by the intrathecal route (Jacobs et al., 1987). With both IFN preparations, some evidence of efficacy was noted primarily in the reduced frequency of new MS attacks.

The first and only trial of gamma IFN as a potential therapy for MS is now viewed as one of the pivotal hallmarks in understanding the pathogenesis of MS. Panitch and co-workers (Panitch et al., 1987) administered recombinant gamma interferon at 3 dose levels to 18 patients by an intravenous route over a one month period and observed that 7 of the 18 had new MS attacks. All of the episodes occurred in anatomic areas which had previously been involved by MS and all were mild in nature. This study identified gamma IFN as a potent stimulus of new MS activity which, in this case, was probably occurring in an area of previous MS involvement. This trial also suggested that any treatment which inhibited gamma IFN and its effects might show efficiency in reducing MS activity. Alpha and beta IFNs were known to have a suppressive effect on gamma IFN.

In 1986, a pilot study in 30 relapsing-remitting (R/R) MS patients was begun to determine the safety and appropriate dose level of a recombinant beta IFN produced in E coli. Within 6 months, it was apparent that the preparation was safe and a subcutaneous dose of 8 miu every other day was identified for further evaluation (Johnson et al., 1990). This led in 1988 to a pivotal Phase 3 trial of 372 R/R ambulatory MS patients randomized to high dose (8 miu), low dose (1.6 miu) or placebo groups who self-administered study medication for 3 years. This trial was analyzed and reported in April of 1993 (The IFNB Multiple Sclerosis Study Group, 1993). In July 1993, the U.S. Federal Food and Drug Administration licensed this preparation (Betaseron®) for the treatment of ambulatory, relapsing-remitting MS patients.

The Betaseron® trial demonstrated that the relapse rate was reduced by 1/3 and the occurrence of serious relapses by 50% when patients received the high dose every other day by subcutaneous injection. The time to first relapse was doubled to 300 days. Importantly, pathologic expansion of the MS burden of disease as measured by yearly MRI, was essentially halted in the group receiving high dose IFN, whereas there was a 20% expansion in the placebo group (Paty et al., 1993).

Side effects of long-term Betaseron® dosing were relatively mild and tolerable for the study population. A flu-like syndrome typical of IFN treatment was common for the first 3 to 8 weeks and consisted of fever, muscle aches, weakness and fatigue. Injection site reactions were frequent. Over time, these adverse reactions modified and then disappeared. Many patients have now been on continuous therapy for 5 to 7 years. Occasionally, intermittent anemia and elevated liver function tests were noted. None of these reactions have been serious enough to limit therapy in most patients.

Mechanisms of interferon action

Gamma IFN has several immunologic functions. Perhaps, most importantly for MS activity, is the induction of MHC Class II cell surface molecules which are essential for antigen presentation within the trimolecular complex (Basham and Merigan, 1983). Gamma IFN also activates macrophages to synthesize proteinases which can degrade myelin. It induces endothelial cells to express adhesion molecules in the CNS which act as homing signals for lymphocytes (Panitch and Bever, 1993). It is also known that these functions of gamma IFN are augmented by another cytokine, tumor necrosis factor, which may also have pathogenic importance in MS demyelination (Sharief and Hentges, 1991).

Alpha and beta IFN are able to inhibit or down-regulate the immunologic effects and synthesis of gamma IFN (Panitch and Bever, 1993). This function has been shown to remain active during prolonged periods of beta IFN therapy. Also, beta IFN enhances suppressor cell activity which has been recognized to be deficient in MS patients (Noronha et al., 1990).

Future considerations

Although beta IFN (Betaseron®) has been approved and is being marketed for R/R MS in the United States, the future course and scope of IFN therapy for demyelinating diseases is far from complete. There are several recombinant beta IFNs available for therapeutic consideration, each of which has somewhat different molecular and possibly therapeutic characteristics (Table 2). The issue of optimal dose and perhaps route of administration and

Table 2. Partial list of commercial human interferon preparations

			Natural	Recombinant
Type 1:	Alpha	Schering Plough		+
		Hoffman LaRoche		+
		Purdue Fredrick	+	
	Beta	Berlex		+
		Biogen		+
		Serono	+	+
		Amgen		+
Type 2:	Gamma	Genentec		+

frequency of treatment has still been only partially explored. There is also preliminary evidence that high doses of recombinant alpha IFN produce very similar immunologic, MRI and clinical effects as Betaseron®, yet little is known about effective dosing for alpha IFN. No therapeutic studies have been done to date on the more advanced chronic-progressive stages of MS with any IFN preparation.

Of great potential therapeutic interest is the concept of combination therapy in which drugs with different mechanisms of action are combined in an effort to more fully control the course and pathology of MS. The next agent most likely to be recognized as a new therapy for MS is Copolymer I, a random polymer of 4 amino acids; alanine, glutamic acid, lysine and tyrosine (Bornstein and Johnson, 1992). Preliminary in vitro studies have shown a synergistic effect between Betaseron® and Copolymer I in controlling immunologic reactions thought to be important in MS.

References

Basham TY, Merigan TC (1983) Recombinant interferon gamma increases HLA-DR synthesis and expression. J Immunol 130: 1492–1494

Bornstein MB, Johnson KP (1992) Treatment of multiple sclerosis with copolymer I. In: Rudick RA, Goodkin DE (eds) Treatment of multiple sclerosis. Trial design, results and future perspectives. Springer, Berlin Heidelberg New York Tokyo

Dhib-Jalbut SS, McFarlin DE (1990) Immunology of multiple sclerosis. Ann Allerg 64: 433–444

IFNB Multiple Sclerosis Study Group (1993) Interferon beta-1b is effective in relapsing-remitting multiple sclerosis. I. Clinical results of a multicenter, randomized, double-blind, placebo-controlled trial. Neurology 43: 655–661

Jacobs L, Salazar AM, et al (1987) Intrathecally administered natural human fibroblast interferon reduces exacerbations of multiple sclerosis: results of a multicenter, double-blinded study. Arch Neurol 44: 589–595

Johnson KP, Knobler RL, et al (1990) Recombinant human interferon beta treatment of relapsing-remitting multiple sclerosis: pilot study results. Neurology 40 [Suppl 1]: 261

Knobler RL, Panitch HS, Braheny SL, et al (1984) Systemic alpha-interferon therapy of multiple sclerosis. Neurology 34: 1273–1279

Noronha A, Toscas A, Jensen MA (1990) Interferon beta augments suppressor cell function in multiple sclerosis. Ann Neurol 27: 207–210

Panitch HS (1992) Interferons in multiple sclerosis. A review of the evidence. Drugs 44 (6): 946–962

Panitch HS, Bever CT jr (1993) Clinical trials of interferons in multiple sclerosis. What have we learned? J Neuroimm 46: 155–164

Panitch HS, Hirsch RL, Schindler J, Johnson KP (1987) Treatment of multiple sclerosis with gamma interferon: exacerbations associated with activation of the immune system. Neurology 37: 1097–1120

Paty DW, Li DKB, et al (1993) Interferon beta-1b is effective in relapsing-remitting multiple sclerosis. II. MRI analysis results of a multicenter, randomized, double-blind, placebo-controlled trial. Neurology 43: 662–667

Sharief MK, Hentges R (1991) Association between tumor necrosis factor-α and disease progression in patients with multiple sclerosis. N Engl J Med 325: 467–472

Author's address: Dr. K. P. Johnson, School of Medicine, Room N4 W46, University of Maryland, 22, South Greene Street, Baltimore, MD 21201–1595, U.S.A.

Role of interferons in demyelinating diseases

B. G. W. Arnason, A. Toscas, A. Dayal, Z. Qu, and **A. Noronha**

Department of Neurology, University of Chicago, Chicago, IL, U.S.A.

Summary. IFNβ-1b reduces the frequency of major multiple sclerosis attacks by 50 percent. Serial MRI scanning over the course of the clinical trial that led to approval of the agent revealed a significant lessening both in disease activity and in accumulating burden of disease in IFNβ-1b-treated patients compared to placebo-treated controls.

The mechanism by which IFNβ-1b exerts its beneficial effect in multiple sclerosis is unknown. T suppressor cell function fails during MS attacks and is persistently subnormal in multiple sclerosis patients with progressive disease. IFNβ-1b partially restores suppressor function in multiple sclerosis patients.

IFNβ-1b also inhibits release of lymphotoxin, tumor necrosis factor, and interferon gamma, at least in vitro. All three cytokines are toxic to oligodendrocytes. In contrast, production of transforming growth factor β-1 (TGFβ1) is increased by IFNβ-1b. TGFβ1 is an immunosuppressive cytokine. All of the above listed actions of IFNβ-1b could contribute to its beneficial effect. Perhaps all do.

Interferon β-1B was approved by the U.S. Food and Drug Administration for the prophylactic treatment of relapsing-remitting Multiple Sclerosis (MS) in July 1993. Many thousands of MS patients are now receiving the agent.

In a double-blind, placebo-controlled trial, IFNβ-1B at a dose of 8 million units (MU) given subcutaneously every other day reduced MS attacks by 35%, major attacks by 50%, and MS-related hospitalizations by 40% (IFNB Multiple Sclerosis Study Group, 1993; Paty and Li, 1993; Arnason, 1993). The decline in attack frequency in the 8MU-treated group held for the entire 5 years of the trial. Time spent experiencing attacks averaged 19 days in the 8MU treatment arm compared to 44 days in the placebo arm over the first 3 years of the trial. Yearly magnetic resonance scans showed a continuing increment in disease burden over 5 years in the placebo group and virtually no change in mean disease burden in the 8MU-treated group. The difference was highly significant at all time points.

For patients who lost at least one point on the EDSS (Kurtzke) disability scale there was a modest difference in favor of the 8MU IFNβ-1B-treated group compared to placebo after 3 years (p < 0.043). This aside, no significant beneficial effect on disability as measured by the EDSS and Scripps rating scales was seen at any time point. On first consideration this failure is surprising and worrisome. One might have anticipated that a difference in accumulating disease burden on MRI would correlate with a beneficial effect on disability score inasmuch as disease burden and disability do correlate with one another when large numbers of patients are evaluated. A correlation between disease burden and disability score was in fact observed when the two were compared for patients entered into all three arms (placebo, 1.6MU IFNβ-1B, 8MU IFNβ-1B) of the IFNβ-1B trial. The failure to demonstrate a significant difference in disability scores when the 8MU IFNβ-1B and placebo subgroups were compared directly to one another could relate to sample size. The study as originally planned was not powered to show a meaningful effect on disability taking into account the known insensitivity of the EDSS rating scale. The placebo group as a whole deteriorated surprisingly little over the course of the trial (0.3 points after 5 years). This may have contributed to the failure to show an effect on disability. In a trial of IFNβ-1A, an IFNβ preparation that differs slightly from IFNβ-1B, a favorable and significant effect on MS disability scores is said to have been shown (Jacobs et al., 1994), but the placebo group in this trial fared considerably worse than the placebo group in the IFNβ-1B trial. Another possibility would be that attack frequency and accumulating disease burden have little to do with ultimate disability, a contention that flies in the face of the literature and clinical experience. A trial of IFNβ-1B in progressive MS is underway at present in Europe and a second trial will commence in the USA and Canada shortly. Since disability advances more rapidly in progressive than in relapsing-remitting MS, and the trials are powered to show an effect on disability if one truly exists, the results of these trials should resolve the disability issue one way or the other.

Given that IFNβ-1B favorably affects the natural history of MS, at least in terms of attack frequency and MRI-determined disease burden, the question as to how it works can be posed. At the outset it should be stated that the mechanism of action of IFNβ-1B in MS is not established. Interferons were so named because of their ability to interfere with viral replication. Viral infections are known to trigger MS attacks (Sibley et al., 1985; Panitch et al., 1991a) and if IFNβ-1B lessened the frequency and/or severity of viral infections this could provide a simple explanation of how and why it works. No evidence that IFNβ-1B lessened the number of viral infections emerged from the pivotal trial which led to its approval by the U.S. Food and Drug Administration. The issue of viral infection severity was not addressed and operationally this would be difficult to assess. Our view is that the basis for the beneficial effect of IFNβ-1B in MS is likely to lie elsewhere.

Interferons are divided into two classes known as Class I and Class II (reviewed in Arnason and Reder, 1994). Class I interferons are comprised of IFNα, of which some 20 types exist, and IFNβ, which exists as a single molecule. IFNα and IFNβ bind to the same receptor and share many, though not all, actions. They also share considerable sequence homology. Class II interferon refers to IFNγ, the sole representative of the class. IFNγ is structurally distinct from Class I IFNs. IFNγ binds to a different receptor and functions chiefly as an immune response modulator. IFNγ also has a very different role in MS from that of IFNβ.

In a brief clinical trial in which IFNγ was administered to MS patients, 7 of 16 subjects promptly developed attacks (Panitch et al., 1987a). The study was quickly abandoned but suggested a role for IFNγ as an initiator of MS disease activity. Other evidence supports this contention. Increased IFNγ can be detected in the circulation at the onset of MS attacks as can increased numbers of IFNγ-secreting cells (Beck et al., 1988; Lu et al., 1993). IFNγ is produced by activated Thl-type T cells pointing to a major role for them in the tissue destruction that characterizes MS attacks. When IFNβ-1B and a stimulating lectin are added to peripheral blood mononuclear cells (PBMC) from MS patients, IFNγ release to the culture supernate is inhibited compared to supernates of PBMCs cultured with lectin alone (Noronha et al., 1993a; Panitch et al., 1987b). At face value these in vitro observations might suggest that IFNβ-1B inhibits MS attacks simply by inhibiting IFNγ production. In vivo, however, the situation is more complex.

IFNγ is the sole inducer of MHC-Class II molecule expression. When patients are started on IFNβ-1B MHC-Class II molecule expression on circulating monocytes rises (Spear et al., 1987). IFNγ is also a potent inducer of neopterin. When patients are started on IFNβ-1B levels of neopterin in the circulation rise (Chiang et al., 1993). The findings suggest that, far from inhibiting IFNγ secretion in vivo, IFNβ-1B may actually induce it. We decided to examine this possibility directly using the immunospot technique to count circulating IFNγ-secreting cells in the circulation of MS patients prior to and after induction with IFNβ-1B. Data are given in Table 1.

It is evident that IFNβ-1B does increase IFNγ-secreting cells in the circulation and that they remain elevated for the first 2 months after starting treatment (Dayal et al., 1995). Prednisone, at a dose of 30 mgm per day, prevents this IFNγ-secreting cell surge.

Systemic side effects are frequent when IFNβ-1B treatment is begun. Fever, chills, muscle aches, headache (i.e., a "flu-like" syndrome) are described by approximately 50 percent of MS patients when they start IFNβ-1B treatment. Fever depends on endogenous pyrogens that act at the level of the

Table 1. Interferon γ-secreting cells (IFNγ-sc) in MS patients treated with IFNβ-1B

Group[a]	Mean number of IFNγ-sc/10^5 cells[b]	Elevated[c]
Controls (44)	98 ± 28	0
MS at baseline (55)	97 ± 58	13%
MS 1st week (27)	154 ± 89	59%
MS 2–10 weeks (12)	173 ± 84	75%
MS > 3 months (9)	78 ± 46	0
MS on prednisone for 1st month (13)	88 ± 37	0

[a]Number of subjects in parentheses; [b]mean ± S.D.; [c]more than 2 S.D. above control mean

hypothalamus. Among the endogenous pyrogens are IL-1 and TNF both of which are made by and released from activated macrophages. Importantly, IFNγ primes macrophages to release IL-1 and TNF so that the systemic side effects observed on starting IFNβ-1B treatment are an expected consequence of IFNγ induction by IFNβ-1B. No effect on attack frequency was noted during the first month on treatment in the pivotal trial of IFNβ-1B in MS. Systemic side effects abate over the first month or two on IFNβ-1B treatment and efficacy surfaces as side effects lessen. Note that the IFNγ-secreting cell surge ends at this time as well. The lesson from the above is that short-term and long-term effects of IFNβ-1B treatment can and do differ. Whether patients on IFNβ-1B for periods of a year or more ultimately show a decline in IFNγ secreting cells to levels below baseline is unknown but under study.

CD8 suppressor function, as measured in vitro, fails during MS attacks and recovers as attacks end (Arnason and Antel, 1978; Antel et al., 1979). CD8 suppressor function is persistently subnormal in secondarily progressive MS. Experimental allergic encephalomyelitis (EAE) provides the best available experimental model for MS. The disease is induced by immunizing experimental animals with myelin or the myelin protein known as myelin basic protein (MBP). In most strains of rats or mice EAE, unlike MS, is a monophasic illness. Recovered animals rarely if ever relapse. Yet, when spleen or lymph node cells from recovered animals are stimulated to proliferate in vitro with a nonspecific lectin and the stimulated cells are then transferred adoptively into a naive animal of the same strain the recipient develops EAE. It follows that recovered animals harbor memory T cells that are fully capable of causing disease, yet in the milieu of the recovered animal they are somehow held in check.

It is possible to generate mice lacking CD8 cells. Such CD8 "knockout" mice develop EAE exactly as controls and recover from the attack exactly as controls as well. Once the first attack has abated the knockout and control mice behave quite differently. Knockouts relapse over and over again whereas controls do not (Koh et al., 1992). The data point to a policing role for CD8 cells once a first attack of EAE has ended.

Table 2. IFNβ-1B augments suppressor function in MS

Group[a]	% suppression[b]
1. Controls (94)	37.2 ± 2.5
2. MS baseline (31)	17.5 ± 4.2
3. MS placebo (41)	20.0 ± 3.3
4. MS IFNβ-1B 1.6MU (45)	24.8 ± 3.8
5. MS IFNβ-1B 8.0MU (40)	32.1 ± 3.6

[a]Number of observations in brackets. 31 MS patients were studied at baseline and then were entered into the treatment or placebo arms. Patients on treatment were studied on multiple occasions. [b]Mean ± S.E.M. Significance: 1 vs. 2,3,4 = all $p < 0.01$; 1 vs. 5 = n.s.; 2 vs 3,4 = n.s.; 2 vs. 5. = $p < 0.02$; 3 vs. 4 = n.s.; 3 vs. 5 = $p < 0.02$

Since CD8 suppressor failure correlates with attacks, we studied it in the MS patients at the University of Chicago entered into the IFNβ-1B clinical trial. CD8 suppressor function was subnormal at time of entry into the trial, an expected finding since the patients were selected for the study on the basis of frequent attacks. The IFNβ-1B-treated patients showed an increase in suppressor function that was not seen in the placebo-treated group (Noronha et al., 1994). The increase was greater for the group treated with 8MU of IFNβ-1B (the currently approved dosage) than for the group treated with 1.6MU of IFNβ-1B.

More recently one of us (A.N.) has revisited this issue, looking to see when after starting IFNβ-1B treatment the increase in suppressor function becomes apparent. The work is ongoing, but preliminary data suggest that suppressor function recovery occurs 2 to 3 months after starting treatment, i.e., as IFNγ-secreting cell numbers fall back into the normal range.

It has been suggested that IFNγ secretion and suppressor function bear a reciprocal relationship to one another (Panitch et al., 1991b). IFNβ-1B added directly to cultured PBMC increases CD8 suppressor cell function promptly and substantially (Noronha et al., 1990). IFNγ added together with IFNβ-1B has no effect on the ability of IFNβ-1B to increase suppressor cell function (Noronha et al., 1992). This finding argues that if there is an interaction between suppressor function and IFNγ the flow is from the suppressor cells to the IFNγ-secreting cells rather than in the other direction.

Since IFNβ-1B augments suppressor function in vitro and since increased suppressor function may tie to the beneficial effect of IFNβ-1B in MS, one strategy to improve the efficacy of IFNβ-1B would be to identify agents that potentiate this action of IFNβ-1B. An example of this approach is given in Table 3. Trans-retinoic acid increases the ability of IFNβ-1B to increase the suppressor cell function of MS patients.

The technique can be applied as a screening mechanism for agents to be considered for adjunctive combination with IFNβ-1B in MS treatment regimens.

Table 3. Trans-retinoic acid (RA) potentiates suppressor function augmentation by IFNβ-1B of PBMC from MS patients

Group[a]	% suppression[d]
A. MS PBMC (14)	9.0 ± 1.8
B. MS PBMC + IFNβ-1B high[b] (14)	25.2 ± 3.2
C. MS PBMC + RA[c] (14)	14.6 ± 2.9
D. MS PBMC + IFNβ-1B high + RA (14)	36.1 ± 3.9
E. MS PBMC + IFNβ-1B low[e] (7)	21.0 ± 4.0
F. MS PBMC + IFNβ-1B low + RA (7)	35.0 ± 4.7

[a]Number of observations in brackets; [b]IFNβ-1B at 1000 μ/ml; [c]RA at 5×10^{-7}M; [d]Mean ± S.E.M.; [e]IFNβ-1B at 100 μ/ml. Significance: A vs. B = $p < 0.001$; A vs. C. = n.s.; A vs. D. = $p < 0.0001$; A vs. E = $p < 0.01$; E vs. F = $p < 0.05$; A vs. F = $p < 0.001$; B vs. C = $p < 0.02$; B vs. D = $p < 0.04$

Several other actions of IFNβ-1B may also be germane to its beneficial effect in MS. Lymphotoxin (LT) and tumor necrosis factor (TNF), the former made by Th1 cells, the latter chiefly by macrophages, are both toxic to oligodendrocytes at least in vitro and may contribute to lesion formation in MS. Both can be detected in active MS plaques. IFNβ-1B lessens release of LT and TNF at least in vitro. Whether it does the same in vivo, either at once or after a delay, is not known. Two other molecules merit comment. IL-10, a product of Th2 cells and of activated macrophages, exerts an inhibitory effect on Th1 cells. Th1 cells it will be recalled make IFNγ and LT and are thought to have a primary role in lesion generation in MS. IFNβ-1B increases IL-10 production by macrophages in vitro, modestly to be sure, but nonetheless reproducibly and significantly. Again, nothing is known about the situation in vivo. Transforming growth factor beta 1 (TGFβ1) is an immunosuppressive cytokine. TGFβ1 inhibits EAE and is being considered for testing in MS. IFNβ-1B augments TGFβ1 production in vitro (Noronha et al., 1993b), but whether it does so in vivo as well is once again not known.

IFNβ-1B exerts an anti-proliferative effect on T cells (Noronha et al., 1993a). Since T cell proliferation is an early event in generation of an immune response, this anti-proliferative action could contribute to the overall effect of IFNβ-1B in MS. Finally, there are circumstances in vitro where IFNβ-1B inhibits the ability of IFNγ to up-regulate MHC-Class II molecule expression. Since MHC molecules present antigen to T cells the overall efficiency of antigen presentation, and hence the magnitude of the immune response generated, could be lessened if IFNβ-1B exerted this effect in vivo. Obviously this does not occur early, but it might well occur at later times.

It is striking that IFNβ-1B has many effects that might be expected to alter the course of MS favorably. At this juncture it is not possible to pick and choose among them for the one or two that are the "key" to its action. Indeed, several may be involved simultaneously.

Acknowledgements

As always we thank Ms. M. Murphy for typing the manuscript and for helpful editorial comments. Supported by PHS-PO1-NINDS-NS 24575, a gift from the Butz Foundation, and a grant from Berlex, Inc.

References

Antel JP, Arnason BGW, Medof ME (1979) Suppressor cell function in multiple sclerosis: correlation with clinical disease activity. Ann Neurol 5: 338–342

Arnason BGW (1993) Interferon beta in multiple sclerosis. Neurology 43: 641–643

Arnason BGW, Antel JP (1978) Suppressor cell function in multiple sclerosis. Ann Immunol (Paris) 129C: 159–170

Arnason BGW, Reder AT (1994) Interferons and multiple sclerosis. Clin Neuropharmacol 17: 495–547

Beck J, Rondot P, Catinot L, Falcoff E, Kirchner H, Wietzerbin J (1988) Increased production of interferon gamma and tumor necrosis factor precedes clinical manifestation in multiple sclerosis: do cytokines trigger off exacerbations? Acta Scand 78: 318–323

Chiang J, Gloff CA, Yoshizawa CN, Williams GJ (1993) Pharmacokinetics of recombinant human interferon-b$_{ser}$ in healthy volunteers and its effect on serum neopterin. Pharm Res 10: 567–572

Dayal AS, Jensen MA, Lledo A, Arnáson BGW (1995) Interferon gamma-secreting cells in multiple sclerosis patients treated with interferon beta-1B. Neurology 45: 2173–2177

IFNB Multiple Sclerosis Study Group (1993) Interferon beta-1b is effective in relapsing-remitting multiple sclerosis I. Clinical results of a multicenter, randomized, double-blind, placebo-controlled trial. Neurology 43: 655–661

Jacobs L, Cookfair D, Rudick R, et al (1994) Results of a phase III trial of intramuscular recombinant beta interferon as treatment for multiple sclerosis. Ann Neurol 36: 259

Koh D-R, Fung-Leung W-P, Ho A, Gray D, Acha-Orbea H, Mak T-W (1992) Less mortality but more relapses in experimental allergic encephalomyelitis in CD8$^{-/-}$ mice. Science 256: 1210–1213

Lu C-Z, Jensen MA, Arnason BGW (1993) Interferon gamma- and interleukin-4-secreting cells in multiple sclerosis. J Neuroimmunol 46: 123–128

Noronha A, Toscas A, Jensen MA (1990) Interferon beta augments suppressor cell function in multiple sclerosis. Ann Neurol 27: 207–210

Noronha A, Toscas A, Jensen MA (1992) Contrasting effects of alpha, beta, and gamma interferons on nonspecific suppressor cell functions in multiple sclerosis. Ann Neurol 31: 103–106

Noronha A, Toscas A, Jensen MA (1993a) Interferon β decreases T cell activation and interferon γ production in multiple sclerosis. J Neuroimmunol 46: 145–154

Noronha A, Jensen M, Toscas A (1993b) TGF-b activity in MS: effect of IFN-b. Neurology 43 [Suppl]: 355

Noronha A, Toscas A, Arnason BGW, Jensen M (1994) IFN-beta augments in vivo suppressor function in MS. Neurology 44 [Suppl]: A212

Panitch HS, Hirsch RL, Schindler J, Johnson KP (1987a) Treatment of multiple sclerosis with gamma interferon: exacerbations associated with activation of the immune system. Neurology 37: 1097–1102

Panitch HS, Folus JS, Johnson KP (1987b) Recombinant beta interferon inhibits gamma interferon production in multiple sclerosis. Ann Neurol 22: 139

Panitch HS, Bever CT, Katz E, Johnson KP (1991a) Upper respiratory tract infections trigger attacks of multiple sclerosis in patients treated with interferon-β [abstract]. J Neuroimmunol 35 [Suppl 1]: 125

Panitch HS, Folus JS, Johnson KP (1991b) Activated suppressor cells inhibit synthesis of interferon-g in patients with multiple sclerosis and normal subjects [abstract]. J Neuroimmunol 35 [Suppl 1]: 186

Paty DW, Li DKB (1993) UBC MS/MRI Study Group, IFNB Multiple Sclerosis Study Group. Interferon beta-1b is effective in relapsing-remitting multiple sclerosis II. MRI analysis results of a multicenter, double-blind, placebo-controlled trial. Neurology 43: 655–661

Spear GT, Paulnock DM, Jordan RL, Meltzer DM, Merritt JA, Borden EC (1987) Enhancement of monocyte class I and II histocompatibility antigen expression in man by in vivo β-interferon. Clin Exp Immunol 69: 107–115

Sibley WA, Bamford CR, Clark KI (1985) Clinical viral infections and multiple sclerosis. Lancet 1: 1313–1315

Authors' address: Dr. B. G. W. Arnason, University of Chicago, Department of Neurology, 5841 S. Maryland Avenue, MC 2030, Chicago, IL 60637, U.S.A.

Mechanism of amyloid β protein induced neuronal cell death: current concepts and future perspectives

C. Behl[1] and Y. Sagara[2]

[1] Clinical Institute, Max-Planck-Institute of Psychiatry, Munich,
Federal Republic of Germany
[2] The Salk Institute for Biological Studies, San Diego, CA, U.S.A.

Summary. Amyloid β protein (Aβ) is a 40 to 43 amino acid peptide which is associated with plaques in the brains of Alzheimer's patients and is cytotoxic to cultured neurons. A number of antioxidants protect both primary central nervous system (CNS) cultures and clonal cell lines from Aβ toxicity, suggesting that one pathway to Aβ cytotoxicity results in free radical damage. Aβ causes increased levels of H_2O_2 and lipid peroxides to accumulate in cells. The H_2O_2 degrading enzyme catalase protects cells from Aβ toxicity. Clonal cell lines selected for their resistance to Aβ toxicity also become resistant to the cytolytic action of H_2O_2. In addition, Aβ induces NF-kB activity, a transcription factor thought to be regulated by oxidative stress. Finally, Aβ induced H_2O_2 production and Aβ toxicity are blocked by reagents which inhibit flavin oxidases, suggesting that Aβ activates a member of this class of enzymes. These results show that the cytotoxic action of Aβ on neurons results from free radical damage to susceptible cells (Behl et al., 1994b).

Introduction

Amyloid beta protein (Aβ) accumulates in the central nervous system plaques which are a characteristic feature of Alzheimer's disease (AD) (Glenner, 1988). Since mutations in the amyloid β precursor protein (AβPP) in some cases of familial AD lead to the overproduction of Aβ (for review see Schubert, 1994), and since Aβ is directly toxic to cultured nerve cells (Yankner et al., 1990; Behl et al., 1992) it can be argued that Aβ toxicity is a major cause of the CNS damage which occurs in AD. It is therefore important to understand the biochemical mechanisms which lead to Aβ induced cell death.

The initial evidence for Aβ toxicity was derived from CNS primary cultures (Koh et al., 1990; Yankner et al., 1990). In addition to primary cultures, some clonal nerve and glial cell lines respond to Aβ. These cells afford homogeneous populations and the ability to select clonal variants for use in dissecting biochemical pathways. Recently, sensitive assays for neurotoxins have been developed which are based upon clonal cell lines (Schubert et al., 1992; Behl et al., 1992). We have used clonal cell lines as well as CNS primary neurons which respond to Aβ to examine the events leading to Aβ induced cell death.

The sympathetic precursor-like cell line PC12 (Greene and Tischler, 1976) and cultured rat cortical neurons are killed by Aβ. Within one hour after exposure to Aβ there is ultrastructural evidence for damage to the Golgi apparatus, mitochondria, and other membrane systems within the cytoplasm, and a breakdown of microtubules in neurites, while the nucleus remains initially intact (Behl et al., 1994a). These data suggest that cytoplasmic membrane damage is an early event in Aβ toxicity. Based both upon biochemical and ultrastructural features of Aβ induced cell death, it was demonstrated that Aβ inducesdeath via a process closely resembling necrosis rather than apoptosis (Behl et al., 1994a).

Since membrane damage can be caused by free radicals (Halliwell and Gutteridge, 1989), and since vitamin E rescues PC12 cells from Aβ toxicity (Behl et al., 1992), it was asked if any molecules specifically associated with oxidative stress could be induced by exposure to Aβ. The following experiments employ PC12 cells, a clonal CNS cell line (B12) and CNS primary cultures to show that Aβ causes the overproduction of H_2O_2 or related peroxides which leads to cell death (Behl et al., 1994b).

Antioxidants protect against Aβ

Exogenously applied Aβ can cause the death of the clonal sympathetic precursor cell line PC12, and Aβ toxicity is blocked by the antioxidants vitamin E and propylgallate (Behl et al., 1992). Of 35 clonal cell lines examined, several are reproducibly very sensitive to Aβ toxicity. One of these cell lines is B12 (Schubert et al., 1974; Schubert, 1984). In addition to PC12 and B12, cultured rat cortical neurons were employed.

Three assays were used to assess Aβ toxicity: (1) MTT assay, (2) LDH-release assay, and (3) trypan blue exclusion/cell counting. The cytotoxic response curves of B12 and CNS primary cultures to Aβ (1–40) are shown in Fig. 1 along with their response to Aβ (25–35), the biologically active fragment of Aβ (Yankner et al., 1990). For the neuronal primary cultures, the IC_{50} for both MTT reduction and LDH release are about the same approximately 2 μM (Fig. 1B). In contrast the clonal cell line is never completely lysed after a 48 hr exposure to Aβ concentrations of up to 10 μM, while the IC_{50} for MTT reduction is 30 nM (Fig. 1A). This difference between clonal cell lines and primary cultures is probably due to the fact that under these assay conditions B12 cells are actively growing while the primary cells are nondividing and are in the G_0 stage of their cell cycle. In all three cell systems the cells are equally sensitive to $Aβ_{25-35}$ and the naturally occurring $Aβ_{1-40}$.

It was previously demonstrated that antioxidants inhibit the toxic action of Aβ on PC12 cells (Behl et al., 1992). To extend these studies to the B12 cells and CNS primary cultures, a wide range of antioxidants were tested for their ability to reduce the toxicity of Ab_{25-35}. Table 1 shows the concentration of each anti-

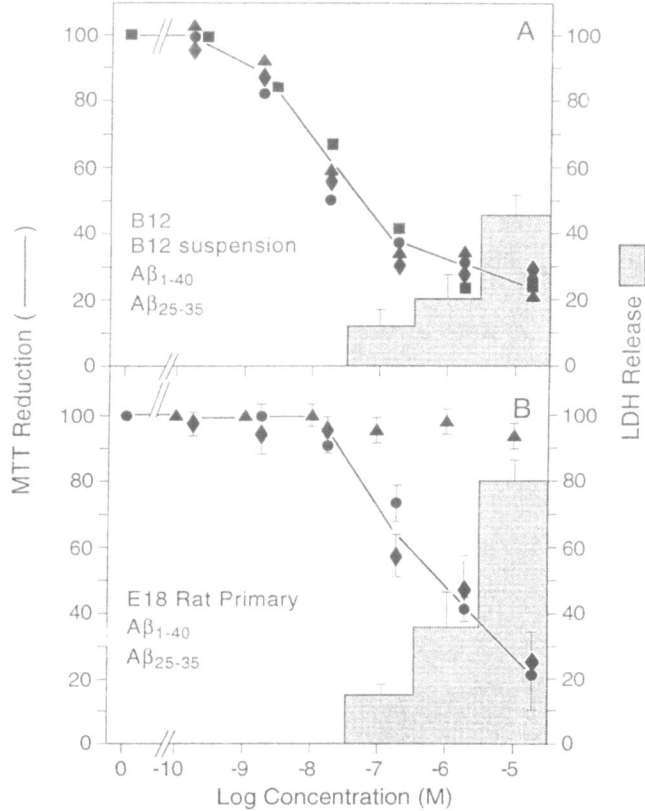

Fig. 1. Aβ toxicity assays. Cells were exposed to the indicated forms of Aβ for 24 hours. Cell viability was assayed using MTT or LDH release. Results are expressed as percent reduction of MTT relative to untreated controls or percent LDH released relative to total LDH in the culture. The data are means plus or minus the standard error of the mean for triplicate cultures. **A** B12: ◆–◆ Aβ$_{1-40}$; ▲–▲ Aβ$_{25-35}$; ▼–▼ scrambled Aβ$_{25-35}$. B12 suspension: ●–● Aβ$_{25-35}$; ■–■ Aβ$_{1-40}$. **B** Rat CNS primary, ◆–◆ Aβ$_{1-40}$; ●–● Aβ$_{25-35}$; ▲–▲ scrambled Aβ$_{25-35}$. For the LDH release assays Aβ$_{1-40}$ was used (Figure taken from Behl et al., 1994b)

oxidant which gave maximum protection from Aβ toxicity. Propylgallate and the lipophilic spin-trapping agent N-t-butyl-phenylnitrone (BPN) are the most effective. Both lipid soluble (e.g. vitamin E) and water soluble (e.g. ascorbate) antioxidants have significant protective effects. Inactive analogues of BPN (t-butyl-phenylcarbonate) and ascorbic acid (dehydroascorbate) do not inhibit. These results suggest that free radical damage may be involved in Aβ toxicity (Behl et al., 1994b).

Aβ causes increased production of peroxide

The major reactive oxidants in cells are superoxide and the more reactive hydroxyl radical, the latter being derived in part from hydrogen peroxide (Halliwell and Gutteridge, 1989). Peroxides can be detected through the use of the

Table 1. Effects of antioxidants on Aβ toxicity and peroxide production

Reagent	Concen-tration	Percent increase (Peroxides in B12)	Percent toxicity in B12 cells (viability Ø Aβ)	Percent toxicity in primary neurons (viability Ø Aβ)
Control Aβ alone	10 μM	100	100	100
Acetylcystine + Aβ	1 mM	23 ± 3	26 ± 5 (93 ± 4)	15 ± 3 (101 ± 5)
Dimethylsulfoxide + Aβ	120 μM	79 ± 11	90 ± 12 (100 ± 8)	95 ± 1 (92 ± 6)
BPN+Aβ	50 μM	17 ± 3	3 ± 5 (107 ± 2)	11 ± 7 (97 ± 8)
BPC + Aβ	50 μM	90 ± 5	97 ± 8 (102 ± 9)	92 ± 6 (105 ± 11)
Vitamin E + Aβ	230 μM	45 ± 6	22 ± 15 (111 ± 5)	9 ± 5 (97 ± 7)
2-Mercaptoethanol + Aβ	12 μM	29 ± 4	37 ± 14 (108 ± 5)	55 ± 6 (101 ± 12)
BHA+Aβ	50 μM	70 ± 1	75 ± 17 (92 ± 1)	83 ± 10 (94 ± 7)
Propylgallate + Aβ	5 μM	10 ± 1	5 ± 7 (97 ± 10)	4 ± 3 (106 ± 5)
Acetylcarnitine + Aβ	10 μM	36 ± 5	23 ± 6 (110 ± 11)	33 ± 7 (93 ± 8)
L-Ascorbate + Aβ	300 μM	42 ± 7	58 ± 20 (102 ± 3)	78 ± 1S (94 ± 3)
Dehydroascorbate + Aβ	300 μM	104 ± 5	98 ± 5 (105 ± 2)	93 ± 4 (91 ± 5)

B12 cells were assayed for the ability of antioxidants to block Aβ induced H_2O_2 accumulation and cell death. Primary cultures were used to examine the effect of antioxidants on Aβ induced cell lysis. For peroxide analyses, B12 suspension cells were exposed to the indicated concentration of antioxidant or both antioxidant and 10 μM Aβ$_{25-35}$ for 24 hrs and intracellular peroxide production determined by FACS analysis. The antioxidant FACS data are normalized to the dye shift in the presence of antioxidant alone and therefore the increase is the fluorescence of the antioxidant plus Aβ compared to the background fluorescence of the antioxidant alone. The increase between untreated control and Aβ alone was 2-fold (100%). The trypan blue based toxicity assays were done after 48 hrs in 2% dialyzed fetal calf serum for B12 and in N2 medium for the primary cultures. In this experiment 63% of the B12 cells were killed by Aβ$_{25-35}$ and 93% of the primaries killed by β$_{1-40}$. The toxicity data are normalized to 63% and 93% reduction in cell number by Aβ; these numbers were called 100% toxicity. BPN is N-t-butyl-phenyl-nitrone, BPC is N-t-butyl-phenylcarbonate and BHA is butylated hydroxyanisole. The numbers in parenthesis after the toxicity data are the number of viable cells in the presence of the antioxidant alone (no Aβ) relative to control cultures containing neither Aβ nor antioxidant (100%) (Table taken from Behl et al., 1994b) non-fluorescent dye 2',7'-dichlorofluorescin (DCF). Intracellular DCF is able to interact with peroxides which convert it to the fluorescent 2',7'-dichlorofluorescein which is readily detected using a fluorescence activated cell sorter (FACS) or a fluorescence microscope (Keston and Brandt, 1965; Royall and Ischiropoulos, 1993; Behl et al., 1994b). The activation of DCF is relatively specific for the detection of H_2O_2 and secondary and tertiary peroxides such as lipid hydroperoxides. Since H_2O_2 is the major peroxide within cells, DCF assays in whole cells primarily measure this molecule (Royall and Ischiropoulos, 1993).

Ab was added to the suspension growing variant of B12 and its effect on intracellular peroxide levels monitored with DCF as a function of time. Aβ causes a rapid increase in intracellular peroxides with a maximum increase of about 3 fold between 3 and 4 hours after exposure. After 24 hrs there was usually a sustained increase of between 1.5 and 2 fold in the surviving cell population (1.8 ± 0.2 averaged over 17 experiments). The reduction of peroxide accumulation by antioxidants supports this conclusion (Table 1).

To extend the role of H_2O_2 to primary cultures two experiments were done: (1) DCF was used to visually assay the extent of H_2O_2 accumulation in cells by fluorescence microscopy. Cortical primary cultures were exposed to Aβ for 23 hrs and then DCF added to the cultures for 1 hr. Cells expressing higher levels of peroxides become more fluorescent as in the FACS assay. Figure 2 shows that Aβ greatly increases the number of cells with elevated peroxides in the population. (2) These data were extended by exposing freshly dissociated E18 rat cortical neurons to Aβ and assaying H_2O_2 production on the FACS 4 hrs later. There is an 86% increase in DCF fluorescence with Aβ compared to untreated control cells. Preexposure of the primary neurons to vitamin E inhibits the Aβ induced increase in peroxide production (Behl et al., 1994b).

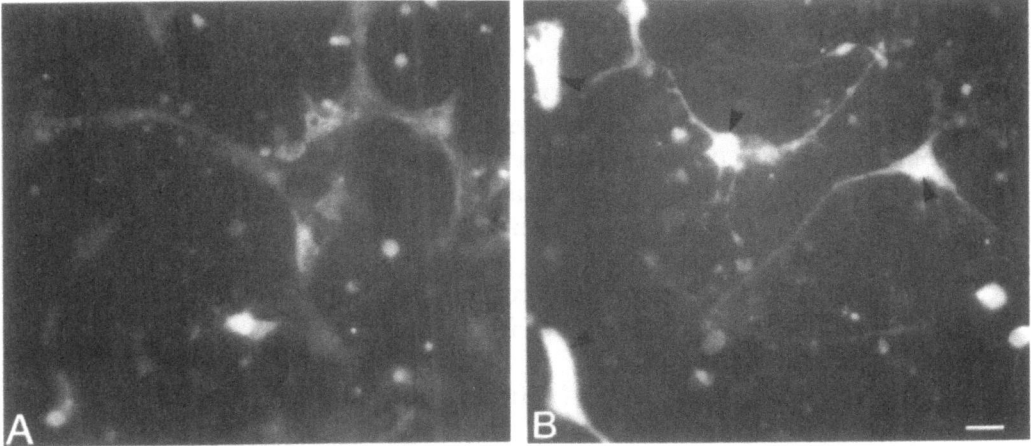

Fig. 2. Aβ induces peroxide accumulation in primary cultures. E18 rat cortical cultures were exposed to 20 µM $A\beta_{1-40}$ for 23 hrs and loaded with 10 µM 2',7'-dichlorofluorescin diacetate for one hour. The cells were washed and viewed with a fluorescence microscope using fluorescein optics. Arrows indicate the intracellularly appearing fluorescence. **A** Control. **B** Cells exposed to Aβ. Bar represents 50 µm (Figure taken from Behl et al., 1994b)

Source of $A\beta$ induced H_2O_2 production

There are numerous cellular mechanisms, both enzymatic and nonenzymatic which can generate superoxide or H_2O_2 (see, for example, Halliwell and Gutteridge, 1989; Coyle and Puttfarcken, 1993; Cross and Jones, 1991 for reviews). In the nervous system H_2O_2 can be produced directly by various oxidases, superoxide dismutase, and xanthine oxidase, and nonenzymatically from the auto-oxidation of catecholamines. The potential sources of superoxide are more numerous and varied, ranging from normal mitochondrial energy metabolism, lipooxygenases and cyclooxygenases, to various flavin oxidases and microsomal cytochrome systems. Although there are no specific inhibitors for most of these radical generating mechanisms, a few reagents inhibit superoxide or H_2O_2 production with varying degrees of specificity, such as allopurinol, indomethacin, deprenyl, and others. These reagents were tested for their ability to reverse the block of MTT reduction and cell lysis caused by $A\beta$ in CNS primary cultures and the block of MTT reduction and H_2O_2 accumulation in B12 cells. Only diphenylene iodonium (DPI), a reagent originally developed as an inhibitor of neutrophil NADPH oxidase (Cross and Jones, 1991) inhibits $A\beta$ toxicity in B12 cells and CNS neurons; it also inhibits H_2O_2 production in B12 suspension cells. Another NADPH oxidase inhibitor, neopterin (Kojima et al., 1993) also inhibits $A\beta$ toxicity. Since DPI also blocks nitric oxide synthase, two specific inhibitors of this enzyme, N^G-methylarginine and N^G-nitroarginine, were also tested and neither one had any effect on $A\beta$ toxicity. Since DPI and neopterin may inhibit the activity of other flavin containing enzymes (Cross, 1990), it can only be concluded that a flavin oxidase such as NADPH oxidase is involved in H_2O_2 generation (Behl et al., 1994b).

Catalase blocks $A\beta$ toxicity

Since H_2O_2 is freely permeable to the cell membrane (Halliwell and Gutteridge, 1989) and the extracellular space constitutes the vast majority of cell culture volume, reagents which lower extracellular H_2O_2 should reduce the overall H_2O_2 concentration in cells and therefore reduce $A\beta$ toxicity. To examine this possibility, B12, PC12, and CNS primary neurons were exposed to $A\beta$ in the presence or absence of catalase, an enzyme which hydrolyses H_2O_2 to O_2 and H_2O (Halliwell and Gutteridge, 1989). For each cell type catalase significantly protects cells from $A\beta$ toxicity. The protection by catalase is dependent upon $A\beta$ concentration, suggesting that the H_2O_2 generating system is activated in an $A\beta$ dose-dependent manner. Of the three cell types, CNS primary cultures are protected from $A\beta_{1-40}$ toxicity most efficiently (Behl et al., 1994b).

$A\beta$ causes lipid peroxidation

To determine if the hydroxyl radical derived from $A\beta$ induced H_2O_2 causes lipid peroxidation, a sensitive fluorimetric assay was used which involves cis-parinaric acid (CPA), a polyunsaturated fatty acid which becomes incorporated into membranes and is subject to peroxidation (Hedley and Chow, 1992). The incorporated fatty acid is fluorescent, and upon peroxidation the double bond structure is lost with a concomitant loss of fluorescence, allowing for a quantitative measure of peroxidation. To assay lipid peroxidation, B12 suspension cells were exposed to Aβ for 23 hrs, and then the cells loaded with CPA and propidium iodide (PI), which is only taken up by dead cells. Between 1 and 2 hrs after loading with CPA the fluorescence intensity of each reagent was determined, and the PI positive dead cells excluded from the quantitation. Within the remaining viable cell population there is a 15–16% decrease in CPA fluorescence between hour 1 and 2 after loading the cells with CPA. This large shift in the mean values was not observed in B12 control cultures. Aβ treatment therefore induces the decay of CPA fluorescence due to enhanced lipid peroxidation. Vitamin E and BPN reagents which suppress peroxide accumulation and protect from Aβ toxicity also suppress the decrease of CPA fluorescence. Primary cultures are much more sensitive to lipid peroxidation than clonal cells and lyse rapidly when lipid peroxidation occurs (Behl et al., 1994b).

Aβ resistant clones are less sensitive to H_2O_2 and exert increased antioxidant enzyme activity

If H_2O_2 toxicity is a mandatory intermediate in Aβ toxicity, then cell lines which are selected for resistance to Aβ toxicity may also become resistant to exogenously applied H_2O_2. Aβ resistant clones of PC12 cells were selected by growth in the presence of 20 μM Aβ$_{25-35}$. After several months, clones of PC12 cells which are able to divide in 20 μM Aβ$_{25-35}$ were isolated. They are all resistant to high levels of Aβ$_{25-35}$ as well as Aβ$_{1-40}$. When these cells are exposed overnight to 125 μM and 250 μM H_2O_2, they are also markedly more resistant to H_2O_2 toxicity than their wild type parental cell line. These data further support the argument that H_2O_2 is involved in Aβ toxicity (Behl et al., 1994b).

More detailed experiments on antioxidant enzyme expression and activity in the Aβ resistant cell clones have revealed that while the enzyme activity of superoxide dismutase is not changed, the mRNA- and protein-levels of catalase and glutathione peroxidase (GSH-Px) as well as the corresponding enzyme activities are highly elevated (Behl et al., 1995a). These elevations correlated with increased survival rates in the presence of Aβ or peroxides. The data suggest a protective effect of antioxidant enzyme activities in these cell clones with respect to H_2O_2 inducing toxic agents.

Amyloid peptides are toxic via a common oxidative mechanism

Aβ is a member of a small group of proteins which accumulate as amyloid deposits in various tissues. Since all of the amyloid diseases are characterized by protein deposited in the structural conformation of the cross β sheet, it was asked if there is a common toxic mechanism. We have recently shown that protein components of other human amyloidoses, including amylin, calcitonin, and atrial natriuretic peptide, are all toxic to clonal and primary neuronal cells (Schubert et al., 1995). The toxic mechanisms induced by these different amyloidogenic peptides are all mediated via a free radical pathway indistinguishable from that of Aβ. Experiments with synthetic peptides and inference about peptide structure clearly suggest that it is the amphiphilic nature of these peptides which causes toxicity.

Free radicals and neuronal disease

Brain cells are at particular risk from damage caused by free radicals. The brain has an extremely high rate of oxygen consumption and CNS neuronal membranes have high contents of polyunsaturated fatty acids which are susceptible to lipid peroxidation. In healthy cells there is a well balanced equilibrium between free radical generation and various enzymatic and non-enzymatic antioxidant defense systems (see Ames et al., 1993; Halliwell and Gutteridge, 1989 for discussion). An imbalance in this status leading to free radical accumulation is defined as oxidative stress (Olanow, 1993). Free radicals have been implicated in different neurological disorders, such as Parkinson's disease, where dopaminergic neurons progressively degenerate, or Amyotrophic Lateral Sclerosis (ALS), where motoneurons are killed (for review see Coyle and Puttfarcken, 1993) (Behl et al., 1994b).

Any explanation of AD must take into account that age is the primary risk factor for the disease. Since normal aged individuals usually acquire the histological manifestations of AD in the form of a limited number plaques as well as some decrease on cognitive ability, it is possible that the mechanisms responsible for AD simply exacerbate those caused by aging. One well documented change in the CNS associated with aging is oxidative damage caused by free radicals (see Ames et al., 1993 for review). On the basis of our observations we conclude that free radicals are also involved in the neurodegeneration in Alzheimer's disease. Specifically, the above data suggest that H$_2$O$_2$ is an important intermediate in Aβ neurotoxicity and that the accumulation of H$_2$O$_2$ related peroxides, and hydroxyl radicals can lead to free radical damage of vital cellular functions. Aβ induced cytotoxicity expressed on a background of already compromised cells may be sufficient to precipitate the clinical and pathological manifestations of the disease.

Future perspectives

As it is clear now that H$_2$O$_2$ is an important mediator of Aβ toxicity, future investigations will focus on the enzymatic source of the H$_2$O$_2$ generation as well as on the downstream signal transduction mechanism. As aging is the primary risk factor for Alzheimer's disease it will also be of importance to investigate the role of other age related changes, such as alterations in the endocrine homeostasis (e.g. hypercorticolism), in neurodegenerative diseases. For example, dramatic neuronal cell death occurs in the hippocampus of AD patients. Since hippocampal neurons express high levels of intracellular receptors for adrenal steroid (MR, mineralocorticoid and GR, glucocorticoid receptors; for review see Reul and DeKloet, 1985), it is of interest to determine whether the activation of these receptors influence the pathway of cell death induced by Aβ. Previously, it has been shown that in cultured hippocampal neurons, corticosterone (the physiological ligand for GR) potentiates the damage produced by a variety of other neurotoxins, such as paraquat, 3-acetylpyridine, and kainic acid (Sapolsky et al., 1988). Therefore the possible role of activated glucocorticoid receptors in Aβ toxicity needs clarification and is currently under intensive investigation. So far we have been able to show that rat primary hippocampal neurons as well as clonal mouse hippocampal cell lines are highly sensitive to Aβ and other amyloidogenic proteins, such as amylin and melittin (Behl et al., 1995b). Therefore these cellular systems offer useful model systems for future investigations of the interplay between glucocorticoid receptors and Aβ induced neurodegeneration.

Note added in proof

The first chapters of this summary on the mechanism of Aβ toxicity were taken from the original publication as cited (Behl et al., 1994b).

References

Ames BN, Shigenaga MK, Hagen TM (1993) Oxidants, antioxidants, and the degenerative diseases of aging. Proc Natl Acad Sci USA 90: 7915–7922

Behl C, Davis J, Cole GM, Schubert D (1992) Vitamin E protects nerve cells from amyloid β protein toxicity. Biochem Biophys Res Com 186: 944–952

Behl C, Davis JB, Klier FG, Schubert D (1994a) Amyloid β protein induces necrosis rather than apoptosis. Brain Res 645: 253–264

Behl C, Davis JB, Lesley R, Schubert D (1994b) Hydrogen peroxide mediates amyloid β protein toxicity. Cell 77: 817–827

Behl C, Lezoualc'h F, Trapp T, Widmann M, Skutella T, Holsboer F (1995) Glucocorticoids enhance oxidative stress-induced cell death in hippocampal neurons in vitro. Endocrinology 138: 101–106

Coyle JT, Puttfarcken P (1993) Oxidative stress, glutamate, and neurodegenerative disorders. Science 262: 689–695

Cross AR (1990) Inhibitors of leukocyte superoxide generating oxidase. Free Rad Biol Med 8: 71–93

Cross AR, Jones OTG (1991) Enzymatic mechanisms of superoxide production. Biochem Biophys Acta 1057: 281–298

Glenner GG (1988) Alzheimer's disease: its proteins and genes. Cell 52: 307–308

Greene LA, Tischler AS (1976) Establishment of a noradrenergic clonal line of rat adrenal pheochromocytoma cells which respond to nerve growth factor. Proc Natl Acad Sci USA 73: 2424–2428

Halliwell B, Gutteridge JMC (1989) Free radicals in biology and medicine. Oxford University Press, Oxford, pp 188–266

Hedley D, Chow S (1992) Fluorocytometric measurement of lipid peroxidation in vital cells using parinaric acid. Cytometry 13: 686–692

Keston AS, Brandt R (1965) The fluorometric analysis of ultramicro quantities of hydrogen peroxide. Anal Biochem 11: 1–5

Koh J, Yang LL, Cotman CW (1990) β amyloid protein increases the vulnerability of cultured cortical neurons to excitotoxic damage. Brain Res 533: 315–320

Kojima S, Nomura T, Icho T, Kajiwara Y, Kitabatake K, Kubota K (1993) Inhibitory effect of neopterin on NADPH-dependent superoxide-generating oxidase of rat peritoneal macrophages. FEBS Lett 329: 125–128

Olanow CW (1993) A radical hypothesis for neurodegeneration. TINS 16: 439–444

Reul JMHM, deKloet ER (1985) Two receptor systems for corticosterone in rat brain: microdistribution and differential accupation. Endocrinology 117: 2505–2511

Royall JA, Ischiropoulos H (1993) Evaluation of 2',7' dichlorofluorescin and dihydro-rhodamine 123 as fluorescent probes for intracellular H_2O_2 in cultured endothelial cells. Arch Biochem Biophys 302: 348–355

Sagara Y, Behl C, Dargusch R, Klier G, Schubert D (1995a) Increased antioxidant enzyme activity in amyloid protein resistant PC12 cells. J Neurosci 16: 497–505

Sapolsky R, Packan D, Vale W (1988) Glucocorticoid toxicity in the hippocampus: in vitro demonstration. Brain Res 453: 367–370

Schubert D (1984) Developmental biology of cultured nerve, muscle, and glia. Wiley & Sons, New York

Schubert D (1994) The structure and function of Alzheimer's amyloid beta proteins. R.G. Landes Company, Austin

Schubert D, Heinemann S, Carlisle W, Tarikas H, Kimes B, Steinbach JH, Culp W, Brandt BL (1974) Clonal cell lines from the rat central nervous system. Nature 249: 224–227

Schubert D, Kimura H, Maher P (1992) Growth factors and vitamin E modify neuronal glutamate toxicity. Proc Natl Acad Sci USA 89: 8264–8268

Schubert D, Behl C, Lesley R, Brack A, Dargusch R, Sagara Y, Kimura H (1995) Amyloid peptides are toxic via a common oxidative mechanism. Proc Natl Acad Sci USA 92: 1989–1993

Yankner BA, Duffy LK, Kirschner DA (1990) Neurotrophic and neurotoxic effects of amyloid β protein: reversal by tachykinin neuropeptides. Science 25: 279–282

Authors' address: Dr. Ch. Behl, Max-Planck-Institute of Psychiatry, Clinical Institute, Kraepelinstrasse 10, D-80804 Munich, Federal Republic of Germany.

From prion diseases to Alzheimer's disease

K. K. Hsiao

Department of Neurology, University of Minnesota, MN, U.S.A.

Summary. Recent advances in the transgenetics of prion diseases and Alzheimer's disease have led to a clearer understanding of the relationship between these two diseases and the pathogenic mechanisms underlying the two disorders. Earlier studies of transgenic mice expressing prion protein (PrP) underscored the importance of PrP levels and PrP primary structure on the resultant phenotype. Three major parameters influencing the phenotypes of mice expressing the Alzheimer amyloid precursor protein (APP) have also been identified: 1) APP levels; 2) APP primary structure; and 3) mouse host strain. The effects and implications of these parameters in transgenic mice expressing APP are discussed.

Prion diseases

Prion diseases are a group of fatal neurodegenerative disorders that includes Creutzfeldt-Jakob disease (CJD), kuru, fatal familial insomnia, familial thalamic dementia, and Gerstmann-Sträussler-Scheinker disease (GSS) in man, as well as scrapie, transmissible mink encephalopathy, and bovine spongiform encephalopathy in animals (reviewed by Prusiner and Hsiao, 1994). These diseases, which in humans are characterized by dementia, cerebellar dysfunction, and extrapyramidal signs, arise in one of three ways: by infection, by inheritance, and sporadically. The infectious etiology is most apparent for kuru, which was transmitted by ritual cannibalism practiced by the Fore tribes in New Guinea, and for iatrogenic CJD, which occurs when patients are exposed to contaminated growth hormone, neurosurgical instruments or cadaveric tissues. Genetic cases result from one of at least eighteen mutations in the host gene that encodes the prion protein. The brains of individuals affected by prion diseases usually display a characteristic spongiform degeneration, accompanied by astroglial proliferation, neuronal loss, and in some cases amyloid plaques containing the prion protein.

Until recently, prion diseases were defined by neuropathologic appearance and infectivity. However, these diagnostic criteria alone failed to identify some patients with a mutant prion protein allele who suffered from fatal familial neurologic conditions. The brains of some such patients displayed minimal or no pathologic changes at death (Collinge et al., 1990; Medori et al., 1992). Furthermore, not all brain tissue from patients with prion diseases transmit disease to experimental animals, and the rate of positive transmissions appears to be lower for brain tissue from patients with inherited prion diseases than from patients with sporadic or iatrogenic prion diseases (Tateishi et al., 1992; Brown et al., 1994). The definition of prion diseases has therefore been broadened to include neurologic disorders associated with mutations in the prion protein, as well as those with the expected neuropathology or infectivity of the brain tissue (Prusiner and Hsiao, 1994).

Prion proteins

Rod-shaped particles were found in highly purified preparations of the infectious agent responsible for experimental scrapie in hamsters. These particles were called *prions*, to distinguish them from conventional viruses (Prusiner, 1982). A glycoprotein of molecular weight 33–35 kD called PrP^{Sc} is the major constituent of scrapie prions (reviewed by Prusiner, 1992). Prions contain little or no nucleic acid, and it is hypothesized that they represent a novel pathogen capable of replicating in the absence of DNA or RNA. Although the claim that PrP^{Sc} itself is infectious is still in dispute, considerable evidence points to an important role for this protein in the disease process. One of the most compelling observations is that mice harboring a mutant PrP transgene spontaneously develop a transmissible disease that resembles GSS (Hsiao et al., 1990, 1994)

Molecular cloning of PrP^{Sc} indicated that the protein is encoded by a host gene (Oesch et al., 1985). This gene, expressed in the brain and at lower levels in other organs, produces a cell-surface protein called PrP^{C}. PrP^{Sc} and PrP^{C} differ in one or more posttranslational modifications that confer infectivity on the scrapie isoform. A covalent modification that distinguishes the isoforms was not found (Stahl et al., 1993). The two isoforms do exhibit distinct conformations (Pan et al., 1993), however, which may explain their biochemical differences, most notably in detergent solubility and protease resistance. PrP^{Sc} can convert PrP^{C} to a protease-resistant form in a cell-free system (Kocisko et al., 1994), but whether PrP^{C} converted in this manner is infectious is unknown.

The normal function of PrP^{C} is unclear. Mice in which the PrP gene has been ablated were normal in several pathological and behavioral tests (Büeler et al., 1992), but were recently shown to exhibit abnormal electrophysiological responses in the hippocampus (Collinge et al., 1994), suggesting that the protein may be involved in synaptic transmission.

Prion disease pathogenesis

Experiments with mice that do not express the PrP gene, or that express transgenes derived from other species, have suggested that prion replication occurs when PrP^{Sc} in the infecting inoculum interacts with homologous host

PrP^C, catalyzing its conversion into PrP^{Sc} (Scott et al., 1989; Prusiner et al., 1990; Büeler et al., 1993). This conversion is postulated to occur spontaneously in PrP^C molecules carrying disease-specific mutations (Hsiao et al., 1994).

While much attention has focussed upon the synthesis of prions, the relationship between prion synthesis and the pathogenesis of prion disease is unclear. In humans and transgenic mice, the clinical and pathological changes are not always accompanied by detectable levels of PrP^{Sc}. In transgenic mice expressing mutant PrP or higher levels of wild-type PrP, central nervous system (CNS) neurodegeneration occurs in the absence of detectable PrP^{Sc} (Hsiao et al., 1990; Westaway et al., 1994). Similarly, a scant amount or absence of detectable PrP^{Sc} in some patients with fatal familial insomnia (Medori et al., 1992) and familial CJD (Brown et al., 1992) has also been reported.

Although brain tissue from transgenic mice expressing mutant PrP transmits disease to recipient hosts, only a fraction of the brain specimens exhibited infectivity and only a proportion of the recipients inoculated with infectious tissue developed neurodegeneration (Hsiao et al., 1994). These results are consistent with the absence of detectable PrP^{Sc}, and indicate a low titer of prions in the brains of these transgenic mice. The discrepancy between the striking clinical and neuropathologic appearance of transgenic mice expressing mutant PrP and both the low prion titer and absence of detectable PrP^{Sc} in their brains raises the possibility that disease pathogenesis in these mice is related to aberrant metabolism of mutant PrP^C rather than the accumulation of PrP^{Sc}. Since the clinical and neuropathological changes in transgenic mice expressing mutant PrP are indistinguishable from those in mice inoculated with prions, it is likely that similar mechanisms of disease pathogenesis pertain to both the inherited and infectious forms of disease.

Alzheimer's disease

A close relationship between prion diseases and Alzheimer's disease has long been proposed (Prusiner, 1985). Both disorders are characterized clinically by dementia, onset in middle or late adulthood, and an inexorable course leading to death. Both conditions are associated with the accumulation or aberrant metabolism of a cell-surface protein, which in some instances can be deposited as amyloid plaques.

However, several salient differences distinguish Alzheimer's disease and prion diseases besides the obvious distinction between the infectious prion protein and the non-infectious amyloid of Alzheimer's disease. Highlighting some of the differences between these two disorders may be as instructive as noting their similarities. Different areas of the brain are affected in the two diseases. Prion diseases can involve a wide variety of gray or white matter structures in the neuraxis, including the cerebellum and spinal cord, but Alzheimer's disease chiefly affects cortico-limbic areas of the brain. Alzheimer's disease is at least 10,000 times more prevalent than prion diseases in elderly individuals. Patients with Alzheimer's disease exhibit to a greater degree the same changes that occur in the brains of many non-demented elderly individuals, suggesting that similar mechanisms are responsible for both aging and Alzheimer's disease, and Alzheimer's disease may be an exacerbated form of the brain's aging process. In contrast, prion diseases resemble nothing that occurs "naturally" in aged human beings.

Alzheimer amyloid precursor protein

Unlike prion diseases, Alzheimer's disease is genetically heterogeneous, involving at least four chromosomal genes. However, disease-specific mutations in the corresponding membrane-associated proteins have been found for both afflictions. Five mutations in the Alzheimer amyloid precursor protein (APP) have been identified in families with Alzheimer's disease (Chartier-Harlin et al., 1991; Goate et al., 1991; Murrell et al., 1991; Hendriks et al., 1992; Mullan et al., 1992), and significant genetic linkage scores have been demonstrated in several kindreds. Interestingly, all five mutations are located within or in the vicinity of the $A\beta$ peptide which is partially embedded in the outer leaflet of the lipid bilayer.

Studies have sought to relate the effects of disease-specific mutations to APP processing and disease pathogenesis. However, analyses of a variety of APP mutants in cultured cells have not revealed any consistently aberrant processing events. APP with M670N-K671L (APP-770 numbering), also known as the Swedish mutation, is associated with increased 1–40 $A\beta$ secretion in cultured cells (Citron et al., 1992; Cai et al., 1993), but APP with the V717I mutation is associated with an increase in the 1–42 $A\beta$ peptide (Suzuki et al., 1994). Whether these mutations exert their pathogenic effects solely through alterations in $A\beta$ production is unknown. However, based upon observations that synthetic $A\beta$ peptides are neurotoxic in vitro and in vivo (reviewed by Price et al., 1993), one prevailing hypothesis is that aberrant processing of APP results in increased $A\beta$ production leading to neuronal dysfunction and death.

The APP gene is comprised of 18 exons spanning > 400 kB on the long arm of chromosome 21. This is the trisomic chromosome in patients with Down's syndrome, nearly all of whom have developed Alzheimer' s disease by the seventh decade, when Alzheimer's disease is just beginning to manifest in the general population. Although many genes on chromosome 21 may be involved in the premature development of Alzheimer's disease in patients with Down's syndrome, it has been postulated that the development of Alzheimer's disease in these patients is related to the overexpression of wild-type APP resulting from the presence of a third APP gene.

Transgenic-models of Alzheimer's disease

One reason to express APP and APP mutants in transgenic mice is to model the human disease. Another important reason is to study the biologic activity of APP in vivo. Such studies must take into consideration the influence of the host on the transgenic phenotype, because there are numerous genetic and biologic differences between humans and mice. Host modification of a transgenic phenotype was clearly demonstrated for GSS. Overexpression in transgenic mice of a mutant PrP genetically linked to GSS produced spongiform neurodegeneration indistinguishable from murine scrapie, but the majority of animals lacked the amyloid plaques that define GSS in humans (Hsiao et al., 1989, 1990, 1994; Prusiner and Hsiao, 1994). Another example of host modification of transgenic phenotypes has been demonstrated in two different strains of mice expressing the same mutant human APP transgene. When high levels of human APP with the Swedish mutation were expressed in FVB mice, 8/9 mice were dead by 100 days of age. In contrast, all 12 C57B6/SJL mice expressing higher levels of the same transgene were alive at 100 days of age (K. Hsiao, unpublished observations). It will not only be important to distinguish the protean clinical and pathological manifestations of transgenic APP expression in different strains of mice, but also to identify features occurring in common. A better understanding of the fundamental biologic activity of aberrant APP expression may be gained by recognizing and focussing attention on the common principles that manifest in different strains and species.

Other variables and parameters that influence the phenotypes of transgenic mice expressing APP include the expression levels and, possibly, the APP genotype. Until overexpression of APP and APP fragments was achieved, there were no reports of robust clinical or pathological phenotypes (Quon et al., 1991; Sandhu et al., 1991; Yamaguchi et al., 1991; Kammesheidt et al., 1992; Buxbaum et al., 1993; Lamb et al., 1993; Pearson and Choi, 1993; Higgins et al., 1994). However, recently, neuronal apoptosis associated with seizures and death was observed in transgenic mice expressing cytoplasmic Aβ (LaFerla et al., 1995), and numerous amyloid plaques were reported in transgenic mice overexpressing mutant HuAPP (Games et al., 1995).

We have also succeeded in overexpressing APP in transgenic mice (Hsiao et al., 1995). Our approach to studying the biological activity of APP in transgenic mice has been to compare their phenotype to that which naturally occurs in aged mice of the same strain. The rationale for this is based upon the observation that age-related CNS dysfunction in rodents can manifest in the absence of Aβ plaque deposition. We reasoned that, since mutant APP is genetically linked to an exacerbation of the cognitive decline and morphological features of aging in humans, then species- and strain-specific senescent changes in the brains of transgenic mice might be induced by expressing APP.

We characterized in FVB mice a naturally occurring central nervous system (CNS) disorder with senescent features including inactivity, agitation, neophobia, seizures, diminished cortico-limbic glucose utilization, cortico-limbic gliosis and death. This condition was accelerated and exacerbated in transgenic FVB mice overexpressing human and mouse APP. In transgenic mice the disorder developed in direct relationship to brain levels of transgenic APP, but mutant human APP conferred the phenotype with higher penetrance than wild-type mouse APP. The disorder occurred in the absence of extracellular amyloid deposition, indicating that some pathogenic activities of APP can be dissociated from amyloid formation.

A summary of two transgenic paradigms overexpressing APPis presented in Table 1. The discrepancy between the clinical, neuropathologic, and functional appearance of our transgenic FVB mice expressing mutant APP transgenes and the amyloid-containing transgenic mice for which no behavioral or functional abnormalities were reported (Games et al., 1995) may be reconciled if mouse strain differences and possibly transgene variations are taken into consideration. It can only be assumed that since no clinical abnormalities were reported in the amyloid-containing mice, they possess a muted behavioral phenotype, if any, relative to our transgenic FVB mice. Levels and targets of APP expression were comparable and cannot explain the different pheno-

Table 1. Transgenic mice overexpressing APP

	Hsiao et al. (1996)	Games et al. (1995)
Transgenic proteins	Human APP Mouse APP	Human APP
Transgene genotype	Wild-type E693Q M670N, K671L V717I, V721A, M722V	V717F
Promoters	PrP	PDGF
APP isoforms	695	695, 751, 770
Highest brain expression	cerebrum cerebellum	cerebrum cerebellum
Expression cell type	mainly neurons less in astrocytes	neurons
APP levels (relative to endogenous mouse APP)	$>8\times$ (by ECL) $4.6\times$ (by ^{35}S)	$\sim 10\times$
Human Aβ	present	present
Behavioral phenotype	agitation inactivity neophobia seizures death	none reported
Functional phenotype	glucose hypoutilization in forebrain (esp. entorhinal cortex, hippocampus, amygdala)	none reported
Pathological phenotype	cortico-limbic gliosis	cortico-limbic amyloid plaques, neuritic changes
Mouse strain	FVB	Swiss Webster $\times$ B6/D2

types. Differences in specific mutations probably do not account for the distinct phenotypes either, because it is difficult to explain why a behavioral phenotype would be suppressed by the particular mutation (V717F) expressed by the amyloid-containing transgenic mice. It is similarly difficult to explain why the presence of two additional APP splice forms would suppress a behavioral phenotype in the amyloid-containing mice. Because host modification of transgenic phenotypes are known to occur, and there is no compelling reason to attribute differences in the phenotypes to variations in transgenes, the distinct mouse strains used to create the transgenic mice most likely explain the different phenotypes. Experiments to test this hypothesis are being conducted.

An alternative perspective

We have shown that transgenic FVB mice overexpressing APP prematurely develop a CNS phenotype that closely resembles the naturally occurring senescent CNS phenotype in non-transgenic FVB mice. The most compelling interpretation of our results would suggest that Alzheimer's disease is not unique to humans. The traditional definition of Alzheimer's disease (with amyloid plaques and neurofibrillary tangles) does not adequately recognize a basic functional disturbance in the aging brain that also occurs in other mammals when they age. Perhaps functional changes, not pathological features, better explain Alzheimer's disease. Just as the definition of prion diseases evolved to encompass disorders sharing molecular genetic features, so too may the definition of Alzheimer's disease broaden as our knowledge of the molecular and biologic basis of this disorder increases. Alzheimer's disease may come to be redefined in humans and other animals as an age-dependent dysfunction of the areas of the brain affecting learning, memory, and emotions, involving specific genes, not all of which have been identified yet. Why these areas of the brain are most vulnerable is unknown. However, it is possible to speculate that because animals must continuously adapt to their environments, these areas of the brain must remain plastic throughout life. This necessity for ongoing plasticity may lead to synaptic instability in the setting of aging and other genetic and environmental insults. The key to understanding Alzheimer's disease may reside in discovering the relationships between perpetual synaptic plasticity, aging, and the Alzheimer's associated genes.

References

Brown P, Goldfarb LG, McCombie WR, Nieto A, Squillacote D, Sheremata W, Little BW, Godec MS, Gibbs CJJ, Gajdusek DC (1992) Atypical Creutzfeldt-Jakob disease in an American family with an insert mutation in the PRNP amyloid precursor gene. Neurology 42: 422–7

Brown P, Gibbs CJJ, Rodgers-Johnson P, Asher DM, Sulima MP, Bacote A, Goldfarb LG, Gajdusek DC (1994) Human spongiform encephalopathy: the National Institutes of Health series of 300 cases of experimentally transmitted disease. Ann Neurol 35: 513–529

Büeler H, Fischer M, Lang Y, Bluethmann H, Lipp HP, DeArmond SJ, Prusiner SB, Aguet M, Weissmann C (1992) Normal development and behaviour of mice lacking the neuronal cell-surface PrP protein. Nature 356: 577-582

Büeler H, Aguzzi A, Sailer A, Greiner RA, Autenried P, Aguet M, Weissmann C (1993) Mice devoid of PrP are resistant to scrapie. Cell 73: 1339–1347

Buxbaum JD, Christensen JL, Ruefli AA, Greengard P, Loring JF (1993) Expression of APP in brains of transgenic mice containing the entire human APP gene. Biochem Biophys Res Commun 197: 639–645

Cai X-D, Golde TE, Younkin SG (1993) Release of excess amyloid β protein from a mutant amyloid β protein precursor. Science 259: 514–516

Chartier-Harlin M-C, Crawford F, Houlden H, Warren A, Hughes D, Fidani L, Goate A, Rossor M, Roques P, Hardy J, Mullan M (1991) Early-onset Alzheimer's disease caused by mutations at codon 717 of the β-amyloid precursor protein gene. Nature 353: 844–846

Citron M, Oltersdorf T, Haass C, McConlogue L, Hung AY, Seubert P, Vigo-Pelfrey C, Lieberburg I, Selkoe DJ (1992) Mutation of the beta-amyloid precursor protein in familial Alzheimer's disease increases beta-protein production. Nature 360: 672–674

Collinge J, Owen F, Poulter M, Leach M, Crow TJ, Rossor MN, Hardy J, Mullan MJ, Janota I, Lantos PL (1990) Prion dementia without characteristic pathology. Lancet 336: 7–9

Collinge J, Whittington MA, Sidle KCL, Smith CJ, Palmer MS, Clarke AR, Jefferys JGR (1994) Prion protein is necessary for normal synaptic function. Nature 370: 295–297

Games D, Adams D, Alessandrini R, Barbour R, Berthelotte P, Blackwell C, Carr T, Clemens J, Donaldson T, Gillespie F, Guido T, Hagoplan S, Johnson-Wood K, Khan K, Lee M, Leibowitz P, Lieberburh I, Little S, Masilah E, McConlogue L, Montoya-Zavaia M, Mucke L, Paganini L, Penalman E, Power M, Schenk D, Peubert P, Snyder B, Soriano F, Tan H, Vitale J, Wadsworth S, Wolozin B, Zhao J (1995) Alzheimer-type neuropathology in transgenic mice overexpressing V717F β-amyloid precursor protein. Nature 373: 523–527

Goate A, Chartier-Harlin MA, Mullan M, Brown J, Crawford F, Fidani L, Giuffra L, Haynes A, Irving N, James L, Mant R, Newton P, Rooke K, Roques P, Talbot C, Pericak-Vances M, Roses A, Williamson R, Rossor M, Owen M, Hardy J (1991) Segregation of a missense mutation in the amyloid precursor protein gene with familial Alzheimer's disease. Nature 349: 704–706

Hendriks L, Duijn CMv, Cras P, Cruts M, Hul WV, Harskamp F v, Warren A, McInnis MG, Antonarakis SE, Martin J-J, Hofman A, Broeckhoven CV (1992) Presenile dementia and cerebral haemorrhage linked to a mutation at codon 692 of the β-amyloid precursor protein gene. Nature Genet 1: 218–221

Higgins LS, Holtzman DM, Rabin J, Mobley WC, Cordell B (1994) Transgenic mouse brain histopathology resembles early Alzheimer's disease. Ann Neurol 35: 598–607

Hsiao K, Baker HF, Crow TJ, Poulter M, Owen F, Terwilliger JD, Westaway D, Ott J, Prusiner SB (1989) Linkage of a prion protein missense variant to Gerstmann-Straussler syndrome. Nature 338: 342–345

Hsiao KK, Scott M, Foster D, Groth DF, DeArmond SJ, Prusiner SB (1990) Spontaneous neurodegeneration in transgenic mice with mutant prion protein. Science 250: 1587–1590

Hsiao KK, Groth D, Scott M, Yang S-L, Serban H, Rapp D, Foster D, Torchia M, DeArmond SJ, Prusiner SB (1994) Serial transmission in rodents of neurodegeneration from transgenic mice expressing mutant prion protein. Proc Natl Acad Sci 91: 9126–9130

Hsiao KK, Borchelt DR, Olson K, Johannsdottir R, Kitt C, Yunis W, Xu S, Eckman C, Younkin S, Price D, Iadecola C, Clark HB, Carlson G (1995) Age-related CNS disorder and early death in transgenic FVB/N mice overexpressing Alzheimer amyloid precursor proteins. Neuron 15: 1203–1218

Hsiao K, Chapman P, Nilsen S, Eckman C, Harigaya Y, Younkin S, Yang F, Cole G (1996) Correlative memory deficits, Aβ elevation, and amyloid plagues in transgenic mice

Kammesheidt A, Boyce FM, Spanoyannis AF, Cummings BJ, Ortegon M, Cotman C, Vaught JL, Neve RL (1992) Deposition of beta/A4 immunoreactivity and neuronal pathology in transgenic mice expressing the carboxyl-terminal fragment of the Alzheimer amyloid precursor in the brain. Proc Natl Acad Sci USA 89: 10857–10861

Kocisko DA, Come JH, Priola SA, Chesebro B, Raymond GJ, Lansbury PT, Caughey B (1994) Cell-free formation of protease-resistant prion protein. Nature 370: 471–474

LaFerla FM, Tinkle BT, Bieberich CJ, Haudenschild CC, Jay G (1995) The Alzheimer's Aβ peptide induces neurodegeneration and apoptotic cell death in transgenic mice. Nature Genet 9: 21–30

Lamb BT, Sisodia SS, Lawler AM, Slunt HH, Kitt CA, Kearns WG, Pearson PL, Price DL, Gearhart JD (1993) Introduction and expression of the 400 kilobase precursor amyloid protein gene in transgenic mice. Nature Genet 5: 22–30

Medori R, Montagna P, Tritschler HJ, LeBlanc A, Cortelli P, Tinuper P, Lugaresi E, Gambetti P (1992) Fatal familial insomnia: a second kindred with mutation of prion protein gene at codon 178. Neurology 42: 669–670

Mullan M, Crawford F, Axelman K, Houlden H, Lilius L, Winblad B, Lannfelt L (1992) A pathogenic mutation for probable Alzheimer's disease in the APP gene at the N-terminus of β-amyloid. Nature Genet 1: 345–347

Murrell J, Farlow M, Ghetti B, Benson MD (1991) A mutation in the amyloid precursor protein associated with hereditary Alzheimer's disease. Science 254: 97–99

Oesch B, Westaway D, Wälchli M, McKinley MP, Kent SBH, Aebersold R, Barry RA, Tempst P, Teplow DB, Hood LE, Prusiner SB, Weissmann C (1985) A cellular gene encodes scrapie PrP 27-30 protein. Cell 40: 735–746

Pan K-M, Baldwin M, Nguyen J, Gasset M, Serban A, Groth D, Mehlhorn I, Huang Z, Fletterick RJ, Cohen FE, Prusiner SB (1993) Conversion of α-helices into β-sheets features in the formation of the scrapie prion proteins. Proc Natl Acad Sci 90: 10962–10966

Pearson BE, Choi TK (1993) Expression of the human β-amyloid precursor protein gene from a yeast artificial chromosome in transgenic mice. Proc Natl Acad Sci USA 90: 10578–10582

Price DL, Borchelt DR, Walker LC, Sisodia SS (1992) Toxicity of synthetic Aβ peptides and modeling of Alzheimer's disease. Neurobiol Aging 13: 623–625

Prusiner SB (1982) Novel proteinaceous infectious particles cause scrapie. Science 216: 136–144

Prusiner SB (1985) Scrapie prions, brain amyloid, and senile dementia. Curr Top Cell Regul 26: 79–95

Prusiner SB (1992) Chemistry and biology of prions. Biochemistry 31: 12277–12288

Prusiner SB, Hsiao KK (1994) Human prion diseases. Ann Neurol 35: 385–395

Prusiner SB, Scott M, Foster D, Pan KM, Groth D, Mirenda C, Torchia M, Yang SL, Serban D, Carlson GA, Hoppe PC, Westaway D, DeArmond SJ (1990) Transgenetic studies implicate interactions between homologous PrP isoforms in scrapie prion replication. Cell 63: 673–686

Quon D, Wang Y, Catalano R, Scardina JM, Murakami K, Cordell B (1991) Formation of beta-amyloid protein deposits in brains of transgenic mice. Nature 352: 239–41

Sandhu FA, Salim M, Zain SB (1991) Expression of the human β-amyloid protein of Alzheimer's disease specifically in the brains of transgenic mice. J Biol Chem 266: 21331–21334

Scott M, Foster D, Mirenda C, Serban D, Coufal F, Wälchli M, Torchia M, Groth D, Carlson G, DeArmond SJ, Westaway D, Prusiner SB (1989) Transgenic mice expressing hamster prion protein produce species-specific scrapie infectivity and amyloid plaques. Cell 59: 847–857

Stahl N, Baldwin MA, Teplow DB, Hood L, Gibson BW, Burlingame AL, Prusiner SB (1993) Structural studies of the scrapie prion protein using mass spectrometry and amino acid sequencing. Biochemistry 32: 1991–2002

Suzuki N, Cheung TT, Cai XD, Okada A, LOtvos J, Eckman C, Golde TE, Younkin SG (1994) An increased percentage of long amyloid beta protein secreted by familial amyloid beta protein precursor (beta APP717) mutants. Science 264: 1335–1340

Tateishi J, Doh-ura K, Kitamoto T, Tranchant C, Steinmetz G, Warter JM, Boellaard JW (1992) Prion protein gene analysis and transmission studies of Creutzfeldt-Jakob disease. In: Prusiner SB, Collinge J, Powell J, Anderton B (eds) Prion diseases of humans and animals. Ellis Horwood, London, pp 129–138

Westaway D, DeArmond SJ, Cayetano-Canlas J, Groth D, Foster D, Yang S-L, Torchia M, Carlson GA, Prusiner SB (1994) Degeneration of skeletal muscle, peripheral nerves, and the central nervous system in transgenic ice overexpressing wild-type prion proteins. Cell 76: 117–129

Yamaguchi F, Richards SJ, Beyreuther K, Salbaum M, Carlson GA, Dunnett SB (1991) Transgenic mice for the amyloid precursor protein 695 isoform have impaired spatial memory. Neuroreport 2: 781–784

Author's address: K. K. Hsiao, M.D., Ph.D., Department of Neurology, Box 295 UMHC, 420 Delaware Street, SE, University of Minnesota, Minneapolis, MN 55455, U.S.A.

A phosphorylation cascade in the basal ganglia of the mammalian brain: regulation by the D-1 dopamine receptor
A mathematical model of known biochemical reactions

J. W. Kebabian

Research Biochemicals International, Natick, MA, U.S.A.

Summary. Stimulation of the dopamine D-1 receptor in the corpus striatum initiates a cascade of biochemical events. These events include: activation of adenylate cyclase, stimulation of cAMP-dependent protein kinase, protein phosphorylation and inhibition of phosphoprotein phosphotase-1. This article presents and discusses a mathematical model of these biochemical events (and their dependence upon the concentration of cytosolic calcium). According to this model, the activity of calcineurin (which is regulated by the concentration of cytosolic calcium ions) counterbalances the activity of the "D-1 cascade". The combined activity of the "D-1 cascade" and calcineurin can regulate the activity of calcium- and calmodulin-dependent protein kinase II.

Introduction

Dopamine receptors are important entities within the mammalian central nervous system. The division of dopamine receptors into two general categories designated as dopamine D-1 receptors and dopamine D-2 receptors is now generally accepted (Kebabian and Calne, 1979). With the application of the techniques of molecular biology, a number of structurally distinct subcategories of dopamine receptors have been identified (Gingrich and Caron, 1993). Although dopamine receptors remain the focus of many research investigators, there are still many unanswered questions.

Indeed, the naggingly simple question, "What does dopamine do in the basal ganglia?" remains open. An all encompassing answer to this question is beyond the scope of this chapter. I will focus on the consequences of stimulating a D-1 dopamine receptor. (In the present discussion, I will not make any differentiation among the subcategories of D-1 dopamine receptors.) I will apply the mathematical approach of John Lisman (originally developed to investigate long term potentiation in the hippocampus) to the biochemical consequences of stimulating a D-1 dopamine receptor (Lisman, 1989). This computational approach towards the enzymology of the basal ganglia has given me some insight into the consequences of stimulating the D-1 dopamine receptor.

The natural history of the D-1 dopamine receptor

The D-1 dopamine receptor occurs in high concentration within the basal ganglia of the mammalian brain. The interaction of dopamine with a D-1 dopamine receptor initiates a cascade of biochemical reactions (for a review see Hemmings et al., 1989). In brief, dopamine stimulates the formation of cyclic AMP; cyclic AMP, in turn, activates a cyclic AMP-dependent protein kinase; the activated kinase phosphorylates serine and threonine residues in Inhibitor 1 and DARPP-32. Either phospho-Inhibitor 1 or phospho-DARPP-32 is a potent inhibitor of phosphoprotein phosphatase-1. Inhibition of this phosphatase may lead to the accumulation of other phosphorylated proteins due to the action(s) of kinases other than cAMP-dependent protein kinase. Either phospho-Inhibitor I or phospho-DARPP-32 is a substrate for calcineurin, a calcium-dependent protein phosphatase. There is extensive experimental evidence related to each of these individual biochemical reactions and substrates.

In his mathematical formalization of a biochemical mechanism underlying long-term potentiation, John Lisman invokes many of the same biochemical events that are known to occur in the basal ganglia. In addition, he presents a series of mathematical equations that model the biochemical reactions. According to Lisman's analysis, changes in intracellular calcium are the critical factor to alter cyclic AMP levels and thereby initiate the cascade of biochemical changes. In the present discussion, I propose that dopamine serves as an endogenous agent that stimulates cAMP formation in a coordinated fashion with calcium. Lisman discusses inhibitor-1 as the substrate for cyclic AMP-dependent protein kinase (PKA), I invoke DARPP-32 to fill this role. These changes seem plausible given the accumulated knowledge about the basal ganglia.

D-1 receptor, adenylate cyclase and phosphodiesterase

Dopamine-sensitive adenylate cyclase occurs within the basal ganglia (Kebabian et al., 1972). This enzyme activity has been used repeatedly as an in vitro model of the functioning of the D-1 dopamine receptor. For the present discussion, no attempt is made to model the biochemical events involving guanyl nucleotides and G-proteins that translate stimulation of the dopamine D-1 receptor into activation o adenylate cyclase. These events are taken as a "given" in the analysis of the phenomena. In the present analysis, I present the effect of dopamine as a step function that can be considered to represent the largest increase that might occur. (Future analysis will investigate the consequences of submaximal effects of dopamine.) Calcium and other divalent cations regulate striatal adenylate cyclase. EGTA, a calcium-specific chelator, diminishes the adenylate cyclase activity of striatal homogenates (Clement-Cormier et al., 1975). Without a calcium chelator, dopamine does not stimulate the adenylate cyclase activity. It is only when enzyme activity has been lowered by the addition of a calcium chelator that the stimulatory effect of dopamine upon adenylate cyclase occurs. Other divalent ions (e.g. Mn++) can stimulate striatal adenylate cyclase to a much greater degree than can dopamine.

In the mathematical model, both dopamine and calcium can increase adenylate cyclase activity, and consequently elevate cAMP levels. This effect of these two agents upon cAMP content is shown in Fig. 1. In my modification of the Lisman model, I impose an arbitrary increase in cyclic AMP formation (+ dopamine). This increase raises cAMP formation to a submaximal level (approximately 35% of the maximal synthetic capacity of the system). Although calcium has the capacity to increase adenylate cyclase activity, at concentrations below 0.1 μM, this effect is minimal. Dopamine is the factor that is critical to increase cyclic AMP at these low concentrations of calcium. These modifications seem reasonable and in accord with experimental data in the open literature.

Cyclic AMP-dependent protein kinase

Protein phosphorylation is an important biological regulatory mechanism. Many of the intracellular actions of cyclic AMP are initiated through a cyclic AMP-dependent protein kinases (PKA). Dopamine, D-1 receptor agonists and other agents that increase cyclic AMP can make PKA activity active in the basal ganglia. Figure 2 presents the mathematical modeling of the effects of calcium and dopamine upon the PKA activity. At concentrations of calcium below 0.1 μM, calcium has no effect upon PKA activity; in contrast, dopamine because of its ability to augment cAMP formation can markedly stimulate

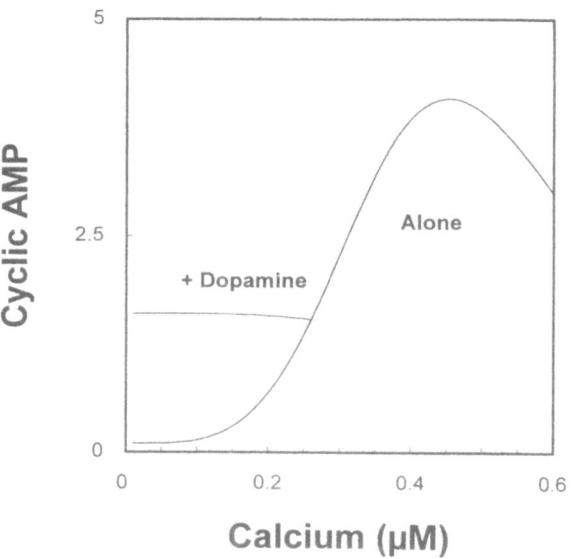

Fig. 1. Mathematical model of the effects of dopamine and calcium on cyclic AMP formation. The equations (for terms C and D in the Appendix of Lisman, 1989) for adenylate cyclase, phosphodiesterase and cyclic AMP levels are used (Alone): an arbitrary increase in cyclic AMP due to dopamine (+ Dopamine) is introduced into the model

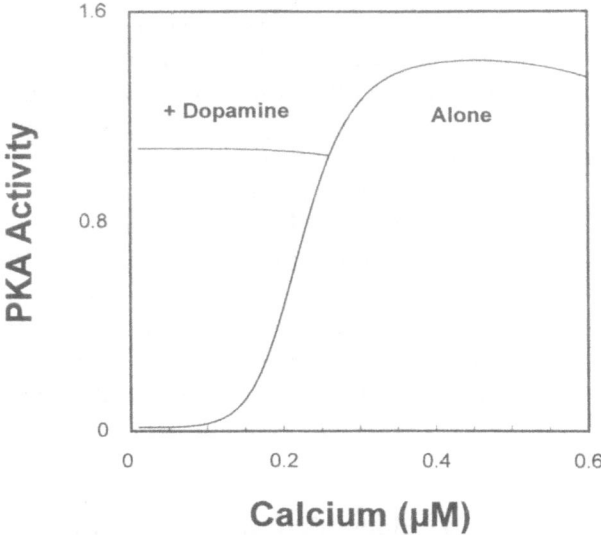

Fig. 2. Mathematical model of the effects of dopamine and calcium on cyclic AMP-dependent protein kinase (PKA) activity. The equation (for the term A in the Appendix of Lisman, 1989) for this enzyme is used (Alone). The arbitrary increase in cyclic AMP (+ Dopamine) introduced in Fig. 1, is applied to PKA activity

PKA activity. There is a slight amplification of the effect of dopamine; the effect of dopamine reaches approximately 70% of the maximal effect of calcium. As happens for cAMP formation, at low concentrations of calcium, dopamine is the important factor that increases protein kinase activity. Lisman in his model, scales the maximal rate of phosphorylation (1.5) of PKA substrates to the maximal rate of dephosphorylation (1.0) of substrates by calcineurin, the calcium-dependent phosphoprotein phosphatase.

Calcineurin is an enzyme in the brain that can dephosphorylate many phosphoproteins in a calcium-dependent manner (Klee, 1991). Klee proposes that this enzyme is sensitive to low, transient changes in intracellular calcium concentration. Figure 3 shows the mathematical model of the phosphatase activity of calcineurin; calcium, but not dopamine, stimulates the enzyme activity. Thus, the two factors regulating the rate of protein phosphorylation would be dopamine that increases phosphorylation through the cAMP cascade and calcium that decreases phosphorylation by activating calcineurin.

Substrates for cyclic AMP-dependent protein kinase

PKA catalyzes the phosphorylation of serine and threonine residues within appropriate substrates. Both inhibitor I and DARPP-32 are substrates for PKA that occur within the striatum (Walaas et al., 1983; Hemmings et al., 1989). DARPP-32 is a particularly attractive focus for the present discussion since it is uniquely associated with the D-1 dopamine receptor throughout the mammalian brain. Figure 4 shows the mathematical modeling of the effects of

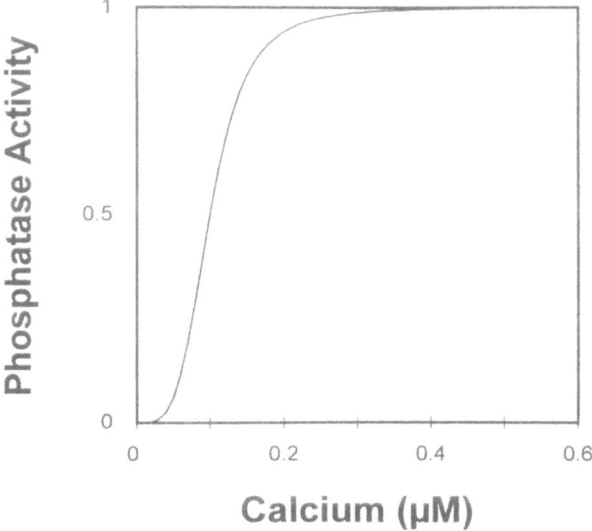

Fig. 3. Mathematical model of the effects of calcium on calcineurin activity. The equation (for the term B in the Appendix of Lisman, 1989) is used. Note that the phosphatase activity of calcineurin (this figure) is more sensitive to calcium than is the kinase activity (alone) shown in Fig. 2

calcium and dopamine upon DARPP-32 phosphorylation. Calcium will dramatically alter protein phosphorylation. In Lisman's model, low concentrations of calcium can stimulate phosphorylation due to kinase activity in the absence of phosphatase activity (caused by low calcium concentration): the present discussion invokes dopamine-sensitive adenylate cyclase as alterna-

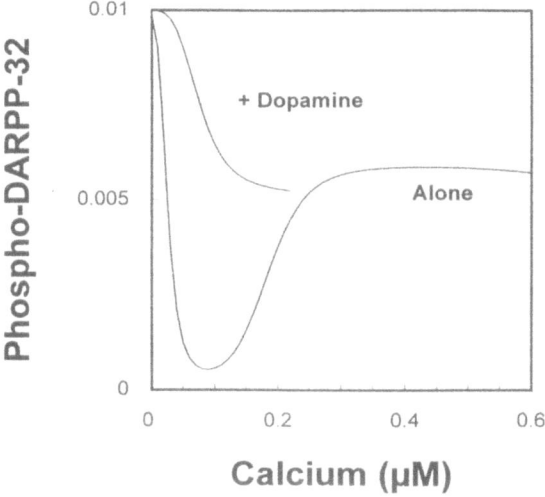

Fig. 4. Mathematical model of the effects of calcium and dopamine on phospho-DARPP-32 formation. The equations (for the term Ip in the Appendix of Lisman, 1989) is used. The effect of the arbitrary increase in cyclic AMP (+ Dopamine) introduced in Fig. 1 and 2 is shown to attenuate the biphasic effect of calcium

tive for increasing phosphorylation. As the concentration of calcium increases so does dephosphorylation; this is due to the stimulation of calcineurin activity. There is some evidence consistent with calcium promoting the dephosphorylation of DARPP (Halpain et al., 1990). The phosphorylation seen at higher concentrations of calcium is a consequence of stimulation by calcium of cyclic AMP formation and protein kinase activity. The effect of dopamine is to maintain the DARPP in a phosphorylated state that is independent of the concentration of cytosolic calcium.

Greengard and his colleagues have shown that phosphorylated DARPP-32 (but not the dephospho-DARPP-32) is a potent inhibitor of phosphoprotein phosphatase 1 (Hemmings et al., 1984). Lisman discusses inhibitor-1 phosphorylation in his model of long-term potentiation; the present discussion will focus on phospho-DARPP as an inhibitor. Figure 5 shows the modeling of the effects of calcium and dopamine on enzyme activity. Phosphatase 1 activity is dramatically stimulated by calcium. This is due to the ability of calcium to alter the amount of phosphatase-1 inhibitor (phospho-DARPP) that is present in the cell; when the concentration of inhibitor is low, phosphatase activity is high. Conversely, dopamine, by virtue of its ability to maintain high levels of phospho-DARPP, is able to attenuate the effect of calcium.

Substrates for phosphatase-1

Calcium and calmodulin dependent protein kinase (CaM-kinase II) is an important enzyme found in postsynaptic complex. Calcium can stimulate this enzyme to phosphoryate other molecules as well as itself. Once activated by calcium, CaM-kinase II maintains high activity and can catalyze phosphoryla-

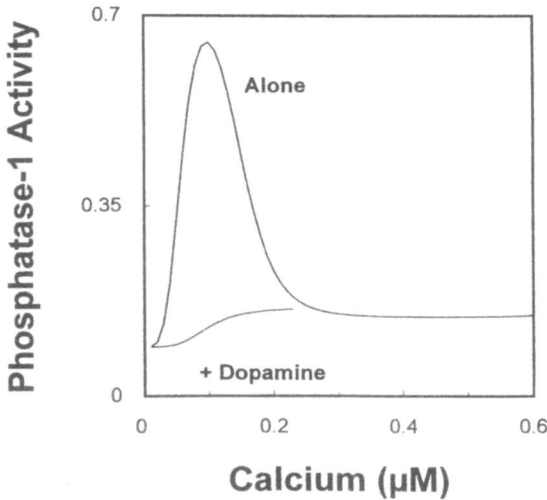

Fig. 5. Mathematical model of the effects of calcium and dopamine on the activity of phosphoprotein phosphatase-1. The equation (for the term f in the Appendix of Lisman, 1989) is used. The inhibitory effect of dopamine (+ dopamine) is a consequence of the arbitrary increase in cyclic AMP introduced in earlier figures

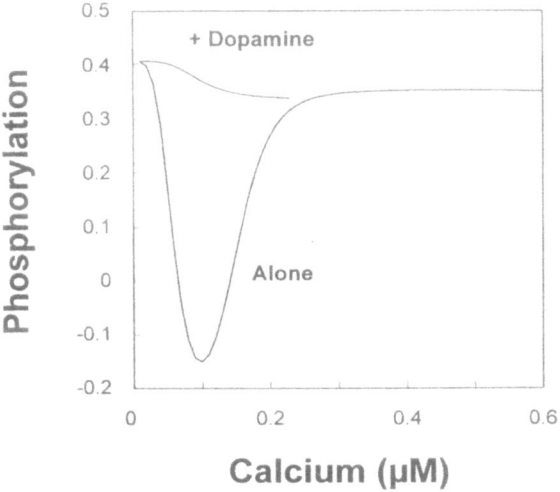

Fig. 6. Mathematical model of the effects of calcium and dopamine on net phosphorylation of "On" CaM-kinase II. The equation (for the term K_{max} and P in the Appendix of Lisman, 1989) is used. The effect of dopamine (+ dopamine) is the consequence of the arbitrary increase in cyclic AMP introduced in earlier figures

tion of its subunits (autophosphorylation). Thus in the absence of calcium, the enzyme is able to maintain a high state of activity once it is turned "on" by a transient rise in calcium. Lisman (1989) proposes that phosphatase-1, by virtue of its ability to dephosphorylate CaM-kinase II, can regulate the activity of the enzyme. Figure 6 shows the mathematical model of the effects of calcium and dopamine on "active" CaM-kinase II. A rise in the concentration of calcium (in the absence of dopamine) will shut off the active (autocatalytic) enzyme and lead to a net dephosphorylation of the enzyme activity. This is due to the rise in phosphatase activity. Dopamine, by virtue of its ability to block the rise in phosphatase activity, would attenuate this ability of calcium to shut off the autocatalyic activity of CaM-kinase II. The large effect of dopamine carried through the mathematical model abolishes completely the effect of calcium on CaM-kinase II phosphorylation. It is intuitively obvious that smaller effects of dopamine would have intermediate actions on CaM-kinase II autophosphorylation. At the present time, I am not aware of any experimental evidence that directly supports such a role for dopamine upon the CaM-kinase II in the striatum.

Implications of the mathematical model

The phenomena that are invoked in the present discussion are, by in large, in accord with the published data from Greengard and his colleagues (see Hemmings et al., 1989). The mathematical model of the phenomena allows an attempt to quantify the biochemical phenomena that Greengard et al. have observed in the phosphorylation cascade in the basal ganglia. Some predictions emerge from the model: they may be useful in thinking about the organization of the striatum. The model predicts that the intracellular calcium concentration "balances" the consequences of stimulating the D-1 dopamine receptor. In addition, the mathematical model provides a series of clear biological targets that might be quantified to test the validity of the model. The model predicts that the phosphorylation state of Cam-kinase II might be affected by dopaminergic agents.

The model also allows *speculation* about physiological events known to occur in the basal ganglia. Removal of the dopaminergic innervation of the striatum introduces profound changes in the properties of this brain region. There is an increased sensitivity or responsiveness to D-1 dopaminergic agonists that can be quantified by tuming behavior, glucose utilization, and c-fos expression (DeNinno et al., 1991; Asin and Wirtshafter, 1993 present examples). According to the model, the absence of dopamine would have dramatic effects on the phosphorylation state of DARPP and CaM-kinase II. These alterations in protein phosphorylation might be important in the phenomena of "priming" that has been studied by DiChiara and his colleagues (Morelli et al., 1990, 1991). Likewise, the biochemical alterations predicted by the model might be involved in the "denervation supersensitivity." In the supersensitive striatum, there are, at best, modest changes in the properties of the D 1 receptors. The advantage of a change in the efficiency of phosphorylation cascade is that it could occur in the absence of a change in the number of D-1 dopamine receptors.

Conclusion

Taken together, the model discussed in this chapter presents a mathematical formalization of the biochemical events that are known to occur in the basal ganglia. The model provides a basis for speculating about the actions of endogenous dopamine within the basal ganglia.

Note added in proof

The interested reader is referred to the interesting and important paper of Rolf Kotter [Kotter R (1994) Postsynaptic integration of glutamatergic and dopaminergic signals in the striatum. Prog Neurobiol 44: 163–196]. Kotter has developed a mathematical model of the dopamine receptor/phosphorylation cascade in medium spiny neurons of the basal ganglia. This model represents the effect of dopamine and calcium upon the activity of CAM kinase.

References

Asin KE, Wirtshafter D (1993) Effects of repeated dopamine D-1 receptor stimulation on rotation and c-fos expression. Eur J Pharmacol 235: 167–168
Clement-Cormier YC, Parrish RG, Petzold GL, Kebabian JW, Greengard P (1975) Characterization of a dopamine-sensitive adenylate cyclase in the rat caudate nucleus. J Neurochem 25: 143–149

DeNinno MP, Schoenleber R, MacKenzie R, Britton DR, Asin KE, Briggs C, Trugman JM, Ackerman M, Artman L, Bednarz L, et al (1991) A68930: a potent agonist selective for the dopamine D1 receptor. Eur J Pharmacol 199: 209–219

Gingrich JA, Caron MG (1993) Recent advances in the molecular biology of dopamine receptors. Annu Rev Neurosci 16: 299–321

Halpain S, Girault JA, Greengard P (1990) Activation of NMDA receptors induces dephosphorylation of DARPP-32 in rat striatal slices. Nature 343: 369–372

Hemmings HC Jr, Greengard P, Tung HY, Cohen P (1984) DARPP-32, a dopamine-regulated neuronal phosphoprotein, is a potent inhibitor of protein phosphatase-1. Nature 310: 503–505

Hemmings HC Jr, Nairn AC, McGuinness TL, Huganir RL, Greengard P (1989) Role of protein phosphorylation in neuronal signal transduction. FASEB J 3: 1583–1592

Kebabian JW, Calne DB (1979) Multiple receptors for dopamine. Nature 277: 93–96

Kebabian JW, Petzold GL, Greengard P (1972) Dopamine-sensitive adenylate cyclase in caudate nucleus of rat brain, and its similarity to the dopamine receptor. Proc Natl Acad Sci USA 69: 2145–2149

Klee CB (1991) Concerted regulation of protein phosphorylation and dephosphorylation by calmodulin. Neurochem Res 16: 1059–1065

Lisman J (1989) A mechanism for the Hebb and the anti-Hebb processes underlying learning and memory. Proc Natl Acad Sci USA 86: 9574–9578

Morelli M, Di Chiara G (1990) Stereospecific blockade of N-methyl-D-aspartate transmission by MK 801 prevents priming of SKF 38393–induced turning. Psychopharmacology (Berl) 101: 287-288

Morelli M, Fenu S, Cozzolino A, Di Chiara G (1991) Positive and negative interactions in the behavioural expression of D1 and D2 receptor stimulation in a model of Parkinsonism: role of priming. Neuroscience 42: 41–48

Walaas SI, Aswad DW, Greengard P (1983) A dopamine- and cyclic AMP-regulated phosphoprotein enriched in dopamine-innervated brain regions. Nature 301: 69–71

Author's address: J. W. Kebabian, Research Biochemicals International, 1 Strathmore Road, Natick, MA 01760, U.S.A.

Molecular heterogeneity of neurotransporters: implications for neurodegeneration

K. P. Lesch*, U. Balling, M. Seemann, A. Teufel, D. Bengel, A. Heils, P. Godeck, and **P. Riederer**

Department of Psychiatry, University of Würzburg, Würzburg, Federal Republic of Germany

Summary. Neurotransporters are high-affinity transport proteins located in the plasma membrane of both presynaptic nerve and glial cells that mediate the removal of neurotransmitters from the synaptic cleft or represent intracellular transport systems that concentrate neurotransmitters in synaptic vesicles. They comprise three subgroups, Na^+/Cl^- or Na^+/K^+-dependent cell surface transporters and H^+-dependent transporters associated with synaptic vesicles. The new insights into neurotransporter diversity provide the means for novel approaches of studying neurotransmitter uptake processes at the molecular level, such as substrate translocation and antagonist binding as well as regulation of gene expression, of intracellular trafficking, and of posttranslational modification. Moreover, modeling neurotransporter-related disorders and therapeutic strategies in genetically engineered animals are now feasible research strategies. Through an improved understanding of the modulation of neurotransporter function in the brain, it may be possible to identify the molecular factors underlying the etiopathogenesis and pathophysiology of neurodegenerative disorders. Due to their specificity for distinct neuronal systems, neurotransporters and their genes are potential targets for novel therapeutic strategies.

Introduction

The structural and functional characterization of neurotransmitter reuptake mechanisms (neurotransporters) is one of the major recent advances in the elucidation of the molecular mechanisms underlying termination of synaptic signaling and fine tuning of neuronal communication. On the basis of their subcellular distribution and pharmacological properties neurotransporters have been placed into three main subgroups, Na^+/Cl^- or Na^+/K^+-dependent cell surface transporters (e.g. serotonin, dopamine and glutamate transporter)

---

* Supported by the Hermann and Lilly Schilling Foundation

and H$^+$-dependent transporter associated with synaptic vesicles (e.g. vesicular monamine transporters). High-affinity reuptake proteins located in the plasma membrane of both presynaptic nerve and glial cells mediate the removal of the neurotransmitter from the synaptic cleft, thereby terminating its action at the receptor. Intracellular transport systems reaccumulate neurotransmitters in synaptic vesicles for additional cycles of release. At both plasma and vesicular membranes, neurotransmitter influx is coupled directly to transmembrane ion gradients that provide the energy for transport against a concentration gradient.

This overview focuses on the molecular biology and pharmacology of neurotransporters and, in particular, their relevance for neurodegeneration. While the majority of cloned neurotransporters correspond to known, functionally identified subfamilies, novel transporters that had escaped detection by radioligand binding have emerged. In all cases, molecular biology is providing a more complete picture of the unexpectedly large number of neurotransporters, their distribution, substrate specificity, drug-binding profiles, molecular structure and gene organization. In addition to defining reuptake mechanisms as candidate genes for neurodegenerative disorders, it specifically emphasizes those neurotransporters that are potential targets for novel drug development.

Sodium/chloride-dependent transporters

A flurry of recent cloning endeavors has led to the molecular characterization of a large number of Na$^+$/Cl$^-$-dependent transporters, including the transporters for the biogenic amines (norepinephrine, dopamine, and serotonin), GABA, glycine, proline, taurine, betaine, and several orphan carriers (for review see Uhl, 1992; Amara and Kuhar, 1993; Blakely et al., 1994; Uhl and Johnson, 1994; Bannon et al., 1995; Barker and Blakely, 1995). Moreover, several subtypes of the GABA and glycine transporters have been identified. Although some variations exist, the structure of proteins in this gene family is best described by a model with 12 transmembrane segments, intracellular

Table 1. Neurotransporter gene families

Na$^+$/Cl$^-$	H$^+$	Na$^+$/K$^+$
Norepinephrine	Vesicular monoamine	Glutamate/aspartate
Dopamine	(1 central/peripheral,	(1 neuronal, 2 glial,
Serotonin	1 peripheral)	1 mixed)
GABA (4)		
Glycine, Proline, Taurine,	Vesicular acetylcholine	More isotypes likely
Creatine, Choline		
Several orphan transporters		

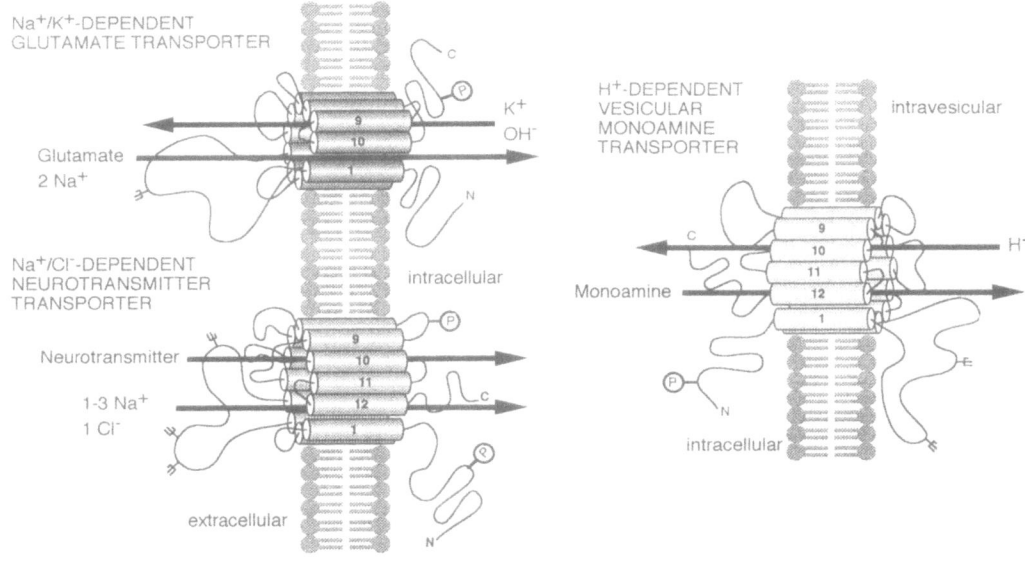

Fig. 1. Structural diversity and ion dependency of neurotransporters

amine and carboxy termini, and an extended extracellular loop with several potential glycosylation sites between transmembrane segments 3 and 4. This loop is most divergent among family members and may therefore participate in substrate recognition and inhibitor binding. On the basis of their remarkable homology of 69–80% and their properties as targets of tri- and heterocyclic antidepressants as well as amphetamine, cocaine, and their analogs, the carriers for the biogenic amines serotonin, dopamine and norepinephrine constitute a subfamily. Several neurotransmitter transporters (e.g. serotonin, dopamine, and GABA transporter) are electrogenic and possess a variety of unexpected ion channel-like properties.

Serotonin transporter

The serotonin (5-HT) transporter plays a central role in the termination of serotonergic neurotransmission by Na^+-driven uptake of 5-HT into the presynaptic neuron and are regarded as initial sites of action of antidepressant drugs and several neurotoxins. Tricyclics, such as imipramine, and the selective 5-HT reuptake inhibitors, fluoxetine and paroxetine, occupy sites overlapping at least partially the substrate binding site and are widely used in the treatment of depression, anxiety and impulse control disorders, as well as substance abuse including alcoholism (Lesch and Beckmann, 1993c). 5-HT transporter (5-HTT) function is altered in the brain of suicide victims and in platelets from patients with depression. Neurotoxins, such as 3,4-methylenedioxymeth-amphetamine (MDMA, "ecstasy"), are concentrated in serotonergic neurons by this transporter. Paralleling the recognition of these clinical links, research has increasingly focused on the structure and function of the human 5-HTT.

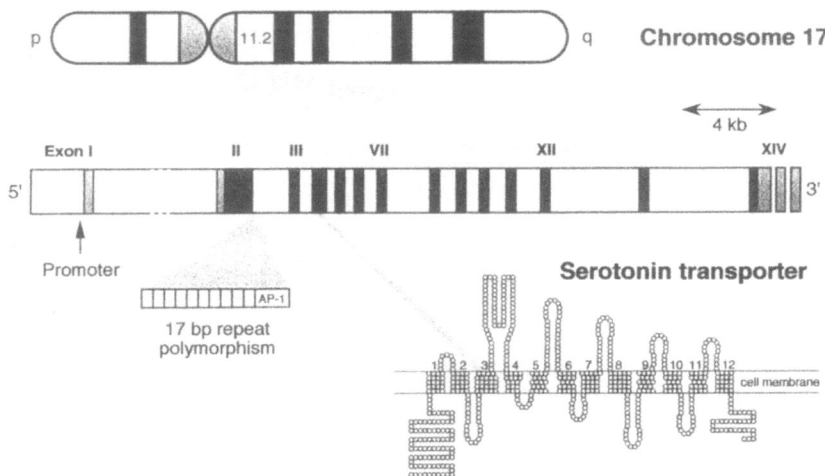

Fig. 2. Organization of the human serotonin transporter gene (hg5-HTT) on chromosome 17q11.2. Positions of exons (shown as black and white boxes for the coding and noncoding regions, respectively) are indicated. The locations of the promoter/enhancer region and of a polymorphic 17 bp repetitive element are also shown

Recent cloning and sequencing of the 5-HTT identified a protein with twelve putative transmembrane domains which is likely to participate in substrate translocation, ion binding and antagonist binding (Blakely et al., 1991; Hoffman et al., 1991). Evidence from studies with deletion mutants indicates that 5-HTT function may be dependent on the formation of quaternary structures, such as dimers and tetramers. For the serotonin transporter several conducting states have been reported: 1) During the gating process a transport-associated current is caused by ions which are not involved in the transport but can pass through the channel, 2) a transient current triggered by extreme negative potentials, and 3) a leakage current (Mager et al., 1994).

In order to facilitate studies of transporter expression and regulation in psychiatric and neurodegenerative disorders as well as the search for polymorphisms and mutations in the human gene, our group has cloned and sequenced a full-length human homolog of the 5-HT transporter cDNA from brain and platelets (Lesch et al., 1993b). The findings demonstrate that the human platelet 5-HT uptake site is identical with the human brain 5-HTT and validate the human platelet 5-HT uptake site as a peripheral model of the brain 5-HTT in molecular pharmacologic and neuropsychiatric research. In support of this concept, the presence of the brain dopamine transporter and vesicular monoamine transporter in human platelets could also be demonstrated (Lesch et al., 1993a, unpublished results). The 5-HTT gene has been mapped on the human chromosome 17q11.2 and is composed of 14 exons spanning > 45 kb (Ramamoorthy et al., 1993; Lesch et al., 1994a; Polymeropoulos and Lesch, unpublished results). Several potential binding sites for transcription factors including a TATA-like motif are present in the promotor region. The role of these transcription factor motifs in the regulation of 5-HTT gene expression are currently being established in gel mobility shift assays and by transfection studies using reporter gene fusion constructs of 5-HTT 5'-flanking sequences. It is anticipated that characterization of the 5-HTT 5'-flanking sequence will identify a promoter/enhancer unit most specific for serotonergic neurons. The fusion of this promoter/enhancer unit to similar or unrelated genes would aid their expression in a specific neuronal system and open the way to novel therapeutic strategies.

Decreased 5-HT uptake and inhibitor binding in patients with depression or bipolar disorders may reflect a structural defect and/or dysregulated expression of the 5-HTT (Ellis and Salmond, 1994). Although screening for abnormalities in the primary structure of the 5-HTT in patients with affective disorders identified several non-coding and coding mutations, the previously observed depression-associated reduction in 5-HTT function is not likely to be caused by a change of amino acids which are thought to be critical to substrate transport (Lesch et al., 1995). Further, structural alterations sufficiently severe to inhibit translation or cell membrane insertion of the 5-HTT protein are not likely to underly the decreased B_{max} in inhibitor binding in platelets and brain of patients with affective disorders. Hence, analysis of the 5-HTT 5'-flanking sequence is required to gain conclusive information on this potential susceptibility locus for affective disorders.

While tri- and heterocyclic antidepressants inhibit monoamine reuptake, several agents increase monoaminergic transmission by facilitating non-exocytic transporter-mediated monamine release. Amphetamines and their highly lipophilic analogs, such as MDMA ("ecstasy"), are either substrates of 5-HT and dopamine transporters or enter nerve terminal by diffusion, displace the biogenic amine from its storage vesicle and ultimately enhance carrier-mediated release (Levi and Raiteri, 1993). Although the dopamine-releasing action of MDMA is likely to be related to the mechanism of reinforcement, it is also a potent releaser of 5-HT and long-term administration results in toxic degeneration of serotonergic terminals. MDMA-induced 5-HT release and neurotoxicity can be prevented by tri- and heterocyclic reuptake inhibitors. In analogy, the 5-HT-releasing effect of the anorectic drug fenfluramine is blocked by clomipramine or fluoxetine.

Dopamine transporter

Following exocytic release into the synaptic cleft, the action of dopamine at its receptors is terminated by neuronal reuptake via a Na^+/Cl^--dependent transporter. Various pharmacologic, structural, and functional aspects of the dopamine transporter (DAT) have recently been discussed in detail (Giros and Caron, 1993; Boja et al., 1994; Bannon et al., 1995). Although the DAT is predominantly associated with dopaminergic neurons, it is also expressed in platelets and several neuroblastoma cell lines (Lesch et al., unpublished results). After its retrieval, dopamine is either repackaged into synaptic vesicles or metabolized by mitochondria-associated monoamine oxidase. Vesicular dopamine uptake (via the vesicular monoamine transporter) is distinct with regard to its ion and energy needs, substrate specificity, and kinetic properties (see also section on proton-dependent transporters). cDNAs for DAT have been cloned from both rat and humans. Molecular studies have shown that the DAT is a 80 kDa glycoprotein with heterogenous forms defined by different carbohydrate side chains in different brain areas. The DAT protein contains degenerate leucine zippers and thus potentially associates to a tetrameric assembly of identical units (Milner et al., 1994).

Inhibition of the DAT (in addition to the blockade of norepinephrine and serotonin transporters) by cocaine and its analogs induces multiple neurochemical and behavioral effects that are thought to represent the molecular basis of reinforcement. Although amphetamine stereoisomers have similar affinities for the norepinephrine transporter, their interaction with the DAT is stereoselective. This stereoselective property at biogenic amine transporters is likely to be involved in psychotomimetic actions of amphetamins. Finally, dopamine reuptake blocking actions may also contribute to the therapeutic efficacy of antidepressants such as nomifensine and bupropion.

Abnormalities in DAT function have been postulated for a variety of neurodegenerative and psychiatric disorders. Postmortem studies indicate that [^{3}H]cocaine and [^{3}H]GBR-12935 binding are decreased in Parkinson's disease (PD). Hemiphere-to-hemisphere differences in the binding of [^{11}C]nomifensine in the striatum and accumulation of [^{18}F]fluorodopa in the putamen have been confirmed by PET studies. Neurodegeneration of the nigrostriatal dopaminergic system induced by 1-methyl-4-phenylpyridine (MPP$^+$), the neurotoxic metabolite of MPTP, has implicated the DAT in the etiology of PD. Since MPP$^+$ enters neurons through the DAT, a dysfunctional transport process may contribute to an increased susceptibility to exogenous MPP$^+$-like neurotoxins. This vulnerability to neurotoxins may be further aggravated by an impaired capacity of the brain vesicular monoamine transporter (VMT), which plays a central role in the sequestration of cytoplasmic toxins and thus in the limitation of mitochondrial damage. With aging, even modest reductions of vesicular uptake would accumulate MPP$^+$-like toxins in the cytoplasmic pool and cause neuronal cell death. In this regard, it is of interest to note that in humans there is evidence for progression of MPP$^+$-induced dopaminergic lesions, thus indicating that transient exposure to a toxin may cause a protracted decline in nigro-striatal dopaminergic function more rapid than in normal aging and similar to PD (Vingerhoets et al., 1994). While the status of VMT expression in PD remains to be investigated, an age-related drastic and sudden onset reduction of DAT mRNA which is contrasted by a more gradual decline in DAT protein has been described in postmortem human substantia nigra (Bannon et al., 1992). Functional and structural analysis of the DAT (and possibly the VMT) using in vitro and in vivo strategies, such as expression in cultured cells and neuroimaging with SPECT and PET, is likely to enhance our understanding of MPP$^+$ neurotoxicity and the role of the DAT in the progression of Parkinson's disease.

Disorders of the schizophrenic spectrum are thought to be associated with an aberrant dopaminergic (and glutamatergic) neurotransmission. Cocaine and amphetamine act as psychostimulants or psychotomimetics. Both amphetamine-evoked psychotic symtomatology in vulnerable individuals or patients with a pre-existing psychotic illness and cocaine-induced kindling and arousal are believed to result from their effects on the mesolimbic dopamine system, which has been implicated in the integrative process of perception, cognition, and emotion. While changes in the kinetics or inhibitor binding of the 5-HTT have been reported in disorders of the schizophrenic spectrum (Joyce et al., 1993), studies on the DAT resulted in conflicting findings (Haberland and Hatey, 1987; Pearce et al., 1989). Moreover, no evidence for a linkage of the DAT gene with schizophrenic disorders in multigenerational pedigrees was found (Byerley et al., 1993). Finally, an increase in DAT sites has been reported in Gilles de la Tourette's syndrome and a deficient DAT-related uptake has been proposed to play a role in compulsive self-mutilating behavior as observed in Lesch-Nyhan syndrome.

Proton-dependent vesicular monoamine transporters

Vesicular monoamine transporters (VMTs) constitute a distinct class of carriers that mediate the transport of neurotransmitters from the cytoplasm into synaptic vesicles (Edwards, 1992). VMTs nonselectively accumulate cytoplasmic biogenic amine neurotransmitters (e.g. serotonin, catecholamines) and neurotoxins (e.g. MPP^+) by ATPase-dependent and H^+-driven uptake into storage vesicles of presynaptic neurons and platelets. They are regarded as initial sites of action of monoamine transport inhibitors, such as reserpine and tetrabenazine (Rudnick et al., 1990). Recent cloning and sequencing of cD-NAs encoding VMTs from rat brain, adrenal medulla, and basophilic leuke-

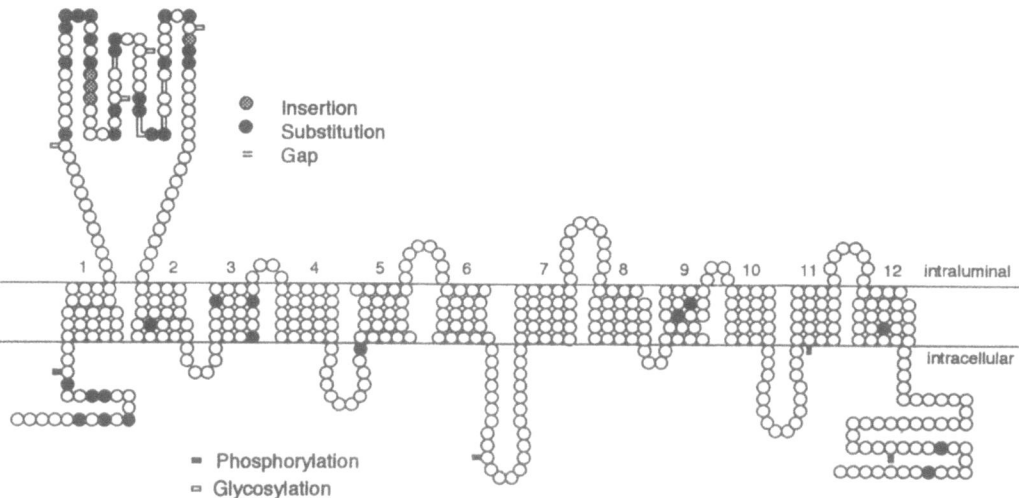

Fig. 3. Topology of the human brain/platelet vesicular monoamine transporter (hVMT). Circles represent individual amino acids. Black and shaded circles or gaps indicate amino acids differing from the rat brain VMT. Open boxes reflect putative glycosylation sites. Potential phosphorylation sites for cAMP-dependent protein kinase and protein kinase C are indicated by black boxes, respectively

mia cells identified two distinct but related proteins (521 and 515 amino acids) with twelve putative transmembrane segments and a large luminal loop located between the first two transmembrane domains which are likely to participate in ion/substrate translocation and antagonist binding (Rudnick et al., 1990; Erickson et al., 1992; Liu et al., 1992). The VMTs show a remote sequence similarity with several bacterial drug resistant genes and have been shown to suppress the neurotoxic effects of MPP$^+$ by subcellular compartmentalization. Sequence divergence between the central and peripheral VMT may play a central role in the differential neurotoxic effects of MPP$^+$ in brain and adrenal medulla.

Recently, our group has isolated and analyzed a VMT cDNA from human midbrain and confirmed its expression in human platelets (Lesch et al., 1993a). Although the human brain VMT is 90.7% homologous to the rat protein, an extensive sequence divergence occurs in the large luminal loop located between the first two transmembrane domains. This region of the polypeptide differs in 23 out of 86 amino acids (26.7%) and contains three additional, two positively and one negatively, charged amino acids. As compared to the rat VMT cDNA, these structural changes result in a reduced overall homology of 64.0% in the large luminal loop of the human VMT, although four putative glycosylation sites are conserved. Analysis of the coding sequence flanking the large luminal loop detected 12 out of 16 conservative amino acid substitutions with an overall homology of 96.3%.

CHO fibroblasts transfected with a cDNA clone encoding the VMT expressed in rat adrenal chromaffin cells confers resistance to MPP$^+$. Although the peripheral VMT is highly homologous to the VMT expressed in rat brain, considerable sequence divergence occurs in the large luminal loop flanked by first two transmembrane segments and, to a lesser extent, in the N-terminal region. Since the large extracellular loop of Na$^+$-dependent plasma membrane transporters and the large luminal loop of H$^+$-driven vesicular transporters have been implicated in ion and/or substrate recognition, sequence divergence in this segment of the VMT expressed in rat brain and adrenal medulla may underlie the tissue-specific differences in MPP$^+$ toxicity. Our finding of an extensive sequence divergence in the large luminal loop of the VMT between the rat and human brain may represent one possible molecular basis for the substantial species differences in the susceptibility to MPP$^+$ demonstrated among humans, nonhuman primates and rodents (for review see Gerlach et al., 1991). The extraordinary extent of variation across species reflecting the process of evolution may indicate that this region of the genome is a "hot spot" for mutational activity and may have intriguing implications for the pathogenesis of idiopathic Parkinson's disease. Structural alterations and/or abnormalities in gene expression as a consequence of an accumulation of somatic mutations could result in a decreased capacity of the brain VMT to remove neurotoxins or regulate cytoplasmic dopamine selectively taken up into nigro-striatal dopaminergic neurons by the DAT (Tanner and Langston, 1990). A dysfunctional VMT may not be able to limit the mitochondrial damage of putative exogenous and endogenous toxins, thus leading to degeneration of dopaminergic neurons and clinical Parkinson's

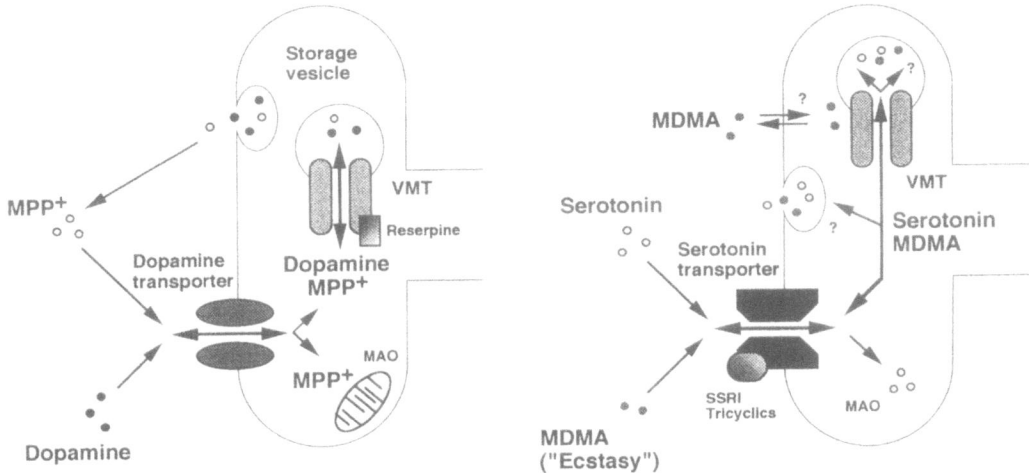

Fig. 4. Neurotransporters and putative mechanisms of MPP⁺ and amphetamine neurotoxicity (for details see text)

disease (also see section on dopamine transporters). Likewise, VMT abnormalities affecting monoamine transporter system function may contribute to
the pathogenesis of affective disorders (Lesch et al., 1994b) and other neuropsychiatric disorders in which these systems have been implicated.

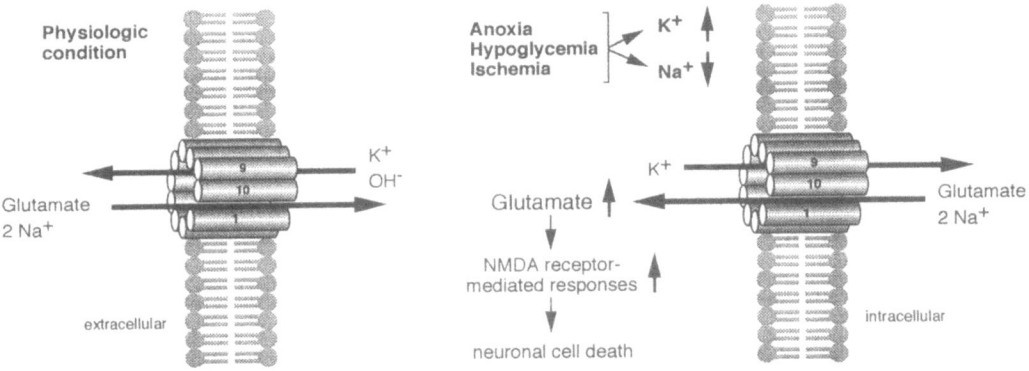

Fig. 5. Reversal of neuronal and glial glutamate transport in pathologic conditions.
Glutamate, a potent excitotoxin, is increased to neurotoxic levels in anoxic, hypoglycae-
mic, and ischaemic states. Following epileptic seizures or stroke, an extracellular increase
of K⁺ and reduction of Na⁺ may favor transport reversal and glutamate release which, in
turn, would prolong NMDA receptor-mediated responses and cause acute and chronic
neuronal cell death

Sodium- and potassium-dependent exitatory amino acid transporters

The joint action of neuronal and glial high-affinity exitatory amino acid transporters plays a critical role in the termination of glutamatergic neurotransmission and the maintenance of extracellular glutatamate concentrations below a neurotoxic level. Four distinct cDNAs encoding structurally related glutamate/aspartate (EAAC1, GLT1, GLAST) have now been identified and characterized (for review see Kanai et al., 1993). Expression of EMC1 appears to be confined to pre- and postsynaptic neurons in the neocortex, hippocampus, and basal ganglia, GLAST is found on certain neurons and astroglia in the cerebellar Purkinje cell layer, and GLT1 is thought to be primarily glial in origin with ubiquitous distribution in the neocortex, hippocampus, and entorhinal cortex (Rothstein et al., 1994). Recently, four human homologs of glutamate transporter cDNAs (EMT1-4) have been cloned from motor cortex (Arriza et al., 1994). Although differences in tissue distributions between species have been demonstrated, the amino acid sequence homologies of EAAT1, EMT2, and EMT3 compared to GLAST, GLT1, and EMC1 are 96%, 95%, and 92%, respectively.

Expression studies with the gene products demonstrated that glutamate transport is an electrogenic process and is characteristically dependent on both extracellular Na^+ and intracellular K^+. During each transport cycle a pH-changing anion (OH^- or HCO_3^-) is also counter-transported. This indicates that the uptake of a glutamate molecule is escorted by the co-transport of two Na^+ and by the counter-transport of one K^+ and one anion, such as OH^- or HCO_3^-. Hence, the glutamate uptake process evokes a depolarizing positive inward current leading to intracellular acidification and extracellular alkalinization. Although there is evidence that this electrogenic uptake is participating in the depolarizing action of glutamate, the functional consequences of the transporter-mediated depolarization and the intra- and extracellular pH changes are not fully understood. Intriguingly, non-vesicular calcium-independent release of glutamate may be induced by reversing the transport process through an extracellular increase in K^+ or a decrease in Na^+ concentrations.

The electrochemical environment of glutamate transport may have important implications for a variety of neuropathological processes. Glutamate is a potent excitotoxin and in vitro and in vivo observations indicate that extracellular glutamate is increased to neurotoxic levels in anoxic, hypoglycemic, and ischemic conditions. An extracellular increase of K^+ and reduction of Na^+ following epileptic seizures or stroke may favor transport reversal and glutamate release (Levi and Raiteri, 1993). Glutamate released in vivo is likely to derive from both neurons and astrocytes. During ischemia and concurrent anoxia/hypoglycemia, the K^+-evoked release of aspartate is increased only in cultured neurons, while the spontaneous release (= leakage) of glutamate and aspartate is enhanced in astrocytes but not in neurons. In line with these findings, microdialysis studies have demonstrated that inhibition of glutamate uptake prolongs NMDA receptor-mediated responses and causes acute and chronic neuronal cell death. The interrelationship of defective excitatory amino acid transporter function and excitotoxic glutamate concentrations in the synaptic cleft is therefore indicative of transporter dysfunction in neurodegenerative disorders that are associated with a disrupted glutamate turnover. There is evidence that concentrations of glutamate transporters are decreased in distinct brain areas of patients with Alzheimer's disease, Parkinson's disease, Huntington's disease, and amyotrophic lateral sklerosis (ALS) (Rothstein et al., 1992), as well as psychiatric disorders of the schizophrenic spectrum (Simpson et al., 1992). Aberrant glutamate transporter function in these conditions would increase synaptic glutamate which would subsequently initiate a cascade of events leading to neuronal death. Whether abnormalities in glutamate uptake are related to the pathogenesis of these disorders or result from as yet unknown neurodegenerative processes is controversial.

An inborn error of glutamate transport may cause dicarboxylic aminoaciduria which is characterized by increased renal excretion of glutamate/aspartate and may also be associated with neurologic abnormalities and mental retardation (Smith et al., 1994). Finally, glutamate transport inhibiting properties of agents such as (S)-β-oxalyl-α,β-diaminopropionic acid (ODAP) and kainic acid may significantly contribute to their neurotoxicity. ODAP is the excitotoxin implicated in lathyrism, a degenerative disorder associated with spastic paraplegia.

Conclusions

There has been considerable progress in the molecular characterization of multiple neurotransporters. The new insights into transporter diversity provide the means for more targeted approaches to studying reuptake processes at the molecular level. Future research strategies will focus on molecular mechanisms of substrate translocation and antagonist binding as well as on posttranslational modification and functional regulation at the neurotransporter protein level. Exciting information will also be derived from the analysis of genomic regulatory elements and from modeling neurotransporter-related disorders in transgenic animals by targeted gene knock-out. Through an improved understanding of the modulation of the function of neurotransporter in the brain, it may be possible to identify the molecular factors underlying both the predisposition to and the pathogenesis of neurodegenerative disorders. Due to their specificity for distinct neuronal systems neurotransporters and their genes are potential targets for novel therapeutic strategies.

References

Arriza JL, Fairman WA, Wadiche JI, Murdoch GH, Kavanaugh MP, Amara SG (1994) Functional comparison of three glutamate transporter subtypes cloned from human motor cortex. J Neurosci 14: 5559–5569

Bannon MJ, Poosch MS, Xia Y, Goebel DJ, Cassin B, Kapatos G (1992) Dopamine transporter mRNA content in human substantia nigra decreases precipitously with age. Proc Natl Acad Sci USA 89: 7095–7099

166   K. P. Lesch et al.

Blakely RD, Berson HE, Fremeau RT, Caron MC, Peek MM, Prince HK, Bradley CC (1991) Cloning and expression of functional serotonin transporter from rat brain. Nature 354: 66–70

Boja JW, Vaughan R, Patel A, Shaya EK, Kuhar MJ (1994) The dopamine transporter. In: Niznik HB (ed) Dopamine receptors and transporters. Marcel Dekker, New York, pp 611–644

Byerley W, Coon H, Hoff M, Holik J, Waldo M, Freedman R, Caron MG, Giros B (1993) Human dopamine transporter gene not linked to schizophrenia in multigenerational pedigrees. Hum Hered 43: 319–322

Edwards RH (1992) Transport of neurotransmitters into synaptic vesicles. Curr Opin Neurobiol 2: 586–594

Ellis PM, Salmond C (1994) Imipramine binding in depression: a meta-analysis. Biol Psychiatry 36: 292–299

Erickson J, Eiden L, Hoffman B (1992) Expression cloning of a reserpine-sensitive vesicular monoamine transporter. Proc Natl Acad Sci USA 89: 10993–10997

Gerlach M, Riederer P, Przuntek H, Youdim MBH (1991) MPTP mechanisms of neurotoxicity and their implications for Parkinson's disease. Eur J Pharmacol Mol Pharmacol Sect 208: 273–286

Giros B, Caron MG (1993) Molecular characterization of the dopamine transporter. Trends Pharmacol Sci 14: 43–49

Haberland N, Hatey L (1987) Studies in postmortem dopamine uptake. II. Alterations of synaptosomal catecholamine uptake in postmortem brain regions in schizophrenia. J Neural Transm 68: 303–313

Hoffman BJ, Mezey E, Brownstein M (1991) Cloning of a serotonin transporter affected by j antidepressants. Science 254: 579–580

Joyce JN, Shane A, Lexow N, Winokur A, Casanova MF, Kleinman JE (1993) Serotonin uptake sites and serotonin receptors are altered in the limbic system of schizophrenics. Neuropsychopharmacology 8: 315–336

Kanai Y, Smith CP, Hediger MA (1993) A new family of neurotransmitter transporters: the high-affinity glutamate transporters. FASEB J 7: 1450–1459

Lesch KP, Beckmann H (1993c) Neurotransporter: Neue Aspekte zum Wirkmechanismus psychotroper Substanzen. Nervenarzt 64: 75–79

Lesch KP, Gross J, Wolozin BL, Murphy DL, Riederer P (1993a) Extensive sequence divergence between the human and rat brain vesicular monoamine transporter: possible evidence for species differences in the susceptibility to MPP+. J Neural Transm 93: 75–82

Lesch KP, Wolozin BL, Murphy DL, Riederer P (1993b) Primary structure of the human platelet serotonin (5-HT) uptake site: identity with the brain 5-HT transporter. J Neurochem 60: 2319–2322

Lesch KP, Balling U, Gross J, Strauss K, Wolozin BL, Murphy DL, Riederer P (1994a) Organization of human serotonin transporter gene. J Neural Transm 95: 157–164

Lesch KP, Gross J, Riederer P, Murphy DL (1994b) Direct sequencing of the reserpine-sensitive vesicular monoamine transporter in unipolar depression and bipolar disorder. Psychiatr Genet 4: 153–160

Lesch KP, Franzek E, Gross J, Wolozin BL, Riederer P, Murphy DL (1995) Primary structure of the serotonin transporter in unipolar depression and bipolar disorder. Biol Psychiatry 37: 215–223

Levi G, Raiteri M (1993) Carrier-mediated release of neurotransmitters. Trends Neurosci 16: 415–419

Liu Y, Peter D, Roghani A, Schuldiner S, Prive GG, Eisenberg D, Brecha N, Edwards RH (1992) A cDNA that suppresses MPP+ toxicity encodes a vesicular amine transporter. Cell 70: 539–551

Mager S, Min C, Henry DJ, Chavkin C, Hoffman BJ, Davidson N, Lester HA (1994) Conducting states of a mammalian serotonin transporter. Neuron 12: 845–859

Milner HE, Beliveau R, Jarvis SM (1994) The in situ size of the dopamine transporter is a tetramer as estimated by radiation inactivation. Biochim Biophys Acta 1190: 185–187

Pearce RKB, Seeman P, Jellinger K, Tourtellotte W (1989) Dopamine uptake sites and dopamine receptors in Parkinson's disease and schizophrenia. Eur Neurol 30: 9–14

Ramamoorthy S, Bauman A, Moore K, Han H, Yang-Feng T, Chang A, Ganapathy V, Blakely R (1993) Antidepressant- and cocaine-sensitive human serotonin transporter: molecular cloning, expression, and chromosomal localization. Proc Natl Acad Sci USA 90: 2542–2546

Rothstein JD, Martin LJ, Kuncl RW (1992) Decreased glutamate transport by the brain and spinal cord in amyotrophic lateral scerosis. N Engl J Med 326: 1464–1468

Rothstein JD, Martin L, Levey Al, Dykes-Hoberg M, Jin L, Wu D, Nash N, Kuncl RW (1994) Localization of neuronal and glial glutamate transporters. Neuron 13: 713–725

Rudnick G, Steiner-Murdoch S, Fishkes H, Stern-Bach Y, Schuldiner S (1990) Energetics of reserpine binding and occlusion by the chromaffin granule biogenic amine transporter. Biochemistry 29: 603–608

Simpson MDC, Slater P, Royston MC, Deakin JFW (1992) Regionally selective deficits in uptake sites for glutamate and gamma-aminobutyric acid in the basal ganglia in schizophrenia. Psychiatry Res 42: 273–282

Smith CP, Weremowicz S, Kanai Y, Stelzner M, Morton CC, Hediger MA (1994) Assignment of the gene coding for the human high-affinity glutamate transporter EMC1 to 9p24: potential role in dicarboxylic aminoaciduria and neurodegenerative disorders. Genomics 20: 335–336

Tanner C, Langston J (1990) Do environmental toxins cause Parkinson's disease? A critical review. Neurology 40: 17-30

Vingerhoets FJG, Snow BJ, Tetrud JW, Langston JW, Schulzer M, Calne DB (1994) Positron emission tomographic evidence for progression of human MPTP-induced dopaminergic lesions. Ann Neurol 36: 765–770

Authors' address: K. P. Lesch, M.D., Department of Psychiatry, University of Würzburg, Füchsleinstrasse 15, D-97080 Würzburg, Federal Republic of Germany.

Central nervous system cytokines and their relevance for neurotoxicity and apoptosis

J. Licinio

Clinical Neuroendocrinology Branch, National Institute of Mental Health, National Institutes of Health, Bethesda, MD, U.S.A.

Summary. Cytokines are molecules that are synthesized not only by the immune system, but also by cells in the central nervous system, including neurons, glia, and brain vascular cells. In the brain, cytokines can be neuroprotective or they can contribute to neurodegeneration. The role of cytokines in the regulation of normal and abnormal brain function represents a rapidly growing frontier in neuroscience. Cytokines are pleiotropic and redundant, and they can modulate the effects of neurotransmitters and neuropeptides; thus, in order to understand the effects of brain cytokines on apoptosis and toxicity, it is necessary to study the temporal and spatial expression of complex networks of cytokines, growth factors, neuropeptides, and neurotransmitters. This effort is currently in progress in many centers. Modulation of cytokine function in the central nervous system represents a new therapeutic strategy for neurodegeneration.

Introduction

Cytokines were first identified as the product of immune cells that function as intercellular signals during the regulation of the immune response. It has been subsequently shown that several cytokines are synthesized not only by immune cells, but also by a variety of cell types in various organs, including the brain. The neurobiology of cytokines is now a rapidly expanding area of neuroscience research. Investigation in this area has had three phases: First, a number of groups, have shown that exogenously administered cytokines affect various functions of the brain. Later, our group and others have examined the pattern of expression and function of endogenous cytokines in brain. The latest development in this field is related to the study of the molecular mechanisms involved in the regulation of cytokine gene expression in brain.

Cytokines in the brain

The genes encoding for a variety of cytokines and cytokine receptors are expressed in brain. Our group has identified in rat brain the mRNA for interleukin 1 (IL-1) receptor antagonist (IL-1ra) (Licinio et al., 1991), IL-1 type I receptor (IL-1RI) (Wong and Licinio, 1994), IL-8 (Licinio et al., 1992) and stem cell factor (SCF) (Wong and Licinio, 1994) Other groups have localized other cytokines in brain, such as IL-1β, IL-1α, IL-2, IL-3, IL-6, IL-6 receptor, and TNF-α, interferon γ (Vanguri, 1995; Pousset, 1994; Schobitz et al., 1993; Lapchack, 1992; Minami et al., 1991; Bandtlow et al., 1990; Farrar et al., 1989). In our initial work we studied constitutive cytokine levels of cytokine/cytokine receptor mRNA using ^{35}S-labeled riboprobes. We have shown cytokine gene expression in glial-like cells, as well as in neuron-like cells. Specifically, we have shown the presence of IL-1ra mRNA in hypothalamus, hippocampus, cerebellum, and choroid plexus (Licinio et al., 1991). We have also found low levels of IL-1RI in hippocampus, cerebellum, and in brain vasculature (Wong and Licinio, 1994).

It has been challenging to demonstrate whether cytokines, acting via cytokine receptors, have a role in the normal functioning of the central nervous system. Under pathological conditions, or experimental situations, cytokine administration results in profound changes in some of the major functions of the brain, including neuroendocrine function, behavior, and the regulation of vegetative functions, such as temperature, food intake, and metabolic rate. Therefore, it is now well established that in pathological states cytokines affect the brain. Whether endogenous cytokines have any effect on the functioning of the brain remains to be demonstrated. That point is hard to demonstrate experimentally. The most logical approach to this question would be to use transgenic animals with a knockout for the gene encoding for a specific cytokine. The recently published interleukin-1β converting enzyme (ICE) knockout mouse, for example, might provide useful insights on the role of endogenous IL-1β in brain (Li et al., 1995). The ICE (-/-) knockout mouse develops normally, and has no overt alterations in brain function. However, that does not rule out a role for IL-1β in normal brain physiology. The hallmark of cytokine biology is that cytokines are pleiotropic and redundant. Thus, creating an organism lacking a specific cytokine may not be very informative, as the effects of the lack one cytokine can be compensated during development by another cytokine with similar effects. For examples several of the actions of IL-1 in the brain can also be accomplished by IL-6. Further studies are certainly needed to determine whether a central cytokine network is involved in the regulation of the normal functions of the central nervous system in physiological conditions. Our lab and others have documented the necessary neuroanatomical substrate for cytokine actions in healthy brain by documenting that the genes encoding for a network of cytokines/cytokine receptors are expressed in normal, unstimulated brain. It remains to be determined whether the product of these genes is actually necessary for the normal functioning of the brain.

Necrosis and apoptosis

The biological processes underlying cell death are being intensively investigated. The ultimate goal in this field is to identify specific genes involved in cell death and to manipulate the expression of those genes or the effects of their

Table 1. General differences between apoptosis and necrosis[a]

Characteristics	Apoptosis	Necrosis
Stimuli	Physiological	Pathological (injury)
Occurrence	Single cells	Groups of cells
Reversibility	No (after morphological changes)	Yes (up to the point of no return)
Adhesions Between cells and to BM	Lost (early)	Lost (late)
Cytoplasmic organelles	Late stage swelling	Very early swelling
Lysosomal enzyme release	Absent	Present
Nucleus	Convolution of nuclear outline and breakdown (karyorrhexis)	Disappearance (karyolysis)
Nuclear chromatin	Compaction in uniformly dense masses	Clumping not sharply defined
DNA breakdown	Internucleosomal	Randomized
Cell	Formation of apoptotic bodies	Swelling and later disintegration
Phagocytosis by other cells	Present	Absent
Exudative inflammation	Absent	Present
Scar formation	Absent	Present

[a]This table summarizes in broad terms the differences between apoptosis and necrosis. There are only a few exceptions to the general concepts presented in it [reproduced from Gerschenson and Rotello (1992) with permission]

products, so as to prevent cell death. This is particularly important in neuroscience, as interventions to prevent both acute and chronic neurodegeneration have enormous therapeutic potential. Briefly, cell death has been traditionally conceptualized to occur via one of two mechanisms: apoptosis or necrosis. Apoptosis [from the Greek for fall (ptosis) off (apo)] is defined as programmed or genetically mediated cell death occurring by fragmentation of a cell into membrane-bound particles that are then eliminated by phagocytosis. Necrosis [from the Greek for deadness] is defined as the sum of the morphological changes indicative of cell death caused by the progressive degradative action of enzymes. Table 1, from Gerschenson and Rotello (1992) summarizes the differences between apoptosis and necrosis. Necrosis is associated with inflammation, and it is logical that cytokines, which function as inflammatory signals, would have a role in necrosis. There is controversy at present as to whether cytokines are involved in apoptosis.

The role of brain cytokines in acute and chronic neurodegeneration

Cytokines are involved in neurodegeneration. IL-1β particularly has been identified in both acute and chronic neurodegeneration and seems to have a role in the pathophysiology of various forms of neurodegeneration. Vascular occlusion is a useful model of acute neurodegeneration. IL-1β gene expression is induced after focal cerebral ischemia caused by experimental middle cerebral artery occlusion (Buttini et al., 1994). Antagonism of the actions of IL-1 by the pure endogenous antagonist to the IL-1 receptor, IL-1ra, causes a 50% decrease in neuronal death (Relton and Rothwel, 1992) and it also decreases edema formation (Yamasaki et al., 1994). IL-1ra also decreases by 70% neuronal cell death associated with excitotoxic damage caused by injection of an NMDA agonist in the striatum (Relton and Rothwel, 1992). We have found IL-1ra mRNA in brain (Licinio et al., 1991), and administration of the translation product of that mRNA decreases acute neuronal cell death (Licinio et al., 1991; Relton and Rothwel, 1992). It is thus possible that IL-1ra might function as an endogenous antiinflammatory cytokine in brain, protecting neurons from death by acute events, such as vascular occlusion. IL-6 mRNA has also been identified in ischemic cortex (Wang et al., 1995). IL-6 mRNA is induced in astrocytes by hypoxia (Maeda et al., 1994). The levels of IL-6 mRNA increase 10 fold after middle cerebral artery occlusion (Wang et al., 1995). TNF-α mRNA and protein is also elevated acutely (1–12 h) after middlecerebral artery occlusion in rats. The neuronal expression of TNF-α appears to facilitate the infiltration of inflammatory cells that can further exacerbate tissue damage in cerebral ischemia and might contribute to increased sensitivity and risk in focal stroke.

In chronic neurodegenerative conditions, such as neuroAIDS (Vitkovic et al., 1994) and Alzheimer's disease (Cacabelos et al., 1994), there is also evidence of increased IL-1β in brain. IL-1β itself is a potent stimulus for the synthesis of amyloid precursor protein (Donnelly et al., 1990). Whether increased IL-1β is an epiphenomenom in those disorders or a component of their intrinsic pathophysiology remains to be determined.

The role of ICE in cell death in brain

IL-1β converting enzyme (ICE) is essential for the cleavage of biologically inactive pro IL-1β into biologically active IL-1β (Thornberry et al., 1992). It has been recently proposed that ICE is the mammalian homologue of the CED-3 gene from C. elegans, which is involved in apoptosis (Yuan et al., 1993). As a result of this homology finding, considerable interest has been centered on the possible of ICE in neurodegeneration. That interest is fueled by the fact that the tri-dimensional structure of ICE has been identified by X-ray cristallography (Wilson et al., 1994), and this enzyme is now the target for the rational development of specific ICE inhibitors.

The initial work in this area seemed to indicate that ICE is indeed involved in neuronal cell death. Transfection of dorsal root gangli (DRG) neurons with crm-A, a vaccinia virus protein that inhibits ICE, is associated with neuronal survival (Gagliardini et al., 1994). That data has been seen as evidence that ICE is indeed involved in neuronal apoptosis. Three more recent pieces of evidence question the concept as an apoptotic signal in mammalian cells. First, neither Keane et al. (1995) nor our own lab have identified ICE in neurons. Second, apoptosis occurs most frequently during development; however, transgenic mice with a knockout (-/-) for ICE develop normally (Li et al., 1995). Third, in a model of apoptosis occurring in T-cells deprived of IL-2, 95% inibition of ICE activity causes no changes in the pattern of apoptosis (Vasilakos et al., 1995). Taken together, these lines of evidence suggest that even if ICE is involved in apoptosis in mammalian cells, it is not required for apoptosis. This is consistent with a reasonable understanding of physiological processes, as it is hard to imagine that apoptosis would occur in all cells cleaving pro-IL-1β into active IL-β. ICE is member of a family of proteases and has a number of homologues, which are potentially inhibited by crm-A. It is thus possible that the finding of crm-A inhibition of neuronal death might be the result of the effects of crm-A on ICE homologues. Shivers et al. have identified ICE in conditions of necrosis. Consequently, even though ICE might not be required for apoptosis, it might be involved in necrosis.

New data indicate that other members of the CED-3 family other than ICE may be involved in clinically-significant apoptosis in the brain. Apoptosis has recently been recognized as a mode of cell death in Huntington's disease (HD). Goldberg et al. (1996) have shown that apopain, a human counterpart of CED-3, has a key role in proteolytic events leading to apoptosis, and that apoptotic extracts and apopain itself specifically cleave the HD gene product, huntingtin. The rate of cleavage increases with the length of the huntingtin polyglutamine tract, providing an explanation for the gain-of-function associated with CAG expansion. These results show that huntingtin is cleaved by a cysteine protease and suggest that HD might be a disorder of inappropriate apoptosis.

Conclusions

Necrosis and apoptosis are two distinct types of cell death. Cytokines might be associated with either process. Cytokines and their receptors exist in brain and are found in elevated levels in conditions of acute and chronic neurodegeneration. The hallmark of cytokine biology is that they are pleiotropic and redundant, so it is not surprising to find elevations of IL-1β, IL-6, and TNF-α in models of acute neurodegeneration. The most convincing evidence for a role of a specific cytokine in neurodegeneration involves IL-1β, which is elevated in transient ischemia, middle cerebral artery occlusion, Alzheimer's disease and AIDS. Inhibition of IL-1 activity decreases neuronal cell death in animal models of middle cerebral artery aocclusion and excitotoxicity. The rate-limiting enzyme in the synthesis of bioactive IL-1β, ICE, has been conceptualized as the mammalian analogue of the C. elegans cell death gene, CED-3. Whether ICE is involved in, or required for, apoptosis particularly in conditions of neurodegeneration is unclear, but ICE may have a role in necrosis. Given the role of cytokines in inflammation, necrosis, and possibly apoptosis, cytokine antagonists have the potential for beneficial effects in the treatment of neurodegenerative conditions.

Acknowledgement

This work was partially supported by the Alma Foster Davis Investigator Award from NARSAD (to J. L.).

References

Bandtlow CE, Meyer M, Lindholm D, Spranger M, Heumann R, Thoenen H (1990) Regional and cellular codistribution of interleukin 1 beta and nerve growth factor mRNA in the adult rat brain: possible relationship to the regulation of nerve growth factor synthesis. J Cell Biol 111: 1701–1711

Buttini M, Sauter A, Boddeke HW (1994) Induction of interleukin-1 beta mRNA after foeal cerebral ischaemia in the rat. Brain Res Mol Brain Res 23: 126–134

Cacabelos R, Alvarez XA, Fernandez-Novoa L, et al (1994) Brain interleukin-1 beta in Alzheimer's disease and vascular dementia. Methods Find Exp Clin Pharmacol 16: 141–151

Donnelly RJ, Friedhoff AJ, Beer B, Blume .J, Vitek MP (1990) Interleukin-1 stimulates the beta-amyloid precursor protein promoter. Cell Mol Neurobiol 10: 485–495

Farrar WL, Vinocour M, Hill JM (1989) In situ hybridization histochemistry localization of interleukin-3 mRNA in mouse brain. Blood 73: 137–140

Gagliardini V, Fernandez PA, Lee RK, et al (1994) Prevention of vertebrate neuronal death by the crmA gene. Science 263: 826–828

Gerschenson LE, Rotello RJ (1992) Apoptosis: a different type of cell death. FASEB J 6: 2450–2455

Goldberg YP, Nicholson DW, Rasper DM, Kalchman MA, Koide HB, Graham RK, Bromm M, Kazemi-Esfarjani P, Thornberry NA, Vaillancourt JP, Hayden MR (1996) Cleavage of huntingtin by apopain, a proapoptotic cysteine protease, is modulated by the polyglutamine tract. Nature Genet 13: 442–449

Keane KM, Giegel DA, Lipinski WJ, Callahan MJ, Shivers BD (1995) Cloning, tissue expressioin, and regulation of rat interleukin-1β converting enzyme. Cytokine 7: 105–110

Lapchak PA (1992) A role for interleukin-2 in the regulation of striatal dopaminergic function. Neuro Report 3: 165–168

Li P, Allen H, Banerjee S, et al (1995) Mice deficient in IL-1β converting enzyme are defective in production of mature IL-1β and resistant to endototoxic shock. Cell 80: 401–411

Licinio J, Wong ML, Gold PW (1991) Localization of interleukin-1 receptor antagonist mRNA in rat brain. Endocrinology 129 (1): 562–564

Licinio J, Wong ML, Gold PW (1992) Neutrophil-activating peptide-1/interleukin-8 mRNA is localized in rat hypothalamus and hippocampus. Neuro Report 3: 753–756

Maeda Y, Matsumoto M, Hori O, et al (1994) Hypoxia/reoxygenation-mediated induction of astrocyte interleukin 6: a paracrine mechanism potentially enhancing neuron survival. J Exp Med 180: 2297–2308

Minami M, Kurashi Y, Yamaguchi T, Nakai S, Hirai Y, Satoh M (1991) Immobilization stress induces interleukin 1-β mRNA in the rat hypothalamus. Neurosci Lett 123: 254–256

Pousset F (1994) Developmental expression of cytokine genes in the cortex and hippocampus of the rat central nervous system. Brain Res Dev Brain Res 81: 143–146

Relton JK, Rothwel NJ (1992) Interleukin-1 receptor antagonist inhibits ischaemic and excitotoxic neuronal damage in the rat. Brain Res Bull 29: 243–246

Schobitz B, de Kloet ER, Sutanto W, Holsboer F (1993) Cellular localization of interleukin 6 mRNA and interleukin 6 receptor mRNA in rat brain. Eur J Neurosci 5: 1426–1435

Thornberry NA, Bull HG, Calaycay JR, et al (1992) A novel heterodimeric cysteine protease is required for interleukin-1 beta processing in monocytes. Nature 356: 768–774

Vanguri P (1995) Interferon-gamma-inducible genes in primary glial cells of the central nervous system: comparisons of astrocytes with microglia and Lewis with brown Noiway rats. J Neuroimmunol 56: 35–43

Vasilakos JP, Ghayur T, Carroll RT, Giegel DA, Saunders JM, Quintal L, Keane KM, Shivers BD (1995) IL-1 beta converting enzyme (ICE) is not required for apoptosis induced by lymphokine deprivation in an IL-2-dependent T cell line. J Immunol 155: 3433–3442

Vitkovic L, da Cunha A, Tyor WR (1994) Cytokine expression and pathogenesis in AIDS brain. Res Publ Assoc Res Nerv Ment Dis 72: 203–222

Wang X, Yue TL, Young PR, Barone FC, Feuerstein GZ (1995) Expression of interleukin-6, c-fos, and zif268 mRNAs in rat ischemic cortex. J Cereb Blood Flow Metab 15: 166–171

Wilson KP, Black JA, Thomson JA, et al (1994) Structure and mechanism of interleukin-1 beta converting enzyme. Nature 370: 270–275

Wong M-L, Licinio J (1994) Localization of interleukin 1 type I receptor mRNA in rat brain. Neuroimmunomodulation 1: 110–115

Wong M-L, Licinio J (1994) Localization of stem cell factor mRNA in adult rat hippocampus. Neuroimmunomodulation 1: 181–187

Yamasaki Y, Shozuhara H, Onodera H, Kogure K (1994) Blocking of interleukin-1 activity is a beneficial approach to ischemia brain edema formation. Acta Neurochir (Wien) [Suppl] 60: 300–302

Yuan J, Shaham S, Ledoux S, Ellis HM, Horvitz HR (1993) The C elegans cell death gene ced-3 encodes a protein similar to mammalian interleukin-1 beta-converting enzyme. Cell 75: 641–652

Author's address: J. Licinio, M.D., CNE, NIMH, NIH Bldg. 10, Rm. 2D46, 10 Center Dr MSC 1284, Bethesda, MD 20892-1284, U.S.A.

Genetics of multiple sclerosis – how could disease-associated HLA-types contribute to pathogenesis?

R. Martin[1, 2]

[1]Department of Neurology, University of Tübingen Medical School, Tübingen, Federal Republic of Germany
[2]Neuroimmunology Branch, NINDS, National Institutes of Health, Bethesda, MD, U.S.A.

Summary. Multiple sclerosis is a chronic demyelinating disease of the central nervous system in young adults. It is considered a T cell-mediated autoimmune disease which is probably triggered by exogenous events, e.g. infectious agents, in susceptible individuals. Population, family and twin studies indicate that genetic factors and most likely several genes are associated with disease, but it is clear from the concordance rates of identical twins (25–30%) that genetic background as well as exogenous or somatic events are required to develop disease. Among many candidate genes which have been analyzed during recent years, the strongest association was shown for genes of the HLA-class II complex, in particular HLA-DR15 Dw2 and -DQw6. At present, it is not clear how the expression of a particular HLA-class II gene translates into susceptibility to develop an organ-specific autoimmune disease. Potential explanations how this could occurr will be discussed.

Introduction

Multiple sclerosis (MS) is the most frequent, inflammatory demyelinating disease in Northern America and Europe. Histopathologically, it is characterized by disseminated lesions with loss of myelin, but relative sparing of axons at least in the early stages of disease. These lesions are well demarcated from the surrounding tissue and primarily occur around small venules in the periventricular white matter (Prineas, 1985). Early lesions are characterized by lymphomonocytic infiltrates and blood brain barrier (BBB) leakage (Raine, 1983). Later, the number of inflammatory cells decreases. Remyelination is often incomplete, and, following frequent inflammatory relapses in one area, fibrous scar tissue composed primarily of astroglial cells may replace the normal cytoarchitecture. The myelin sheath produced by oligodendrocytes is the primary target of the inflammatory process (Prineas, 1985; Raine, 1983). It serves as a capacitator insulating stretches of the axon between nodes of

Ranvier and thus facilitates saltatory nerve conduction from node to node. Following incomplete or complete damage of the myelin sheath, nerve conduction may be impaired or completely blocked resulting in neurological deficits such as impairment or loss of sensation, vision or paresis. Clinically, MS most often affects young adults between the ages of 20 to 40 years (McFarlin and McFarland, 1982a, b). Inflammatory lesions can occur everywhere in the central nervous system (CNS) white matter and, accordingly, almost any clinical sign attributable to those functional CNS systems can occur. About 80% of MS patients present with a relapsing-remitting course (RR-MS) that starts out with an isolated syndrome such as optic neuritis or sensory problems, often with complete remission. Later on, relapses tend to occur more frequently and remit incompletely until disease starts to take a secondary chronic progressive or relapsing progressive course (RP-MS). RR-MS begins at a younger age (20–40) affects women more often than men (female to male ratio = 2:1) and generally runs a more benign course than the major other form, primary chronic progressive MS (CP-MS). CP-MS starts later in life (mean age of onset around 40 years of age), is characterized by an insidious onset with steady progression of neurological deficits, shows an equal sex distribution and worse prognosis than RR-MS. Since there is no clinical sign specific for MS, the diagnosis is still based on the clinical syndrome and its course. The diagnosis, definite MS, can only be made, when at least two separate exacerbations occurred that can be attributed to two different functional systems of the CNS, for example problems of vision and sensory deficits (McFarlin and McFarland, 1982a, b). Laboratory and electrophysiological examinations can lend support to the diagnosis if only one exacerbation occurred in a young adult or only one functional system is affected. Supportive measures include increased immunoglobulin G (IgG) levels and oligoclonal Ig bands in the cerebrospinal fluid (CSF) (positive in > 90% of MS patients), delayed visual, somatosensory or auditory evoked potentials (if combined, they are positive in about 85% of patients), and magnetic resonance imaging (MRI) (positive in > 90% of MS patients). In recent years, especially MRI of the brain and spinal cord has gained importance since it allows direct visualization of early inflammatory changes by showing the leakage of paramagnetic contrast materials (gadolinium-DTPA) through the blood brain barrier (BBB) (McFarland et al., 1992). In addition, by using T2-weighted images, it allows the quantitation of the total amount of white matter damaged during the demyelinating process. Furthermore, studies with serial MRI imaging demonstrated that fresh inflammatory CNS lesions occur every month and even at times when there is no clinical sign of disease activity (McFarland et al., 1992). These studies have completely changed our view of the course of RR-MS. It is now clear that disease activity was most often already present considerable time before the clinical onset and that the inflammatory process is active all the time. In addition, the lesion load shown by MRI has allowed to make some prognostic predictions since the presence and size of MRI lesions during the time of isolated CNS syndromes such as optic neuritis was highly predictive of future development of MS (Filippi et al., 1994). In conclusion from this information, MRI is now accepted as a secondary outcome measure for clinical trials and as a means that has some prognostic value if several and large MRI lesions can be shown at first onset of CNS symptoms (McFarland et al., 1992; Filippi et al., 1994).

The etiology of MS is as yet unknown, but it is now widely accepted that a T cell-mediated autoimmune response against myelin antigens such as myelin basic protein (MBP), proteolipid protein (PLP) or others contributes to the pathogenesis (Martin et al., 1992). Evidence supporting this concept stems from the nature of the CNS infiltrates which are composed of inflammatory cells, mainly lymphocytes and macrophages, from improvement of disease by treatment with immunomodulatory substances such as interferon-β (IFN-β) (Paty et al., 1993) and from parallels with an experimental animal model for demyelinating diseases, experimental allergic encephalomyelitis (EAE) (Wekerle et al., 1986). Finally, it has been shown that genes that are relevant for immune responses against foreign and self antigens such as major histocompatibility complex (MHC; HLA in humans and H-2 in mice) contribute to disease susceptibility.

This short review will briefly summarize the evidence for genetic influences on disease susceptibility, but focus on the functional consequences of the presence of certain HLA genes and how these could contribute to disease susceptibility (for a more detailed review of the genetics of MS see Ebers and Dessa Sadovnick, 1994).

Genetic factors associated with multiple sclerosis

Family and twin studies

A number of observations provide evidence suggesting that genetic factors are associated with MS. These include population-, family- and twin studies as well as linkage analysis. It has long been known from population and family studies that the risk of acquiring MS is increased about 10–20 fold if one family member is affected (Sadovnick et al., 1988, 1991). Up to 20% of the family members of an affected individual show either clinical or subclinical findings consistent with MS (Sadovnick et al., 1988, 1991). Interestingly, daughters of affected mothers carry the highest risk whereas sons of affected fathers almost never develop disease (Sadovnick et al., 1991). Futher support for genetic influences comes from twin studies. Although sometimes flawed by ascertainment bias, a number of twin studies have compared twin pairs from large populations and consistently shown that concordance rates of monozygotic twins are 5–10 times higher than those observed in dizygotic twins (25–30% in monozygotic as opposed to 2.5–5% in dizygotic twins) (Sadovnick et al., 1993). Together with population and family studies the data derived from examining twins allow the following conclusions. Firstly, there is a considerable genetic component to susceptibility even if it is taken into account that monozygotic twins are slightly more likely to share environmental influences than dizygotic twins. Furthermore, the concordance rate of about 25% in monozygotic twings suggests that genetic susceptibility is not determined by a single dominant or recessive gene, but rather by two or more genes. Third, the large percentage of discordant identical twins even though these individuals share genetic background, argues for the influence of environmental factors that are required for induction of disease in genetically susceptible individuals. There are, however, other possibilities that could in part explain the high rate of discordance in monozygotic twins which include somatic mutations and alterations of genes by deletions and translocations or other random events.

Population studies

The data mentioned above argues for both genetic and environmental factors to be associated with disease susceptibility. When the distribution of MS was examined in entire populations different prevalence rates for MS have been reported in various ethnic groups (Bulman and Ebers, 1992; Hammond et al., 1988). Populations with Caucasian origin in Northern America and Northern Europe showed prevalence rates between 10-130/100,000 with a mean of approximately 60/100,000 (Kurtzke, 1985; Bulman and Ebers, 1992). A considerably higher prevalence rate ranging from between 137 and 178/100,000 respectively is found in areas such as Northern Ireland or northeast Scotland (Francis et al., 1987). In general, the rates decrease with latitude on both hemispheres (Kurtzke, 1985; Hammond et al., 1988) ; this has been interpreted as an argument favoring environmental influences. In a number of ethnic groups, exceptions show, however, that genetic predisposition as well as exogenous factors play a role. A low prevalence rate is, for example, seen not only in Japan (2/100,000), but also in Japanese living in Hawaii or at the Pacific coast of the United States. Similarly, the rate found in Hungarians of caucasian descent is 37/100,000, but only 2/100,000 in Hungarian gypsies (Palffy, 1982). A low prevalence rate is also documented for other distinct ethnic groups like Hutterites, Yakuts, Inuit or Bantu (Waksman and Reynolds, 1984). Although these differences are striking, they may be at least partly caused by differing lifestyles such as nutrition or by other environmental factors. Further support for environmental factors stems from the description of epidemics of disease such as the one on the Faroe islands (Kurtzke, 1985).

Taken together, population, family- and twin studies document genetic influences probably exerted by several genes. In addition, environmental factors appear to be involved in inducing disease or exacerbations. Unless the genes associated with disease are identified it will be difficult to sort out to what extent genetic makeup and environment contribute to the risk to develop MS.

Linkage analysis for non-HLA genes that are associated with MS

Since an autoimmune pathogenesis has been postulated for MS and susceptibility to develop EAE was found to be linked to MHC-class II genes, attention in MS also focused on genes that are involved in immune reactivity, particularly those of the HLA- and TCR gene complexes. During the last two decades,

Table 1. Gene loci potentially associated with multiple sclerosis (for details see text)

Gene complex	Associated Allele/RFLP	Chromosome	Confirmed
HLA-class I	HLA-A3, -B7	6	+
HLA-class II	DRB1*1501, DRB5*0101	6	+
HLA-class II	DQA1*0101, DQB1*0602	6	+
TCR β-chain	Vβ8, Vβ11, Cβ	7	±
TCR α-chain	Vα12.1, Cα	14	–
Immunoglobulin complex	Constant and variable region (VH2)	14	N.D.
Peptide transporter	TAP1/TAP2	6	–
Proteasome	LMP2/LMP7	6	–
Complement	C3F (third component)	19	–
Alpha-1 Antitrypsin	Alpha-1	14	–
Myelin basic protein	5'-untranslated region	18	±

a number of genes was found to be associated with MS. These include genes of the HLA-class I and -class II complex (Bertram and Kuwert, 1982; Tiwari and Terasaki, 1985) (see below), restriction fragment length polymorphisms (RFLP) of the TCR α- and β-chain region (Beall et al., 1989; Oksenberg et al., 1989; Seboun et al., 1989), genes of the immunoglobulin constant and variable region (Gaisner et al., 1987; Walter et al., 1991) as well as complement- (Bulman et al., 1991), alpha-1 antitrypsin- (McCombe et al., 1985) and myelin basic protein genes (Boylan et al., 1990) (for summary see Table 1). Other genes such as those coding for antigen processing- (LMP2 and LMP7) and transporter proteins (TAP1 and TAP2) are located in close proximity of HLA genes and have only recently been described (Brown et al., 1991; Momburg et al., 1994). Although first reports are discouraging (Liblau et al., 1993; Vandevyer et al., 1994), it is probably too early to decide whether they are associated with disease or not. Several of the above associations could not be confirmed by later reports (Hillert et al., 1991), but it should be kept in mind that most studies had been performed in MS populations from different areas and with small numbers of patients. It is therefore difficult to accept or discard some of these associations before larger studies in well-matched and -stratified (for ethnic background, sex, disease type, age, severity and length of disease) groups of patients and controls have been performed.

Human leukocyte antigens (HLA) and multiple sclerosis

The human HLA gene complex is located on chromosome 6 and consists of several loci coding for heterodimeric membrane proteins which are divided into HLA class I (HLA-A,-B,-C) and -class II (HLA-DP, DQ,-DR) genes. HLA class I antigens are composed of β2 microglobulin and an HLA-encoded polymorphic a chain glycoprotein, HLA class II antigens of an α chain and a more polymorphic b chain. HLA-A,-B,-C,-DR and -DQ antigens are typed by serological techniques usually with sera from multiparous women. In recent years, conventional serological and cellular typing techniques are more and more replaced by typing with sequence-specific oligonucleotides which allow the exact determination of each of the alleles (for the last nomenclature of HLA molecules and genes see Bodmer et al., 1994).

If one is to understand how MHC- or HLA genes contribute to disease susceptibility we need to briefly review their influence on immune responses. In 1974, Zinkernagel and Doherty (Zinkernagel and Doherty, 1974) made a landmark observation in immunology and demonstrated that T cells are not able to recognize "free" antigen, but only respond to it, if presented on antigen-presenting cells (APC) in the context of self MHC molecules. Since then, this joint recognition was coined MHC-restricted antigen recognition by T cells. Consequently, the highly polymorphic MHC antigens which had been recognized before by their importance in graft rejection are thus not only involved in shaping the T cell repertoire during thymic maturation, but also in any T cell response to foreign and self antigens. It was therefore no surprise when MHC antigens were found to be associated with a number of autoimmune diseases including type I diabetes, rheumatoid arthritis or celiac disease to name only a few (Tiwari and Terasaki, 1985). Reports in the early 80s documented a weak association of the class I antigens HLA-A3 and -B7 with MS (Bertram and Kuwert, 1982; Tiwari and Terasaki, 1985), but later a much stronger one was shown with HLA-DR2 Dw2 (DR15 Dw2 according to a recent HLA workshop) in Caucasian MS patients (Tiwari and Terasaki, 1985). In ethnic groups with lower prevalence rates of MS, associations with DR4 [Sardinian and Arab (Kurdi et al., 1977; Marrosu et al., 1988)] and DR6 [Mexican and Japanese MS patients (Gorodezky et al., 1986; Naito et al., 1978)] were observed. The recent use of molecular HLA-typing techniques disclosed MS-associated HLA-genes at the molecular level. With respect to MS, it has been shown that the relative risk conferred by certain DQ genes is higher than that conferred by DR2 (Vartdal et al., 1989; Sollid et al., 1989; Spurkland et al., 1991). One study of 61 MS patients that were positive for either DR15 Dw2, DR4 or DR6 demonstrated that 97% of these individuals expressed HLA-DQβ genes that shared polymorphic stretches of the membrane distal domain (Vartdal et al., 1989). This part of HLA-class II molecules forms the peptide binding groove, and, based on this knowledge, it was concluded that MS-associated HLA-DQβ chains are presenting a single or a set of autoantigen/s that is/are related to the pathogenesis of MS. Subsequent studies of DQα chain RFLP and DQα chain genes suggested that both the combination of certain DQβ (encoded by DQB1*0602) and DQα (encoded by DQA1*0102) chains may contribute to disease either alone or in combination. In one group of 69 MS patients, 99% carried DQα alleles encoding glutamine at residue 34 supporting the abovementioned concept that a specific DQ heterodimer may be related to disease via binding of a relevant autoantigen (Spurkland et al., 1991). In a Canadian study (Haegert and Francis, 1993), disease association in French Canadians with the common DR2 haplotype (DRB1*1501-DQA1*0102-DQB1*0602) was stronger than that with shared

DQB1 sequences and a glutamine in position 34 of DQA1. In the same group of patients, a leucine at position 26 of DQB1 appeared to be associated as strongly with MS as the DR2 haplotype. In mixed ethnic whites, an association with the DR haplotype, but not with the DQ alleles has been shown. Consequently, none of the DQ polymorphisms fully explained the disease association with HLA alleles in either group. The results of this study stress the importance to choose comparable patient and control groups (see above). It will otherwise be hard to delineate the relative contribution of each allele. Finally, a relatively lower and independent association between MS and HLA-DPw4 has been described in two Scandinavian studies (Moen et al., 1984; Odum et al., 1988), but was not confirmed by another (Begovich et al., 1990).

The analysis of the relative importance of HLA-DR and -DQ molecules was obscured for a long time by the close physical proximity of their genes on chromosome 6. HLA-DR and -DQ genes are therefore very rarely separated by crossing-over events and usually inherited together by an individual's offspring, a mechanism that is referred to as linkage dysequilibrium. Due to the linkage of HLA-DR15 Dw2 and -DQw6 (encoded by DQA1*0102 and DQB1*0602) and segregation of these genes in a complex haplotype that includes HLA-class I genes and non-MHC elements such as complement genes, TNF genes and genes of peptide transporters, the question as to which gene is most relevant for disease susceptibility is not yet answered. The issue is complicated further by disease heterogeneity. It has been mentioned before that the etiology of MS is still unclear and there is no MS-specific diagnostic or pathogenetic finding. The two major forms, RR-MS versus CP-MS manifest at different ages, differ with respect to sex distribution and clinical findings. It was therefore not surprising that recent studies indicate that the two forms could be distinguished based on different HLAassociations. One report (Hammond et al., 1988) found RR-MS to be associated with DR2, HLA-A3 and -B7 and CP-MS with HLA-A1, -B8 and -DR3. These findings were based on serotyping and later extended by two Scandinavian studies (Olerup et al., 1989; Hillert et al., 1992) analyzing 100 MS patients (26 with CP-MS and 74 with RR-MS) by RFLP analysis. Both forms were associated with the DRw15 DQw6 haplotype, but additional risk for CP-MS was conferred by a DQB1 RFLP seen in DR4, DQw8, DR7, DQw9 and DRw8, DQw4 haplotypes, whereas a DQB1 allelic pattern corresponding to the DQw7 was negatively associated (Olerup et al., 1989). In contrast, additional risk for RR-MS was associated with a DQB1 RFLP observed in the DRw17, DQw2 haplotype (Olerup et al., 1989). These three DQB1 alleles are in strong negative linkage with the DR15 haplotype. It was concluded that DQB1 alleles may confer either additional risk or resistance for the two forms of MS, and that RR-MS and CP-MS may therefore represent immunogenetically distinct entities. A more recent report failed to show an association with shared polymorphic DQB1 sequences or the presence of glutamine at position 34 of DQA1 (Hillert et al., 1992), but confirmed the above findings.

These partially controversial observations raise the question whether it is a single HLA allele, a combination of two or a complex haplotype that confers increased risk. Findings from other autoimmune diseases such as insulin-dependent diabetes (IDDM) (Nepom and Erlich, 1991), rheumatoid arthritis (RA) (Seyfried et al., 1988) or celiac disease (CD) (Bugawan et al., 1989; Sollid et al., 1989) indicate that the influence of HLA may differ in different diseases. In IDDM, for example, the association with DR4 has been localized to a DQB1 gene (Baisch et al., 1990), whereas the exact locus within DR3 haplotypes that contributes to susceptibility is not clear (Nepom, 1993). Linkage studies documented that only HLA-DR4 and -DQ3.2 containing haplotypes are associated with IDDM, whereas those carrying other DQ alleles are not. It should be stressed, however, that these observations are only valid for DR4-positive haplotypes. In contrast to IDDM, susceptibility genes have been mapped to polymorphic stretches of the DRB1 genes coding for DR4 Dw4 (DRB1*0401) and DR4 Dw14 (DRB1*0404) in rheumatoid arthritis (Nepom, 1990, 1993). In other diseases such as CD, different loci may be involved in conferring risk. Here, the strongest association is observed with the DR3 Dw3 DQ2 haplotype (Kagnoff et al., 1989). Among DR3-negative patients, most patients are heterozygous for the DR5 Dw11 DQ3 and DR7 Dw7 DQ2 haplotypes (Kagnoff et al., 1989; Bugawan et al., 1989; Sollid et al., 1989). The latter haplotypes do not share the same combination of linked loci that is found in the DR3 Dw3 DQ2 haplotype. The DQA1 and DQB1 genes that are present in *cis* in the DR3 Dw3 DQ2 haplotype are found in *trans* in a DR5-DR7 heterozygote. Besides single disease-associated loci or a combination of two genes that, when present in either *cis* or *trans,* code for a disease-related HLA molecule, susceptibility may also be linked to an extended haplotype. As for example in the IDDM-related DR3 haplotype, in which a specific locus has not yet been identified (Nepom, 1993), one or more classical HLA genes or even non-HLA genes (TNF, peptide transporter, heat-shock protein 70) within such a haplotype may be associated with disease.

According to the above data the association of HLA genes with MS appears to be complex. There is evidence that single loci or shared polymorphic stretches present in DQB1 alleles are associated with disease, but that a locus found on DQA1 alleles may also contribute. From other studies, it seems less clear whether the contribution of HLA-DQB1 alleles is stronger than that of HLA-DRB1 and additional risk was found to be conferred by other haplotypes than DR15 Dw2 DQw6 in RR-MS versus CP-MS. We do therefore not know yet how many individual loci or genes within a complex haplotype are associated with disease. Future studies should extend the previous findings and take special care to characterize the clinical form of MS.

How could a disease-associated HLA molecule be related to pathogenesis?

Besides linkage analysis, several concepts have been proposed of how the presence of certain HLA genes or membrane heterodimers could contribute to an autoimmune response. HLA-class II molecules may influence immune responses in a number of ways:

1. HLA-DR or-DQ molecules may preferentially bind and present autoantigens to disease-related T helper 1 cells. This mechanism has elegantly been shown in EAE.
2. HLA-DQ may serve as restriction element for CD4+ "suppressorinducer" cells that could be involved in activating CD8+ regulatory cells. Evidence for this concept comes from studies of various human immune responses against foreign antigens by Sasazuki and colleagues (Hirayama et al., 1987) as well as from TCR vaccination studies in EAE (Offner et al., 1991) and MS (Zhang et al., 1993).
3. Expression of HLA-DQ may be important for organ-specific immune response through differential regulation (Andersen et al., 1991) in the tissue and through shaping the TCR repertoire during thymic selection. Shared amino acid stretches of disease-related DR- or DQ molecules may be more important for contacting TCR residues than binding of autoantigens and may thus participate in shaping the TCR repertoire.
4. As shown for IDDM, HLA-DQ alleles may also confer resistance for disease (Thomson et al., 1988).
5. Genes encoded in the MHC region (TNF, TAP transporters, Proteasome complex, LMP) could be involved in upregulation of inflammatory responses (TNF). Presentation of different sets of self peptides on MHC class I molecules (LMP, TAP transporters) (Powis et al., 1992) could result in differences in T cell repertoire selection during thymic maturation.

With respect to MS and the influence of disease-associated HLA-class II antigens, possibility number 1 is currently being favored. Similar to the situation in EAE, it is believed that MS-associated HLA-DR molecules (i.e. DR15 in Caucasians) preferentially bind and present one or several antigenic peptides derived from myelin proteins. We will outline below, how this can be envisioned. We can only speculate about possibilities 2–5. HLA-DQ molecules have been found to serve as restriction elements only in a few in vitro studies, and it is difficult to relate these to autoimmune diseases. The reasons for this may be both the low level of HLA-DQ expression on most class II-positive cell types and also our tissue culture techniques that are probably optimized for expansion of HLA-DR-restricted rather than DQ-restricted T cells. At present, we do not understand why so many autoimmune disease, such as type I diabetes, rheumatoid arthritis and others, are associated with HLA-DQ molecules. Direct evidence for influences of HLA-DQ molecules on suppressive mechanisms is lacking. It is also unlikely that shaping of the TCR repertoire in MS patients is largely under the control of DQ, because the skewed TCR Vα repertoire that has been observed in identical twins discordant for disease (Utz et al., 1993), is probably due to exogenous factors or somatic events since monozygotic twins by definition share the same genetic background. With respect to possibility 5, loading of MHC-class I antigens with different sets of self peptides could result in differences in selecting T cells responding to self-, foreign- or allo-antigens. Experimental evidence that such a mechanism is operative in MS is, however, not yet available.

Interaction between MS-associated HLA-DR antigens, immunodominant MBP peptides and MBP-specific T cells

Disease susceptibility in EAE is under the control of immune response genes (I-A- or I-E-molecules) (Fritz et al., 1985). Depending on the diseaseassociated MHC-class II type (for example I-A^S in SJL/J mice), a certain peptide, i.e. MBP peptide 89–101, is immunogenic and encephalitogenic in this particular mouse strain (Fritz and McFarlin, 1989). T cell assays using truncated peptides and peptides with L-alanine substitutions and, in particular, studies of the three-dimensional structure of MHC-class I and -class II molecules allowed to delineate the exact interactions between antigenic peptide and MHC-binding groove at the molecular level (Brown et al., 1993). It is now clear that structural constraints of the peptide binding groove favor high affinity and stable binding of defined amino acid stretches within a given antigen, whereas others bind only poorly or not at all. Although oversimplified, these rules that are responsible for high affinity binding are at least one important factor that determines the formation of a stable peptide/MHC complex, its proper conformation and targeting through the intracellular pathways as well as the stability on the surface membrane and thus also the time it is available for interaction with T cells. In theory, peptides with such properties would be expected to be immunogenic, and, in case of an autoantigenic myelin peptide, also to be encephalitogenic.

Based on this knowledge and hoping that a T cell response against myelin antigens that it is unique to MS would be found, several investigators examined T cell reactivity against myelin antigens such as MBP and PLP. It was expected that mechanisms that are responsible for maintaining tolerance against self antigens such as MBP would be fully operative in healthy individuals, but less stringent in MS. If MS resembles EAE in that certain MHC class II genes predispose for disease and disease is mediated by T cells recognizing a well defined myelin component, it would be expected that tolerance in normal individuals results from any combination of deletion of self-reactive T cells in the thymus, sequestration of antigen, suppression of myelin-specific T cells or lack of costimulation of these cells. In recent years, a number of studies have characterized the MBP-specific T cell response in MS patients and controls with respect to 1. presence and frequency, 2. phenotype and function, 3. fine specificity, 4. HLA-restriction and 5. TCR usage of MBP-specific T cells (Burns et al., 1983; Richert et al., 1988; Martin et al., 1990; Pette et al., 1990; Ota et al., 1990; Martin et al., 1992). These studies made the following observations: MBP-specific T cells can not only be expanded from MS patients, but also from healthy individuals indicating that they are not deleted, but rather part of the normal T cell repertoire. The question as to whether the frequency of myelin-speciific T cells is higher in controls or MS patients is still controversial, some investigators have found similar frequencies in unrelated individuals (Pette et al., 1990), and even in family members (Voskuhl et al., 1993) and identical twins of MS patients (Martin et al., 1993), others documented higher numbers of MBP- and PLP-specific in peripheral blood and CSF (Olsson et al., 1990, 1992; Ota et al., 1990) and also in the fraction of activated (IL-2-receptor-positive) T cells (Zhang et al., 1994). Later on, MBP-specific T cells were characterized as being primarily CD4+, T helper 1-like (se-

crete IFN-γ and TNF-α/β) (Voskuhl et al., 1993) T cells that are usually restricted by HLA-DR antigens, often show cytotoxic activity and respond to a whole
range of different MBP peptides. Further analysis revealed that certain MBP
peptides are immunodominant in the context of MS-associated HLA-DR antigens. Immunodominant regions were identified in the middle (84–102 and 87–
106 respectively) (Ota et al., 1990; Martin et al., 1991) and C-terminus (143–168
and 154–172) (Ota et al., 1990; Martin et al., 1990) respectively, but probably also
the N-terminus (residues 1–21 and 40–60) of the molecule (Martin et al., 1990;
Pette et al., 1990; Meinl et al., 1993). The latter observation of immunodominant
regions, for example in the context of HLA-DR15, was interesting with respect
to the situation in EAE since similar areas of MBP are encephalitogenic in different animals strains (Ac1–9 in PUJ and B10.PL mice, 89–100 in SJL/J mice, 69–
86 in Lewis rats and 153–165 in rhesus monkeys) (Fritz and McFarlin, 1989;
Martin et al., 1992). The vast majority of human MBP-specific T cell lines has
been restricted by HLA-DR. HLA-DQ and -DP rarely served as restriction elements, and therefore information about their influence is sparse. Interestingly,
those HLA-DR molecules that have been found to be associated with MS in the
various ethnic groups (DR15, DR4 and DR6) restricted most of the MBP-specific T cell lines with DR15 being the most important HLA-DR molecule. The
immunodominant MBP peptide 87–106 was recognized by different T cell lines
(TCL) in the context of at least four MS-associated HLA-DR molecules including DR15 (Martin et al., 1991), and a few TCL responded to a similar peptide
(84–102) in the context of HLA-DQw6 (Ota et al., 1990). Besides promiscuity of
some MBP peptides in binding to several HLA-DR molecules, HLA-DR2,
-DR4 and DR6 were each able to bind a number of different MBP peptides.
Together with the demonstration of TCL responding to nested epitopes in the
87–106 peptide (Martin et al., 1992), it is concluded from these studies that the
MBP-specific T cell response is extremely complex and clearly shows parallels
to EAE. The question as to whether MBP-specific T cells or T cells found in CNS
lesion express a restricted TCR usage is still controversial. Some studies documented restricted TCR usage (Kotzin et al., 1991; Wucherpfennig et al., 1990),
whereas others failed to confirm these findings. The detection of common TCR
V-D-J sequences in brain lesions of patients positive for the DR15 Dw2 DQw6
haplotype (Oksenberg et al., 1993), in MBP-specific, cytotoxic, DR15-restricted
T cells specific for the immunodominant epitope (Martin et al., 1991) and in
encephalitogenic T cells in rodents (Allegretta et al., 1994) argues for the importance of certain TCR heterodimers for disease. It was discussed before that identical twins discordant for MS show a skewed TCR usage (Utz et al., 1993), another indication that autoreactive T cells or at least differences in T cell repertoires
compared to healthy individuals are involved in the disease process.

Recognition of the immunodominant MBP-peptide 87–106 in the context of MS-associated HLA-DR molecules

As mentioned before, MBP peptide 87–106 is interesting for a number of
reasons: 1. Although TCL specific for MBP 87–106 had been established from

both MS patients and controls, it was shown to be immunodominant in the context of the MS-associated HLA-DR molecules DR15, DR4 and DR6. 2. MBP 89–101 and 87–99 are encephalitogenic in SJL mice (I-As) and Lewis rats (I-E) (Vandenbark et al., 1989) respectively. 3. TCR CDR3 sequences similar to that of a MBP 87–106-specific and DR15 Dw2-restricted, cytotoxic TCL have been isolated from brains of MS patients, but also from encephalitogenic T cells and in vivo activated MBP-specific T cells from MS patients (Allegretta et al., 1994). Based on these observations, we have studied the interaction between MBP peptide 87–106 and MS-associated HLA-DR antigens at the molecular level and later extended these experiments to analyze the recognition of the MBP 87–99/HLA-DR complex by MBP 87–99-specific T cells. We will present this as an example of how molecular interactions between autoantigenic peptide and disease-associated HLA molecules could be related to the pathogenesis of MS.

When MBP 87–106-specific T cells were tested with truncated and L-Ala-substituted peptides for fine specificity, peptide 87–99 was found to represent the core sequence (Martin et al., 1991, 1992). Further analysis aimed at the question which of the two DR2 molecules (DRα paired with DRB1*1501 = DR2b; DRα paired with DRB5*0101 = DR2a) that are isotypically expressed in the DR15 Dw2 DQw6 haplotype (DR15 is a split of the former serologic specificity DR2) is responsible for restricting 87–106-specific T cells. In this context, it is important to know that the DR2 haplotype is unique in that both DRβ chains are polymorphic and that the DRB1 gene product (DRB1*1501) is less polymorphic than the DRB5 (DRB5*0101) gene product. The amino acids differences in the membrane-distal domain of the DRβ chains which account for the differences between the two DR2 molecules are involved in formation of the peptide binding groove. This implies that both molecules have distinct requirements for peptide binding. Experiments using murine L cells or human fibroblasts transfected with cDNAs for DRα and DRB1*1501 (DR2b) or DRB5*0101 (DR2a) as antigen presenting cells (APC) demonstrated that both DR2a- and DR2b-restricted T cells with specificity for peptide 87–99 existed (Jaraquemada et al., 1990; Pette et al., 1990; Wucherpfennig et al., 1994). In addition, DR2a-restricted TCL were found to recognize a number of other MBP peptides (1–44, 43–60, 76–91, 131–145 or 135–153) (Jaraquemada et al., 1990), whereas DR2b-restricted TCL only responded to 80–99 or 148–162 (Pette et al., 1990). From these results, it became not only clear that both DR molecules within the DR2 Dw2 haplotype are able to bind MBP peptide 87–99, but also that this region is immunodominant for TCL with either restriction element. It is still controversial whether higher frequencies of DR2a- or DR2b-restricted TCL are found in MS patients and whether one or both alleles are disease-associated via restricting MBP-specific T cells. In contrast to two other reports that described an overrepresentation of DR2b as restriction element in the DR2-restricted, 87–99-specific T cell response (Pette et al., 1990; Wucherpfennig et al., 1994), our own recent results argue for a relatively even distribution of 87–99-specific TCL restricted by either DR2a or DR2b (Vergelli et al., manuscript in preparation). Besides the above mentioned T cell data, recent experiments delineated the structural interac-

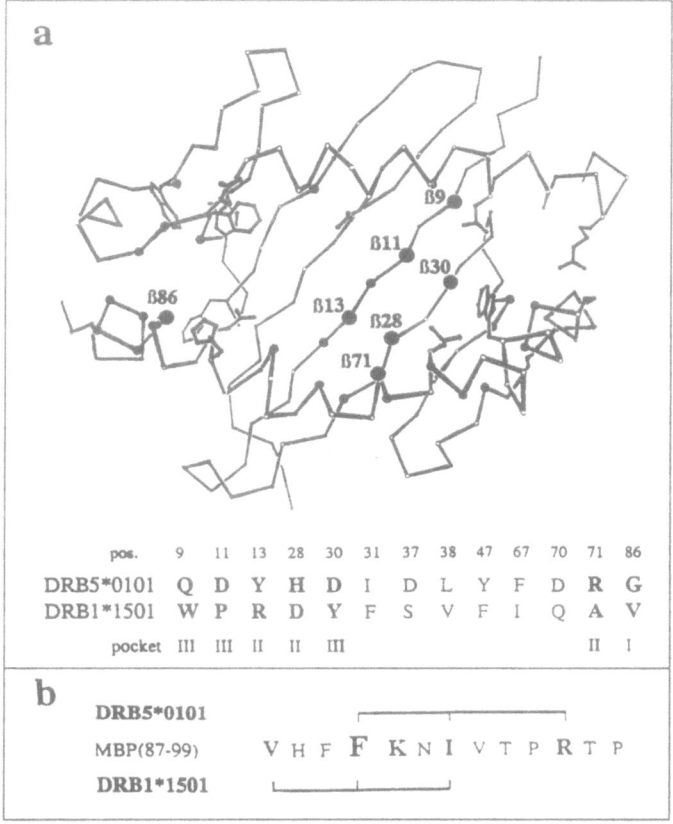

Fig. 1. Schematic model of the peptide binding groove of HLA-DR2 according to the X-ray crystal structure of DR1 (Brown et al., 1993) (adopted from Vogt et al., 1994; kind permission of The Journal of Immunology). Polymorphic residues of DRB1*1501 and DRB5*0101 are indicated by dark circles, and different residues are listed below. The immunodominant MBP peptide 87–99 contains both anchor motifs overlapping each other

tions between MBP 87–99 or 82–100 (human MBP sequence numbering) and DR2a or DR2b by in vitro binding assays using isolated DR heterodimers or by aligning foreign- and self peptides released from isolated DR molecules in order to determine the amino acids responsible for anchoring the peptide in the DR binding groove (Wucherpfennig et al., 1994; Vogt et al., 1994). When data from T cell assays, sequencing of self peptides and in vitro binding studies were analyzed the following anchor motifs were established for DR2a and DR2b (Fig. 1). According to the amino acid (AA) in position 86 of the DRβ chain (G in DR2a and V in DR2b), either the bulky, aromatic AA phenylalanine (F; single letter AA code used from here on) binds at position i in pocket I of the antigen binding groove [based on the X-ray crystal structure of DR1 (Brown et al., 1993)] in DR2a or the considerably larger, aliphatic AA V in DR2b. AA I and R serve as secondary anchors at relative positions i+3 and i+7/i+8 in DR2a. A bulky, hydrophobic F is required in position i+3 in DR2b and appears equally important for binding as V in position i. An AA with a hydrophobic side chain, I, is observed at position i+6 and serves as secondary anchor. Without going into further detail (see Fig. 1 and Vogt et al., 1994), it becomes obvious through this analysis that the anchor motifs for both MS-associated DR2 alleles are present in the immunodominant MBP peptide 87–99 and overlap each other (Fig. 1). Interaction with DR2b occurs via residues V(87), F(90) and I(93), residues F(90), I(93) and R(97), and to a lesser extent also K(91) are important for binding to DR2a. AA F(90) and I(93) are MHC-contact residues for both DR2 subtypes.

Similar analyses have been performed for DR4 Dw4 and DR6. In conclusion from these experiments, we are now able to understand why MBP peptide 87–99 is immunodominant in the context of various MS-associated DR alleles. With respect to the two DR2 subtypes, MBP 87–99 contains ideal binding motifs that overlap each other and contribute to high affinity binding to DR2a and DR2b. Probably due to the stability of this complex on the surface of presenting cells, strong T cell responses against 87–99 are found when compared to other MBP peptides which bind with lower affinity. This notion has been confirmed in the meantime by comparing large numbers of MBP-specific T cells with different specificities and HLA-restrictions with the binding affinities of overlapping sets of MBP peptides to the various HLA-DR alleles (Kalbus et al., manuscript in preparation). Furthermore, the understanding of the molecular interaction of an autoantigenic peptide with disease-associated HLA-class II alleles and the definition of MHC-anchor residues helped to determine which M contact the TCR, the third ligand in the trimolecular complex. The knowledge of the MHC- and TCR contact sites of MBP 87–99 may help to design altered peptides that retain binding to the MHC molecule, but are modified in TCR contact positions in a way that they antagonize T cells specific for 87–99 or modify their function.

Conclusions

According to currently available data, multiple genes are associated with susceptibility to develop MS, but among these, only the association between HLA-class II genes and disease can be accepted. Although our understanding about the interaction between HLA molecules, antigenic peptides and T cells has rapidly increased during recent years, we still do not know how exactly the presence of a certain HLA gene contributes to disease. The analysis of the association between HLA and disease is complicated by ethnic diversity and disease heterogeneity. Despite these problems, there is hope that the influence of genetic as well as environmental factors on the pathogenesis of MS will become clearer and that the analysis of the immunologic as well as the biochemical events leading to demyelination will allow specific immune intervention.

Acknowledgement

R. Martin is a Heisenberg Fellow of the Deutsche Forschungsgemeinschaft (DFG Ma 965/4-1).

References

Allegretta M, Albertini RJ, et al (1994) Homologies between T cell receptor junctional sequences unique to multiple sclerosis and T cells mediating experimental allergic encephalomyelitis. J Clin Invest 94: 105–109

Andersen LC, Beaty JS, et al (1991) Allelic polymorphism in transcriptional regulatory regions of HLA-DQβ genes. J Exp Med 173: 181–192

Baisch JM, Weeks T, et al (1990) Analysis of HLA-DQ genotypes and susceptibility in insulin-dependent diabetes mellitus. N Engl J Med 322: 1836–1841

Beall SS, Concannon P, et al (1989) The germline repertoire of T cell receptor β-chain genes in patients with chronic progressive multiple sclerosis. J Neuroimmunol 21: 59–66

Begovich AB, Helmuth RC, et al (1990) HLA-DP beta and susceptibility to multiple sclerosis: an analysis of caucasoid and Japanese patient populations. Hum Immunol 28: 365–372

Bertram J, Kuwert E (1982) HLA antigen frequencies in multiple sclerosis. Eur J Neurol 7: 74–79

Bodmer JG, Marsh SGE, et al (1994) Nomenclature for factors of the HLA system, 1994. Hum Immunol 41: 1–20

Boylan KB, Takahashi N, et al (1990) DNA length polymorphism 5' to the myelin basic protein gene is associated with multiple sclerosis. Ann Neurol 27: 291–297

Brown JH, Jardetzky TS, et al (1993) Three-dimensional structure of the human class II histocompatibility antigen HLA-DR1. Nature 364: 33–39

Brown MG, Driscoll J, et al (1991) Structural and serological similarity of MHC-linked LMP and proteasome (multicatalytic proteinase) complexes. Nature 353: 355–357

Bugawan TL, Angelini G, et al (1989) A combination of a particular HLA-DPb allele and an HLA-DQ heterodimer confers susceptibility to coeliac disease. Nature 339: 470–473

Bulman DE, Ebers GC (1992) The geography of multiple sclerosis reflects genetic susceptibility. J Trop Geograph Neurol 2: 66–72

Bulman DE, Armstrong H, et al (1991) Allele frequencies of the third component of complement (C3) in MS patients. J Neurol Neurosurg Psychiatry 54: 554–555

Burns J, Rosenzweig A, et al (1983) Isolation of myelin basic protein-reactive T cell lines from normal human blood. Cell Immunol 81: 435–440

Ebers GC, Dessa Sadovnick A (1994) The role of genetic factors in multiple sclerosis susceptibility. J Neuroimmunol 54: 1–17

Filippi M, Horsfield MA, et al (1994) Quantitative brain MRI lesion load predicts the course of clinically isolated syndromes suggestive of multiple sclerosis. Neurology 44: 635–641

Francis DA, Batchelor JR, et al (1987) Multiple sclerosis in Northeast Scotland. An association with HLA-DQw1. Brain 110: 181–196

Fritz RB, McFarlin DE (1989) Encephalitogenic epitopes of myelin basic protein. In: Sercarz EE (ed) Antigenic determinants and immune response. Karger, Basel, pp 101–125 (Chem Immunol 46)

Fritz RB, Skeen MJ, et al (1985) Major histocompatibility complex-linked control of the murine immune response to myelin basic protein. J Immunol 134: 2328–2332

Gaisner CN, Johnson MJ, et al (1987) Susceptibility to multiple sclerosis associated with immunoglobulin gamma 3 restriction fragment length polymorphism. J Clin Invest 79: 309–313

Gorodezky C, Najera R, et al (1986) Immunogenetic profile of multiple sclerosis in Mexicans. Hum Immunol 16: 364–374

Haegert DG, Francis GS (1993) HLA-DQ polymorphisms do not explain HLA class II associations with multiple sclerosis in two Canadian patient groups. Neurology 43: 1207–1210

Hammond SR, English D, et al (1988) The clinical profile of MS in Australia. A comparison between medium-frequency and high-frequency prevalence zones. Neurology 38: 980–986

Hillert J, Leng C, et al (1991) No association with germline T cell receptor betachain gene alleles or haplotypes in Swedish patients with multiple sclerosis. J Neuroimmunol 32: 141–147

Hillert J, Grönning M, et al (1992) An immunogenetic heterogeneity in multiple sclerosis. J Neurol Neurosurg Psychiatry 55: 887–890

Hirayama K, Matsushita S, et al (1987) HLA-DQ is epistatic to HLA-DR in controlling the immune response to schistosomal antigen in humans. Nature 327: 426–430

Jaraquemada D, Martin R, et al (1990) HLA-DR2a is the dominant restriction molecule for the cytotoxic T cell response to myelin basic protein in DR2Dw2 individuals. J Immunol 145: 2880–2885

Kagnoff MF, Harwood JI, et al (1989) Structural analysis of the HLA-DR, -DQ, and -DP alleles on the celiac disease-associated HLA-DR3 (DRw17) haplotype. Proc Natl Acad Sci USA 86: 6274–6278

Kotzin BL, Karuturi S, et al (1991) Preferential T-cell receptor Vβ-chain variable gene use in myelin basic protein-reactive T-cell clones from patients with multiple sclerosis. Proc Natl Acad Sci USA 88: 9161–9165

Kurdi A, Ayesh I, et al (1977) Different B-lymphocyte alloantigens associated with multiple sclerosis in Arabs and Northern Europeans. Lancet i: 1123–1125

Kurtzke JF (1985) Epidemiology of multiple sclerosis. In: Vinken PJ, Bruyn GW, Klawans HL, Koetsier JC (eds) Handbook of clinical neurology. Demyelinating diseases. Elsevier, Amsterdam New York, pp 259–287

Liblau R, van Endert PM, et al (1993) Antigen processing gene polymorphisms in HLA-DR2 multiple sclerosis. Neurology 43: 1192–1197

Marrosu HG, Muntoni F, et al (1988) Sardinian multiple sclerosis is associated with HLA-DR4: a serological and molecular analysis. Neurology 38: 1749–1753

Martin R, Jaraquemada D, et al (1990) Fine specificity and HLA restriction of myelin basic protein-specific cytotoxic T cell lines from multiple sclerosis patients and healthy individuals. J Immunol 145: 540–548

Martin R, Howell MD, et al (1991) A myelin basic protein peptide is recognized by cytotoxic T cells in the context of four HLA-DR types associated with multiple sclerosis. J Exp Med 173 (1): 19–24

Martin R, McFarland HF, et al (1992) Immunological aspects of demyelinating diseases. Annu Rev Immunol 10: 153–187

Martin R, Utz U, et al (1992) Diversity in fine specificity and T cell receptor usage of the human CD4+ cytotoxic T cell response specific for the immunodominant myelin basic protein peptide 87–106. J Immunol 148: 1359–1366

Martin R, Voskuhl R, et al (1993) Myelin basic protein-specific T-cell responses in identical twins discordant or concordant for multiple sclerosis. Ann Neurol 34: 524–535

McCombe PA, Clark P, et al (1985) Alpha-1 antitrypsin phenotypes in demyelinating diseases: An association between demyelinating disease and the allele PiM3. Ann Neurol 18: 291–297

McFarland HF, Frank JA, et al (1992) Using gadolinium-enhanced magnetic resonance imaging lesions to monitor disease activity in multiple sclerosis. Ann Neurol 32: 758–766

McFarlin DE, McFarland HF (1982a) Multiple sclerosis, part 1. N Engl J Med 307: 1183–1188

McFarlin DE, McFarland HF (1982b) Multiple sclerosis, part 2. N Engl J Med 307: 1246–1251

Meinl E, Weber F, et al (1993) Myelin basic protein-specific T lymphocyte repertoire in multiple sclerosis. Complexity of the response and dominance of nested epitopes due to recruitment of multiple T cell clones. J Clin Invest 92: 2633–2643

Moen T, Stein R, et al (1984) Distribution of HLA-SB antigens in multiple sclerosis. Tissue Antigens 4: 126–127

Momburg F, Roelse J, et al (1994) Selectivity of MHC-encoded peptide transporters from human, mouse and rat. Nature 367: 648–651

Naito S, Kuroiwa Y, et al (1978) HLA and Japanese MS. Tissue Antigens 12: 19–24

Nepom GT (1990) The HLA genetic contribution to rheumatoid arthritis. Clin Immunol 10: 127–131

Nepom GT (1993) MHC and autoimmune diseases. In: Bach J-F (ed) Monoclonal antibodies and peptide therapy in autoimmune diseases. Marcel Dekker, New York, pp 143–164

Nepom GG, Erlich H (1991) MHC class-II molecules and autoimmunity. Annu Rev Immunol 9: 493–526

Odum N, Hyldig-Nielsen JJ, et al (1988) HLA-DP antigens are involved in the susceptibility to multiple sclerosis. Tissue Antigens 31: 235–237

Offner H, Hashim GA, et al (1991) T cell receptor peptide therapy triggers autoregulation of experimental encephalomyelitis. Science 251: 430–432

Oksenberg JR, Sherritt M, et al (1989) T-Cell receptor V_α and C_β, alleles associated with multiple sclerosis and myasthenia gravis. Proc Natl Acad Sci USA 86: 988–992

Oksenberg JR, Panzara MA, et al (1993) Selection for T-cell receptor Vβ-Dβ-Jβ gene rearrangements with specificity for a myelin basic protein peptide in brain lesions of multiple sclerosis. Nature 362: 68–70

Olerup O, Hillert J, et al (1989) Primarily chronic progressive and relapsing/remitting multiple sclerosis: two immunogenetically distinct disease entities. Proc Natl Acad Sci USA 86: 7113–7117

Olsson T, Wei Zhi W, et al (1990) Autoreactive T lymphocytes in multiple sclerosis determined by antigen-induced secretion of interferon-γ. J Clin Invest 86: 981–985

Olsson T, Sun J, et al (1992) Increased numbers of T cells recognizing multiple myelin basic protein epitopes in multiple sclerosis. Eur J Immunol 22: 1083–1087

Ota K, Matsui M, et al (1990) T-cell recognition of an immunodominant myelin basic protein epitope in multiple sclerosis. Nature 346: 183–187

Palffy G (1982) MS in Hungary, including Gypsy population. In: Kuroiwa Y, Kurland LT (eds) Multiple sclerosis east and west. Karger, Basel, pp 149–157

Paty DW, Li DKB, et al (1993) Interferon beta-1 b is effective in relapsing-remitting multiple sclerosis. II. MRI analysis results of a multicenter, randomized, double-blind, placebo-controlled trial. Neurology 43: 662–667

Pette M, Fujita K, et al (1990) Myelin basic protein-specific T lymphocyte lines from MS patients and healthy individuals. Neurology 40: 1770–1776

Pette M, Fujita K, et al (1990) Myelin autoreactivity in multiple sclerosis: recognition of myelin basic protein in the context of HLA-DR2 products by T lymphocytes of multiple sclerosis patients and healthy donors. Proc Natl Acad Sci USA 87: 7968–7972

Powis SJ, Deverson EV, et al (1992) Effect of polymorphism of an MHC-linked transporter on the peptides assembled in a class I molecule. Nature 357: 211–215

Prineas JW (1985) The neuropathology of multiple sclerosis. In: Vinken PJ, Bruyn GW, Klawans HL, Koetsier JC (eds) Handbook of clinical neurology. Demyelinating diseases 3 (47). Elsevier, Amsterdam New York, pp 213–257

Raine CS (1983) Multiple sclerosis and chronic relapsing EAE: comparative ultrastructural neuropathology. In: Hallpike JF, Adams CW, Tourtellotte WW (eds) Multiple sclerosis. Williams and Wilkins, Baltimore, pp 413–478

Richert JR, Reuben-Burnside CA, et al (1988) Peptide specificities of myelin basic protein-reactive human T-cell clones. Neurology 38: 739–742

Sadovnick AD, Baird PA, et al (1988) Multiple sclerosis: updated risks for relatives. Am J Med Genet 29: 533–541

Sadovnick AD, Bulman D, et al (1991) Parent-child concordance in multiple sclerosis. Ann Neurol 29: 252–255

Sadovnick AD, Armstrong H, et al (1993) A population-based study of multiple sclerosis in twins: update. Ann Neurol 33: 281–285

Seboun E, Robinson MA, et al (1989) A susceptibility locus for multiple sclerosis is linked to T cell receptor β chain complex. Cell 57: 1095–1100

Seyfried CE, Mickelson E, et al (1988) A specific nucleotide sequence defines a functional T cell recognition epitope shared by diverse HLA-DR specificities. Hum Immunol 21: 289–299

Sollid LM, Markussen G, et al (1989) Evidence for a primary association of celiac disease to a particular HLA-DQ α/β heterodimer. J Exp Med 169: 345–350

Spurkland A, Rønningen KS, et al (1991) HLA-DQA1 and HLA-DQB1 genes may jointly determine susceptibility to develop multiple sclerosis. Hum Immunol 30: 69–75

Thomson G, Robinson WP, et al (1988) Genetic heterogeneity, modes of inheritance, and risk estimates for a joint study of Caucasians with insulin-dependent diabetes mellitus. Am J Hum Genet 43: 799–816

Tiwari JL, Terasaki PI (1985) HLA and disease associations. Springer, Berlin Heidelberg New York Tokyo, pp 152–167

Utz U, Biddison WE, et al (1993) Skewed T cell receptor repertoire in genetically identical twins with multiple sclerosis correlates with disease. Nature 364: 243–247

Vandenbark AA, Hashim GA, et al (1989) Determinants of human myelin basic protein that induce encephalitogenic T cells in Lewis rats. J Immunol 143: 3512–3516

Vandevyer C, Stinissen P, et al (1994) TAP 1 and TAP 2 transporter gene polymorphisms in multiple sclerosis: no evidence for disease association with TAP. J Neuroimmunol 54: 35–40

Vartdal F, Sollid LM, et al (1989) Patients with multiple sclerosis carry DQB1 genes which encode shared polymorphic aminoacid sequences. Hum Immunol 25: 103–110

Vogt AB, Kropshofer H, et al (1994) Ligand motifs of HLA-DRB5*0101 and DRB1*1501 molecules delineated from self-peptides. J Immunol 153: 1665–1673

Voskuhl RR, Martin R, et al (1993) T helper 1 (TH1) functional phenotype of human myelin basic protein-specific T lymphocytes. Autoimmunity 15: 137–143

Voskuhl RR, Martin R, et al (1993) A functional basis for the association of HLA class II genes and susceptibility to multiple sclerosis: cellular immune responses to myelin basic protein in a multiplex family. J Neuroimmunol 42: 199–208

Waksman BH, Reynolds WE (1984) Minireview: multiple sclerosis as a disease of immune regulation. Proc Soc Exp Biol Med 175: 282–294

Walter MW, Gibson WT, et al (1991) Susceptibility to multiple sclerosis is associated with the proximal immunoglobulin heavy chain variable region. J Clin Invest 87: 1266–1273

Wekerle H, Linington C, et al (1986) Cellular immune reactivity within the CNS. Trends Neuro Sci 9: 271–277

Wucherpfennig KW, Ota K, et al (1990) Shared human T cell receptor V beta usage to immunodominant regions of myelin basic protein. Science 248: 1016–1019

Wucherpfennig KW, Sette A, et al (1994) Structural requirements for binding of an immunodominant myelin basic protein peptide to DR2 isotypes and for its recognition by human T cell clones. J Exp Med 179: 279–290

Zhang J, Medaer R, et al (1993) MHC-restricted depletion of human myelin basic protein-reactive T cells by T cell vaccination. Science 261: 1451–1454

Zhang J, Markovic-Plese S, et al (1994) Increased frequency of interleukin 2-responsive T cells specific for myelin basic protein in peripheral blood and cerebrospinal fluid of patients with multiple sclerosis. J Exp Med 179: 973–984

Zinkernagel RM, Doherty PC (1974) Restriction of in vitro T cell-mediated cytotoxicity in lymphocytic choriomeningitis within a syngeneic or semiallogeneic system. Nature 248: 701–702

Author's address: Dr. R. Martin, Department of Neurology, University of Tübingen Medical School, Hoppe-Seyler-Strasse 3, D-72076 Tübingen, Federal Republic of Germany.

Modulation of control mechanisms of dopamine-induced apoptosis – a future approach to the treatment of Parkinson's disease?

I. Ziv[1], **D. Offen**[1], **A. Barzilai**[2], **R. Haviv**[2], **R. Stein**[2], **R. Zilkha-Falb**[2], **A. Shirvan**[1], and **E. Melamed**[1]

[1]Department of Neurology, Beilinson Medical Center, Petah-Tiqva, and
[2]Department of Biochemistry, The George Wise Faculty of Life Sciences, Neuroscience, Tel-Aviv University, Tel-Aviv, Israel

Summary. The cause for the progressive and selective degeneration of the dopaminergic (DA) nigrostriatal neurons in Parkinson's disease (PD) is still unknown. We suggest a novel approach, that links this neuronal degenerative process to inappropriate triggering of apoptosis, an active, controlled program of cellular self destruction, by excess oxidative stress mediated by DA metabolism. In support of this concept, we found that DA, the endogenous neurotransmitter, is capable of initiating apoptosis in cultured, postmitotic chick sympathetic neurons, an observation further extended to other cellular systems (PC-12 cells, cerebellar granular cells, thymocytes, splenocytes). In comparing the relative apoptosis-triggering potency of other mononamine neurotransmitters, DA was found to be the most active, whereas norepinephrine and serotonin had a moderate and a mild effects, respectively. This grading can be correlated with the relative involvement of the relevant neuronal systems (i.e., substantia nigra, locus ceruleus and raphe nuclei) in PD.

We therefore hypothesize that neuronal degeneration in PD may be caused, at least in part, by a failure, either inherited or acquired, in cellular control systems of apoptosis, that may normally restrain the lethal potential of these endogenous neuro-transmitters and their potentially-toxic oxidation products. We therefore point at apoptosis-control systems as a critical scene of events, where the fate of nigrostriatal neurons is ultimatly determined, and whose modulation may yield attenuation of the neuronal degenerative process. In support of this concept, we found that vector-driven stable expression of the proto-oncogene bcl-2, an inhibitor of apoptosis, can exert powerful cellular protection against DA toxicity in rat pheochromocytoma PC-12 cells. Furthermore, cell extracts from bcl-2-expressing cells were found to markedly inhibit in vitro oxidation of DA and production of DA-melanin. We also found that expression of bcl-2 can inhibit the decrease in intracellular reduced thiol (-SH) groups which we observed following exposure to DA. Research of the bcl-2 system and associated control mechanisms of apoptosis, possibly acting in association with intra-cellular anti-oxidant pathways, may therefore lead to novel therapeutic approaches for neuroprotection in PD.

Parkinson's disease (PD) results from selective degeneration of the dopaminergic (DA) nigrostriatal neurons of still unknown etiology. Currently, there exists no therapeutic approach which may affect and slow down this degenerative process. Levodopa preparations, the cornerstone of current treatment for PD, basically provide a symptomatic relief, and moreover, their long-term use is associated with severe and disabling adverse effects.

It is currently suggested that the etiology of the neuronal loss in PD is linked to DA-induced excessive oxidant stress produced during oxidation of DA. It is assumed that the generation of toxic free radical species leads to lipid peroxidation, dysfunction and rupture of cellular membranes, resulting in neuronal disintegration (Olanow, 1993). However, such mode of cell death by necrosis would be expected to cause rapid nigral degeneration, with cell swelling, rupture and uncontrolled spillage of its contents into the extracellular space (Boobis et al., 1989), with significant local inflammatory reaction. In contrast, nigral histopathology in PD is characterized by a slow, protracted degeneration of individual neurons, lack of marked inflammation and shrunken, condensed appearance of some of the remaining neurons. Apoptosis is a mode of genetically-controlled, active cellular death process, with a highly characteristic chain of events, including loss of contacts among neighbouring cells, their shrinkage and condensation, "blebbing" of cellular membranes, nuclear fragmentation at internucleosomal sites, and degradation of cells to membrane-bound particles (apoptotic bodies) that are rapidly phagocytized by macrophages without a significant inflammatory response. It occurs in various cells (Boobis et al., 1989; Wyllie et al., 1980), including neurons (Martin et al., 1988) and seems to have a major physiological role in the differentiation and organization of the nervous system during its development (Barinaga, 1993). However, since this "suicide" program exists in every cell in a "restrained" form throughout life, it is theoretically possible that its inappropriate activation may also be involved in the pathogenesis of neurodegeneration in older age.

DA, the endogenous neurotransmitter, has been shown to exert antitumor effects and is toxic to cultures of melanoma and neuroblastoma cells (Wick, 1978; Graham et al., 1978). Its antitumor action is probably related to its genotoxicity, since DA causes both DNA damage and inhibition of its repair, and its effect has been shown to be synergistic to that of irradiation (Moldeus et al., 1983; Wick, 1989). After reaching a certain critical level, DNA damage is one of the triggers of apoptosis. Therefore, many anti-cancer drugs act through induction of apoptosis (Eastman, 1990).

Recently, we were the first to show that DA, in physiological concentrations [estimated as 0.1–1 mM within neuronal cell bodies (Michel and Hefti, 1990; Johnson, 1971)], is capable of initiating apoptosis in cultured, postmitotic chick sympathetic neurons (Ziv et al., 1994). Exposure to DA initiated the various morphological and flow-cytometric changes pathognomonic of apoptosis. DA caused extensive axonal thinning and disintegration, condensation and shrinkage of cell bodies (Fig. 1). Scanning electron microscopic (SEM) studies revealed extensive membrane blebbing, with actual cell transformation to clusters of apoptotic bodies, ready to be engulfed by macrophages. Furthermore, exposure to DA caused the characteristic apoptotic nuclear condensation and fragmentation, evident in flow cytometric analysis of purified, propidium-iodide-stained cell nuclei, (Fig. 2) and flourescent microscopy of DAPI (4,6-diamidino-2-phenylindole)-stained cell nuclei (Fig. 3). These marked changes were similar to the effect of nerve growth factor (NGF) deprivation, a well-known model for apoptosis in sympathetic neurons (Martin et al., 1988). We later extended these observations to other cellular systems, including mouse thymocytes (Offen et al., submitted), cerebellar granullar cells and rat pheochromocytoma PC-12 cells (Ziv et al., submitted), and also showed the inhibition of this DA-induced apoptosis by thiol (-SH) - group- containing antioxidants (e.g., N-acetylcysteine, reduced gluthatione and dithiothreitol) (Offen et al., submitted), thus implying a role for free radical species in the observed death process.

In Parkinson's disease, in addition to nigrostriatal neuronal degeneration, there is also involvement, though to lesser extent, of noradrenergic and serotoninergic pathways (Agid et al., 1990). We therefore examined the relative apoptosis-triggering potency of other monoamine neurotransmitters, including norepinephrine and serotonin (Zilka et al., submitted). DA showed the highest degree of toxicity, whereas norepinephrine and serotonin had a moderate and a mild effect, respectively. Hypothetically, this graded lethality of monoamines, i.e., DA > norepinephrine ≫ serotonin, can be correlated with the relative involvement of dopaminergic (nigrostriatal) > noradrenergic (locus ceruleus) ≫ serotoninergic (raphe nuclei) pathways in PD, and thus may be compatible with the concept of a common, underlying multisystem abnormality in apoptosis-control mechanisms.

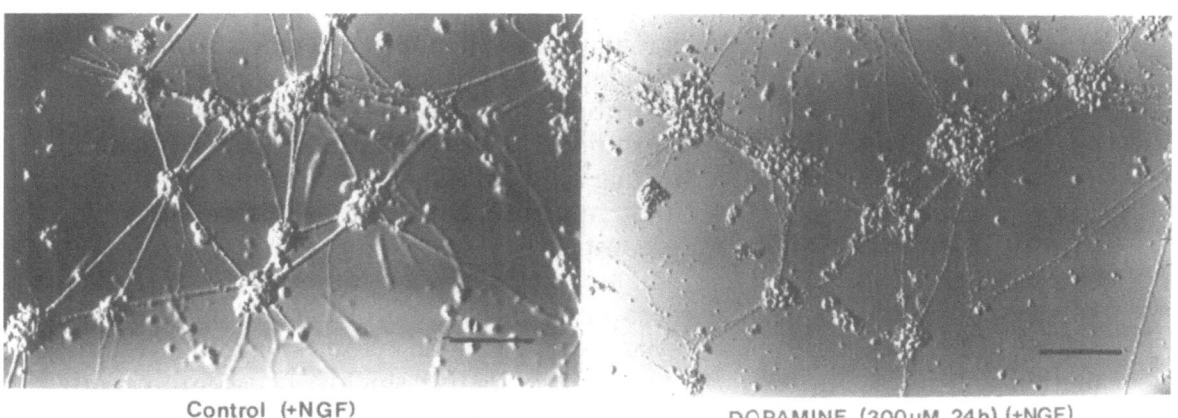

Control (+NGF)   DOPAMINE (300μM, 24h), (+NGF)

Fig. 1. Apoptotic morphological alterations, induced by dopamine, in cultured, postmitotic, chick sympathetic neurons (Nomarski optics). **a** Control: Well developed neuronal network, bright cell bodies. **b** DA-effect: Exposure to DA, 0.3 mM for 24 hours: extensive neurite thinning and disruption, severe shrinkage and condensation of cell bodies (× 300, bar = 100 nm) (Neurosci Lett 170: 136–140)

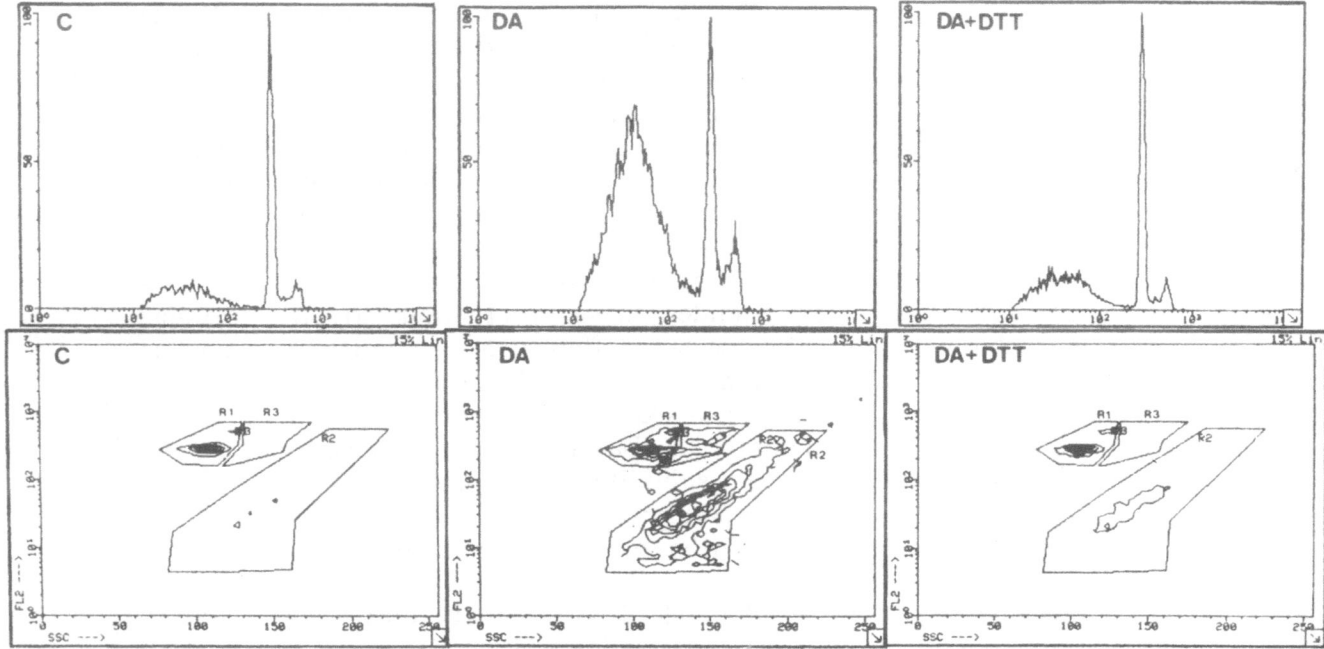

Fig. 2. Apoptotic, dopamine-induced alterations in cell nuclei of cultured, postmitotic, chick sympathetic neurons: flow cytometric analysis of propidium-iodide (PI)-stained purified cell nuclei. *C* control, *DA* dopamine (0.3 mM, 24 h), *DA+DTT* dopamine (0.3 mM) + dithiotreitol (DTT, 0.5 mM, 24 h). *Upper figures*: PI fluorescence (FL^2): The emergence of a large, distinct, subdiploid apoptotic peak, which reflects apoptotic nuclear fragmentation, and the inhibition of its formation by the antioxidant DTT. *Lower figures*: FL^2 vs. side scatter (SSC): *Region 1 (R1)*: normal diploid nuclei. *Region 2 (R2)*: apoptotic nuclei. *Region 3 (R3)*: nuclei with increased granularity but normal DNA content, probably in the beginning of the apoptotic process (Neurosci Lett 170: 136–140)

These observations suggest that nigrostriatal neuronal degeneration in PD may not be merely due to an oxidative-induced membranal damage, but rather reflects an active apoptotic process, governed, at least in part, at the nuclear level. Our studies therefore point at the cellular control systems of apoptosis as the possible critical scene of events where nigrostriatal neuronal fate is finally determined. DA, other monoamine neurotransmitters, and their toxic oxidative derivatives may constitute a continuous challenge for the competence of these cellular defense systems, that normally restrain the apoptosis-triggering effect of these endogenous neurotransmitters. Among others, these natural protective systems may include cytoplasmatic DA-vesiculation and intra- and extra-nuclear anti-oxidative shields. In addition (and not previously explored in PD research), specific, strictly monitored, apoptosis-inhibiting mechanisms may also be extremly important. These systems are only beginning to be unfolded, and may hold the key for the determination of the fate the neuronal cells to survival or degeneration. Their dysregulation, either acquired or inherited, may lead to premature activation of apoptosis, and may thus contribute, at least in part, to the pathogenesis of PD. On the

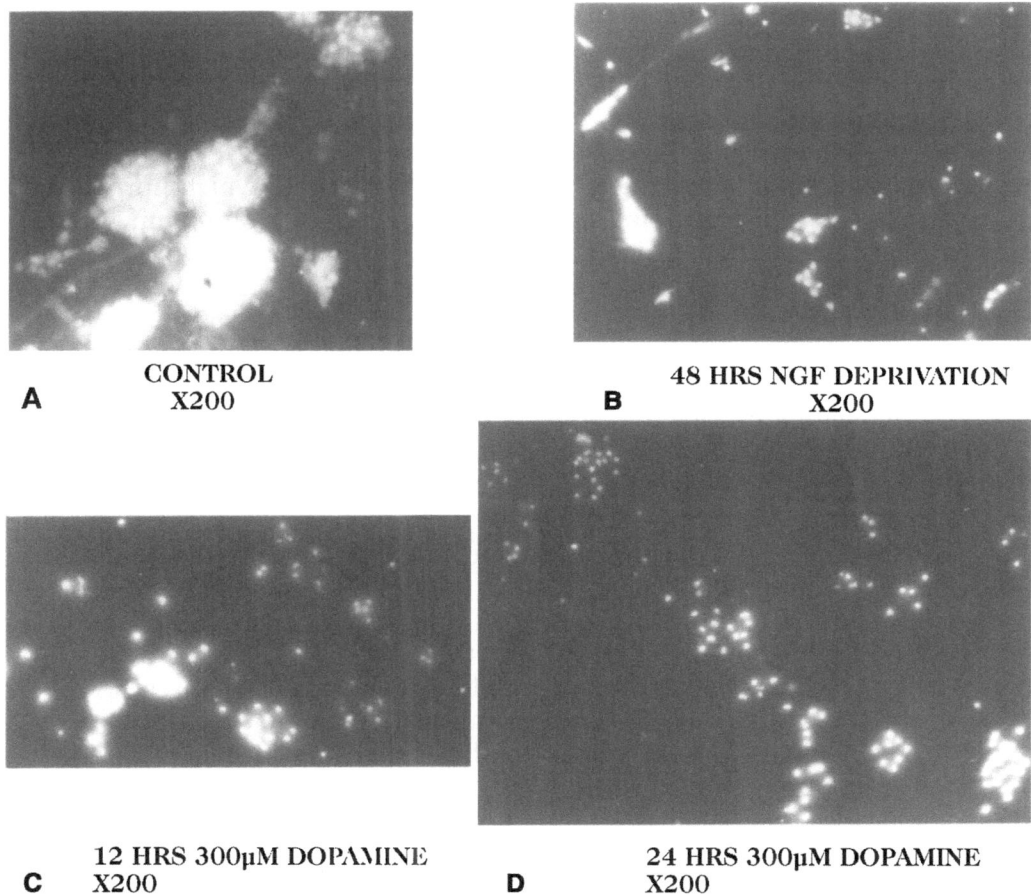

Fig. 3. Fluorescent microscopy of DAPI (4,6-diamidino-2-phenylindol)-stained nuclei of cultured chick sympathetic neurons, following exposure to dopamine (0.3 mM, 24 h). Cells were grown with nerve growth factor (NGF). **A** Control cells. **B** Apoptotic alterations induced by NGF-deprivation (treatment with anti-NGF-ab 0.01% for 48 hours, a well-established model of neuronal apoptosis). **C, D** Effect of exposure to dopamine (0.3 mM) for 12 and 24 hours, respectively. Dopamine caused the characteristic apoptotic nuclear condensation and fragmentation, similar to the effect of NGF-deprivation

other hand, it is possible, that their manipulation, pharmacologically or by gene therapy, may hold a promise for novel therapeutic approaches for neuroprotection in this degenerative neurological disorder.

Testing this line of thought, we examined the capability of the proto-oncogene bcl-2 to block DA-induced cellular death process. Bcl-2 (acronym for the B-cell lymphoma/leukemia-2 gene) is a proto-oncogene that was first discovered in association with B-cell malignancies, where its overexpression, induced by 18–14 chromosomal translocation, has a role in promoting tumor growth (Reed, 1994). Bcl-2 was later shown to be a potent inhibitor of apoptosis in various systems. Among others, it can rescue cultured postmitotic neuronal cells destined for apoptosis by growth factor deprivation (Garcia et al., 1992). Neuronal cells are among the several cell types were bcl-2 is normally found (Hockenberry et al., 1991), and it has been shown to be expressed in pigmented dopaminergic nigral neurons (F.Javoy-Agid, personal communication). Bcl-2 may therefore act as one of the safeguards, which maintain the longevity of these postmitotic cells.

We induced stable expression of bcl-2 in PC-12 cells (which normally do not express this onco-protein) by transfection with recombinant pCMV5 expression vector (Andersson et al., 1989), containing mouse bcl-2 coding-sequence cDNA. Differential expression in transfected cells [Bcl-2(+) cells] vs. control [bcl-2(–) cells] was verified at the mRNA and protein levels by Northern and Western blot analyses, respectively. In addition, function of the expressed protein was verified by confirming the relative resistance of bcl-2 (+) cells against apoptosis induced by serum deprivation.

DA (0.3 mM for 24 hours) was lethal to all control, bcl-2(–) cells. In contrast, bcl-2 was found to confer marked resistance against DA toxicity, and bcl-2-expressing cultures manifested 90% viable cells after same exposure to DA, as measured by the trypan blue-exclusion and 3H-thymidine-incorporation assays. Scanning electron microscopic studies revealed extensive apoptotic morphological alterations in the control cells, alterations which were inhibited in the bcl-2(+) cultures. Bcl-2 was therefore found to be a powerful inhibitor of DA-induced apoptosis, capable of dramatically rescuing PC-12 cells from the toxicity of this neurotransmitter (Ziv et al., submitted).

The mechanisms by which bcl-2 inhibits apoptosis in general, and that triggered by exposure to DA in particular, are not fully elucidated. The bcl-2 gene encodes a 25kD protein structurally and functionally analogous to the ced-9 gene of the roundworm Caenorhabditis Elegans, thus implying conservation of this "anti-death" gene throughout evolution (Reed, 1994). Bcl-2 appears to block a final common pathway for apoptosis. One theory suggests that bcl-2 may prevent the translation of presence of damaged DNA into a signal for activation of apoptosis-associated genes, and it should be noted at this point, that DA has been shown to exert substantial genotoxic effects (Moldeus et al., 1983). Alternatively, bcl-2 may act through blocking the action of products of apoptosis-related genes once induced (Reed, 1994).

The intracellular localization of this onco-protein may provide other clues to its mode of action. Bcl-2 resides in the nuclear envelope, [where it is concentrated in patches in relation to the nuclear pore complexes (NPCs)], parts of the endoplasmic reticulum (ER) and mitochondrial membrane. Interestingly, bcl-2 was not identified in the cellular plasma membrane (Reed, 1994). This selective subcellular distribution may suggest its role in the control of traffic of apoptosis mediators such as Ca^{++} from stores in the ER to the nucleus. Alternatively, the mitochondrial location of bcl-2 suggests its association with oxidative metabolism. Indeed, there is evidence for the role of free radical species in the triggering of apoptosis, and moreover, it has been demonstrated that bcl-2 may act to inhibit formation and/or activity of free radicals (Kane et al., 1993).

These considerations, and our previously-mentioned findings of inhibition of DA toxicity by (-SH)-containing antioxidant agents prompted us to further clarify the association between the protective effect of bcl-2 against DA-toxicity and cellular antioxidant capacity. We therefore examined the effect of cellular lysate from bcl-2(+) PC-12 cells on the in vitro auto-oxidation of DA and formation of DA-melanin, evaluated by spectrophotometry. We found that the extract from bcl-2-expressing cells significantly inhibited the production of DA-melanin, as compared to that of the control, bcl-2(−) cells. Furthermore, by measuring the free thiol groups [by the DTNB (5,5'-dithiobis-2-nitrobenzoic acid 3-carboxy-4-nitrophenyl disulfide) method], we have been able to show that expression of bcl-2 inhibited the reduction in intracellular reduced-thiol groups encountered following exposure to DA (Offen et al., submitted). Our findings therefore support an association between the protective cellular effect of bcl-2 against DA toxicity and antioxidant pathways, and suggest a role for this onco-protein in the monitoring and enhancement the anti-oxidant state of the cell.

Recently, several bcl-2-related proteins have been discovered, manifesting the complexity of this apoptosis-control system. These include Bax, which heterodimerizes with bcl-2 and opposes its action and the bcl-x proteins, which constitute a similar competing pair with opposite effects on apoptosis. This complexity possibly reflects the importance of fine tuning of this system as a vital checkpoint of cell survival. Delicate balance between these related proteins is considered to have a key role in the control of the inherent genetic death program by bcl-2 (Oltavi et al., 1993; Boise et al., 1993).

Bcl-2 and associated control systems of apoptosis therefore constitute a previously-unexplored field in PD research. Hypothetically, modulation of this system in neuronal cells may enhance their intrinsic antioxidant capacity, inhibit premature activation of apoptosis in these cells, and possibly slow-down the degenerative neuronal process.

Acknowledgements

Supported, in part, by the National Parkinson Foundation, Miami, Florida, U.S.A., Teva Pharmaceutical Industries, Ltd., Israel, The Prevot Foundation, Switzerland, and The American Physicians Fellowship for Medicine in Israel, New York, NY, U.S.A.

References

Agid Y, Graybiel AM, Ruberg M, Hirsch E, Blin J, Dubois B, Javoy-Agid F (1990) The efficacy of levodopa treatment declines in the course of Parkinson's disease: do nondopaminergic lesions play a role? Adv Neurol 53: 83–100

Andersson S, Davis DL, Dahlback H, Jornvall H, Russell DW (1989) Cloning, structure and expression of the mitochondrial cytochrome p-450 sterol 26-hydroxylase, a bile acid biosynthetic enzyme. J Biol Chem 264: 8222–8229

Barinaga M (1993) Death gives birth to the nervous system. But how? Science 259: 762–763

Boise LH, Gonzalez-Garcia MZ, Thompson CB, Postema CE, Ding L, Lindsten T, Turka LA, Mao X, Nuñez G (1993) Bcl-x, a bcl-2-related gene that functions as a dominant regulator of apoptotic cell death. Cell 74: 597–608

Boobis AR, Fawthrop DJ, Davies DS (1989) Mechanisms of cell death. Trends Pharmacol Sci 10: 275–280

Eastman A (1990) Activation of programmed cell death by anticancer agents: cisplatin as a model system. Cancer Cells 2: 275–280

Garcia I, Martinou I, Tsujimoto Y, Martinou JC (1992) Prevention of programmed cell death of sympathetic neurons by the bcl-2 proto-oncogene. Science 258: 302–30

Graham DG, Tiffany SM, Bell WR, Gutknecht WF (1978) Autoxidation versus covalent binding of quinones as the mechanism of toxicity of dopamine, 6-hydroxydopamine, and related compounds toward c1300 neuroblastoma cells in vitro. Mol Pharmacol 14: 644–653

Hockenbery DM, Zutter M, Hickey W, Nahm M, Korsmeyer SJ (1991) Bcl-2 protein is topographically restricted in tissues characterized by apoptotic cell death. Proc Natl Acad Sci USA 88: 6961–6965

Johnson G (1971) Quantitation of fluorescence of biogenic monoamines. Prog Hystochem Cytochem 2: 299–344

Kane DJ, Sarafian TA, Anton R, Hahn H, Gralla EB, Valentino JS, Örd T, Bredesen DE (1993) Bcl-2 inhibition of neural death: decreased generation of reactive oxygen species. Science 262: 1274–1277

Martin DP, Schmidt RE, Distefano PS, Lowry OH, Carter JG, Johnson EM Jr (1988) Inhibitors of protein synthesis and RNA synthesis prevent neuronal death caused by nerve growth factor deprivation. J Cell Biol 106: 829–844

Michel PP, Hefti F (1990) Toxicity of 6-hydroxydopamine and dopamine for dopaminergic neurons in culture. J Neurosci Res 26: 428–435

Moldeus P, Nordenskjold M, Bolcsfoldi G, Eiche A, Haglund U, Lambert B (1983) Genetic toxicity of dopamine. Mut Res 124: 9–24

Olanow CW (1993) A radical hypothesis for neurodegeneration. Trends Neurosci 16: 439–444

Oltvai ZN, Milliman CL, Korsmeyer SJ (1993) Bcl-2 heterodimerizes in vivo with a conserved homolog, Bax, that accelerates programmed cell death. Cell 74: 609–619

Reed JC (1994) Bcl-2 and the regulation of programmed cell death. J Cell Biol 124: 1–6

Wyllie AH, Kerr JFR, Currie AR (1980) Cell death: the significance of apoptosis. Int Rev Cyto 68: 251–306

Wick MM (1978) Dopamine: a novel antitumor agent active against B-16 melanoma in vivo. J Invest Dermatol 71: 163–164

Wick MM (1989) Levodopa/dopamine analogs as inhibitors of DNA synthesis in human melanoma cells. J Invest Dermatol 92: 329S–331S

Ziv I, Melamed E, Nardi N, Luria D, Achiron A, Offen D, Barzilai A (1994) Dopamine induces apoptosis-like cell death in cultured chick sympathetic neurons – a possible novel pathogenetic mechanism in Parkinson's disease. Neurosci Lett 170: 136–140

Authors' address: E. Melamed, M.D., Department of Neurology, Beilinson Medical Center, 49100, Petah-Tiqva, Israel.

GTP cyclohydrolase I gene, dystonia, juvenile parkinsonism, and Parkinson's disease

T. Nagatsu and **H. Ichinose**

Institute for Comprehensive Medical Science, School of Medicine,
Fujita Health University, Toyoake, Aichi, Japan

Summary. GTP cyclohydrolase I is the rate-limiting enzyme for the biosynthesis of tetrahydrobiopterin, which is the cofactor for tyrosine hydroxylase, the rate-limiting enzyme for dopamine biosynthesis. We found that dominantly inherited, hereditary progressive dystonia (HPD), first described by Segawa and also called dopa responsive dystonia (DRD), is caused by the mutations of GTP cyclohydrolase I gene, the partial decrease in the enzyme activity, and probably in striatal dopamine level, to less than 20% of the normal values. Juvenile parkinsonism and Parkinson's disease are also striatal dopamine deficiency, but no mutation in the enzyme has not been found, and they are supposed to be different from HPD/DRD in which no cell death of the nigrostriatal dopamine neurons occurs.

Dystonia, juvenile parinsonism (JP) and Parkinson's disease (PD) are the movement disorders accompanied by the dopamine deficiency in the basal ganglia, and are responsive to L-dopa treatment with varying degrees of efficacy.

Dopamine is synthesized from L-tyrosine: L-tyrosine → L-dopa → dopamine. Two dopamine-synthesizing enzymes are required: (1) tyrosine hydroxylase (TH, Nagatsu et al., 1964), aromatic L-amino acid decarboxylase (AADC) (also called dopa decarboxylase, DDC). TH requires a tetrahydropterin as a cofactor, and (6R)-L-*erythro*-5, 6, 7, 8,-tetrahydrobiopterin (6R-BH4) is the natural cofactor (Kaufman, 1963). 6R-BH4 is synthesized in the TH-containing dopamine neurons from GTP: GTP → 7, 8-dihydroneopterin triphosphate → 6-pyruvoyltetrahydropterin → 6R-BH4. Three enzymes are required for 6R-BH4 biosynthesis from GTP: (1) GTP cyclohydrolase I, (2) 6-pyruvoyltetrahydropterin synthase, and (3) sepiapterin reductase. The first enzyme, GTP cyclohydrolase I is assumed to be the rate-limiting enzyme (Nichol et al., 1985).

In vivo TH activity is considered to be partly regulated by the concentration of 6R-BH4 (Kettler et al., 1974; Niwa et al., 1985; Matsuura et al., 1986; Nagatsu et al., 1994), and thus GTP cyclohydrolase I activity may also regulate in vivo TH activity via regulation of 6R-BH4 biosynthesis, and ultimately dopamine biosynthesis. Therefore, GTP cyclohydrolase I activity may be important for the molecular mechanism of dystonia, JP, PD. We have been studying biochemistry and molecular biology of BH4-synthesizing enzymes (Nagatsu et al., 1985; Ichinose et al., 1991; Togari et al., 1992; Nomura et al., 1993). We attempt to examine whether or not BH4-synthesizing enzymes, especially GTP cyclohydrolase I, are related to the dopamine deficiency in the basal ganglia of dystonia, JP, and PD.

GTP cyclohydrolase I gene as the causative gene of hereditary progressive dystonia with marked diurnal fluctuation (HPD) / Dopa responsive dystonia (DRD)

Dystonia refers to a heterogeneous group of movement disorders of unknown etiology. The primary dystonias can be divided into a number of genetic and clinical subtypes. Hereditary progressive dystonia with marked diurnal fluctuation (HPD) is a dystonia, originally described by Segawa et al. (1971, 1986). Clinical onset is mostly in the first decade. The clinical course shows progression in the first two decades, which subsides from the third decade and becomes almost static from the 3rd to 4th decades. Low doses of L-DOPA are essentially curative over the life of the individual without apparent unfavorable side effects. Dopa responsive dystonia (DRD) is a term first proposed by Nygaard et al. (1989, 1991) to describe dystonia responding to L-dopa including dystonia with heterogeneous etiologies. DRD is now considered to be identical to HPD when DRD is diagnosed with strictly defined criteria. HPD/ DRD is inherited as autosomal dominant traits with reduced penetrance. Recently the gene for DRD was mapped to chromosome 14q (Nygaard et al., 1993). We have determined the chromosomal localization of GTP cyclohydrolase I to 14q 22.1–q22.2 within the DRD/HPD locus (Ichinose et al., 1994). The GTP cyclohydrolase I gene has six exons (Ichinose et al., 1995a). As we already know the sequences around the exons, we amplified exons, including splicing junctions, from genomic DNA, using polymerase chain reaction (PCR). The antisence primer for the amplification of exon 3 was set in the coding region (19 nucleotides). Amplified DNA fragments were directly sequenced with an automated DNA sequencer. Using our primer sets, we could have determined 97.5% of the nucleotides in the coding region sequence (731 out of 750 nucleotides). We examined four HPD families and a sporadic case and discovered four variations as compared to the control sequence: three single-base changes that predict nonconservative amino acid substituents (Arg 88 Trp, Asp 134 Val, and Gly 201 Glu) and a two-base insertion (ATG GAG → ATG GG GAG) that shifts the reading frame just after the translational starting methionine (Table 1). All patients were heterozygous in terms of these mutations, and no other mutations were found in the coding region of GTP cyclohydrolase I gene. None of the mutations was found on 108 other chromosomes from unrelated normal Japanese individuals. Both the Arg 88 Trp and Gly 201 Glu substitutions abolished the increase in the GTP cyclohydrolase I activity shown in the wild-type cDNA, demonstrating that these mutations are not simply polymorphisms.

Table 1. Mutations of GTP cyclohydrolase I in HPD/DRD and GTP cyclohydrolase I deficiency

	Base-pair chage	Amino-acid change
HPD/DRD[a]	CGG → TGG	Arg 88 Trp
	GAC → GTC	Asp 134 Val
	ATG GAG → ATG GG GAG	frame shift
	GGA → GAA	Gly 201 Glu
GTP cuclohydrolase I deficiency	CGC → CAC[b]	Arg 184 His
	GTA → ATA[b, c]	Met 211 Ile

[a]Ichinose et al. (1994); [b]Ichinose et al. (1995a); [c]Blau et al. (1995)

We measured the GTP cyclohydrolase I activity stimulated with phyto-hemagglutinin (PHA) to induce the enzyme in mononuclear cells isolated from blood of HPD patients. All patients with HPD phenotype showed very low GTP cyclohydrolase I activities, about 2–20% of normal values (Table 2). These results confirm that GTP cyclohydrolase I gene is a causative gene for HPD/DRD. This is the first report of a causative gene for the inherited dystonia.

We found two men without any symptoms who had the same mutations as the patients. The presence of asymptomatic male carriers in HPD/DRD has been expected from pedigree analyses which indicates incomplete penetrance, especially in males. GTP cyclohydrolase I activity in mononuclear blood cells from patients with HPD was decreased to less than 20% of the mean value of healthy controls in spite of the presence of a normal allele (heterozygous

Table 2. GTP cyclohydrolase I activity in mononuclear blood cells and postmortem striatum in patients with HPD/DRD, JP, and PD and healthy controls

			GTP cyclohydrolase I activity (mean ± SEM, pmol/h/mg protein)
Mononuclear blood cells	controls	(18)[b]	18.7 ± 2.3 (9.0–46.1)
	HPD/DRD	(7)[a]	1.9 ± 0.4 (0.3–3.6)
	Jp	(7)[b]	20.5 ± 4.8 (7.8–35.8)
Postmortem caudate nucleus[c]	controls	(10)	4.2 ± 0.4
	PD	(6)	0.8 ± 0.6
	JP	(1)	4.5

[a]From data by Ichinose et al. (1994); [b]Ichinose et al. (1995b); [c]from data by Nagatsu et al. (1986). The numbers in parentheses indicate the range of values

state), whereas those in unaffected carriers were 36 and 37%. Neopterin content in cerebrospinal fluid, which is thought to reflect the GTP cyclohydrolase I activity in the brain, was reported to be less than about 20% in normal levels in HPD/DRD patients (Fink et al., 1988; Furukawa et al., 1993), while an unaffected carrier of HPD/DRD showed 35% neopterin content compared to normals (Takahashi et al., 1994). These results suggest that the level of the enzyme activity is critical in this disease. It should be noted that GTP cyclohydrolase I activity is slightly higher in men than in women in healthy controls and that there is marked (4:1) female predominance in HPD/DRD (Ichinose et al., 1994). Higher GTP cyclohydrolase I activity may protect men from the appearance of HPD/DRD symptoms.

Autosomal recessive GTP cyclohydrolase I deficiency

There have been several reports of an atypical hyperphenylalaninemia caused by a deficiency in GTP cyclohydrolase I with autosomal recessive inheritance (Niederwieser, 1984; Blau et al., 1985). These patients show severe retardation in development, severe muscular hypotonia of the trunk and hypertonia of the extremities, convulsions and frequent episodes of hyperthermia in the absence of infections, probably due to lack of both catecholamines (dopamine, noradrenalin, and adrenalin) and serotonin. GTP cyclohydrolase I activity was not detectable in PHA-stimulated mononuclear blood cells or in liver biopsies of these patients (Niederwieser, 1984; Blau et al., 1985). Heterozygous carriers of this recessive deficiency state do not have dystonia symptoms and have been reported to have enzyme activity in the range of 30–36% of normal levels (Blau et al., 1985).

A single-base change that causes an amino acid substitution (Met 211 Ile, Arg 184 His) from patients with GTP cyclohydrolase I defect has been found, and the mutated enzyme expressed in *Escherichia coli* had no activity (Blau et al., 1995; Ichinose et al., 1995a) (Table 1). HPD patients have normal levels of phenylalanine in their urine and blood, probably owing to their low but never the less substantial GTP cyclohydrolase I activity in the liver. Expression of dystonic symptoms may reflect the relative decrease in enzyme activity in the basal ganglia, whereas complete defect in enzyme activity may result in several neurologic phenotypes which may be produced by the deficiency of catecholamines and serotonin.

GTP cyclohydrolase I and juvenile parkinsonism (JP)

Juvenile parkinsonism (JP) is defined as parkinsonism manifesting clinically below the age of 40, in contrast to idiopathic Parkinson's disease (PD) with onset after middle age. JP is distingusished from PD in the following characteristics; higher familiar incidence, relatively slow progression and benign prognosis, and dramatic or marked responses to L-dopa. These features of JP are similar to those of dystonia, HPD/DRD (Yokochi, 1993).

In order to study the difference between JP and HPD/DRD, we have examined GTP cyclohydrolase I activity in mononuclear blood cells isolated from blood of 7 patients with JP after PHA stimulation. The mean enzyme activity in mononuclear blood cells in JP was similar to that in healthy controls (Table 2). This agrees with our previous result that a case of JP had normal enzyme activity in the postmortem striatum tissue (Nagatsu et al., 1986), suggesting that the enzyme activity in mononuclear blood cells can represent the activity in the striatum. The result also indicates that the GTP cyclohydrolase I gene in JP does not have any significant mutations as the gene in HPD/DRD, and JP is distinct from HPD/DRD in terms of normal GTP cyclohydrolase I (Ichinose et al., 1995b). Measurement of GTP cyclohydrolase I activity in mononuclear blood cells can be a useful method for biochemical differential diagnosis between JP and HPD/DRD.

GTP cyclohydrolase activity in Parkinson's disease (PD)

We have reported that Parkinson's disease (PD) is characterized by decreases in activity (Nagatsu et al., 1979), protein (Mogi et al., 1988) and mRNA (Ichinose et al., 1994) of TH in the nigrostriatal dopamine regions. We have also reported decreased GTP cyclohydrolase I activity in the striatum of postmortem brains from patients with PD (Nagatsu et al., 1986). The decreases in TH and GTP cyclohydrolase I in the striatum seem to be due to the results of cell death. Thus the real cause of the decreases in TH and GTP cyclohydrolase I should be the cause of cell death in PD. This is in contrast to HPD/DRD in which the dopaminergic neurons are intact and only dopamine biosynthesis is impaired due to mutation of GTP cyclohydrolase I (Ichinose et al., 1994).

Conclusion

GTP cyclohydrolase I regulates dopamine biosynthesis via regulation of TH activity. The results on GTP cyclohydrolase I activity in mononuclear blood cells in autosomal dominant HPD/DRD (Ichinose et al., 1994) and in autosomal recessive GTP cyclohydrolase I deficiency (Blau et al., 1985) suggest interesting relation between the activity and phenotypes. No phenotype change may appear up to 30% normal levels, activity in the range of 2–20% normal levels may produce HPD/DRD owing to the specific decrease in TH activity in the basal ganglia. No activity of GTP cyclohydrolase I in homozygote patients with autosomal recessive GTP cyclohydrolase I deficiency may produce the dysfunction of the phenylalanine, tyrosine, and tryptophan hydroxylases in the liver and brain with hyperphenylalaninemia and severe neurologic symptoms. Our study suggests high susceptibity of the nigrostriatal dopaminergic neurons to a dificiency of BH4. GTP cyclohydrolase I activity in mononuclear blood cells can be a useful biochemical marker for differential dianosis of HPD/DRD and JP or PD.

References

Blau N, Joller P, Atarés M, Cardesa-Garcia J, Niederwieser A (1985) Increase of GTP cyclohydrolase I activity in mononuclear blood cells by stimulation: detection of heterozygotes of GTP cyclohydrolase I deficiency. Clin Chim Acta 148: 47–52

Blau N, Ichinose H, Nagatsu T, Heizmann CW, Zacchello F, Burlina AB (1995) A missense mutation in a patient with guanosine triphosphate cyclohydrolase I deficiency missed in the newborn screening program. J Pediatr 126: 401–405

Fink JK, Barton N, Cohen W, Lovenberg W, Burns RS, Hallett M (1988) Dystonia with marked diurnal variation associated with biopterin deficiency. Neurology 38: 707–711

Furukawa Y, Nishi K, Kondo T, Mizuno Y, Narabayashi H (1993) CSF biopterine levels and clinical features of patients with juvenile parkinsonism. In: Narabayashi H, Nagatsu T, Yanagisawa N, Mizuno Y (eds) Advances in neurology, vol 60. Raven Press, New York, pp 562–567

Ichinose H, Katoh S, Sueoka T, Titani K, Fujita K, Nagatsu T (1991) Cloning and sequencing of cDNA encoding human sepiapterin reductase – an enzyme involved in tetrahydrobiopterin biosynthesis. Biochem Biophys Res Commun 179: 183–189

Ichinose H, Ohye T, Fujita K, Pantucek F, Lange K, Riederer P, Nagatsu T (1994) Quantification of mRNA of tyrosine hydroxylase and aromatic L-amino acid decarboxylase in the substantia nigra in Parkinson's disease and schizophrenia. J Neural Transm [P-D Sect] 8: 149–158

Ichinose H, Ohye T, Takahashi E, Seki N, Hori T, Segawa M, Nomura Y, Endo K, Tanaka H, Tsuji S, Fujita K, Nagatsu T (1994) Hereditary progressive dystonia with marked diurnal fluctuation caused by mutations in the GTP cyclohydrolase I gene. Nature Genet 8: 236–242

Ichinose H, Ohye T, Matsuda Y, Hori T, Blau N, Burlinov A, Rouse B, Matalon R, Fujita K, Nagatsu T (1995a) Characterization of mouse and human GTP cyclohydrolase I genes. Mutations in patients with GTP cyclohydrolase I deficiency. J Biol Chem 270: 10062–10071

Ichinose H, Ohye T, Yokochi M, Fujita K, Nagatsu T (1995b) GTP cyclohydrolase I activity in mononuclear blood cells in juvenile parkinsonism. Neurosci Lett 190: 140–142

Kaufman S (1963) The structure of the phenylalanine hydroxylase cofactor. Proc Natl Acad Sci USA 50: 1085–1093

Kettler R, Bartholini G, Pletscher A (1974) In vivo enhancement of tyrosine hydroxylation in rat striatum by tetrahydrobiopterin. Nature 249: 476–478

Matsuura S, Murata S, Sugimoto T, Sawada M, Nagatsu T (1986) Preparation and cofactor activity of (6S)-tetrahydrobiopterin. Chem Express 1: 403–406

Miwa S, Watanabe Y, Hayaishi O (1985) 6R-L-Erythro-5, 6, 7, 8-tetrahydrobiopterin as a regulator of dopamine and serotonin biosynthesis in the rat brain. Arch Biochem Biophys 239: 234–241

Mogi M, Harada M, Kikuchi K, Kojima K, Kondo T, Narabayashi H, Rausch D, Riederer P, Jellinger K, Nagatsu T (1988) Homospecific activity (activity per enzyme protein) of tyrosine hydroxylase increases in parkinsonian brain. J Neural Transm 72: 77–82

Nagatsu T (1985) Biopterin cofactor and monoamine-synthesizing monooxygenases. In: Osborne NN (ed) Selected topics in neurochemistry. Pergamon Press, Oxford, pp 325–340

Nagatsu T, Levitt M, Udenfriend S (1964) Tyrosine hydroxylase: the initial step in norepinephrine biosynthesis. J Biol Chem 239: 2910–2917

Nagatsu T, Kato T, Nagatsu I, Kondo Y, Inagaki S, Iizuka R, Narabayashi H (1979) Catecholamine-related enzymes in the brain of patients with parkinsonism and Wilson's disease. In: Poirier LJP, Sourkes TL, Bédard PJ (eds) Advances in neurology. Raven Press, New York, pp 283–242

Nagatsu T, Horikoshi T, Sawada M, Nagatsu I, Kondo T, Iizuka R, Narabayashi H (1986) Biosynthesis of tetrahydrobiopterin in parkinsonian human brain. In: Yahr MD, Bergmann KJ (eds) Advances in neurology. Raven Press, New York, pp 223–226

Nagatsu T, Nakahara D, Kobayashi K, Morita S, Sawada H, Mizuguchi T, Kiuchi K (1994) Peripherally administered (6R)-tetrahydrobiopterin increases in vivo tyrosine hydroxylase activity in the striatum measured by microdialysis both in normal mice and in transgenic mice carrying human tyrosine hydroxylase. Neurosci Lett 182: 44–46

Nichol CA, Smith GK, Duch DS (1985) Biosynthesis and metabolism of tetrahydrobiopterin and molybdopterin. Ann Rev Biochem 54: 729–764

Niederwieser A, Blau N, Wang M, Joller P, Atares M, Cardesa-Garcia J (1984) GTP cyclohydrolase I deficiency, a new enzyme defect causing hyperphenylalaninemia with neopterin, biopterin, dopamine, and serotonin dificiencies and muscular hypotonia. Eur J Pediatr 141: 208–214

Nomura T, Ichinose H, Sumi-Ichinose C, Nomura H, Hagino Y, Fujita K, Nagatsu T (1993) Cloning and sequencing of cDNA encoding mouse GTP cyclohydrolase I. Biochem Biophys Res Commun 191: 523–527

Nygaard TG, Marsden CD, Duvoisin RC (1989) Dopa-responsive dystonia. Neurology 50: 377–384

Nygaard TG, Marsden CD, Fahn S (1991) Dopa-responsive dystonia: long-term treatment response and prognosis. Neurology 41: 174–181

Nygaard TG, Wilhelmsen KC, Risch NJ, Brown DL, Trugman JM, Gilliam TC, Fahn S, Weeks DE (1993) Linkage mapping of dopa-responsive dystonia (DRD) to chromosome 14q. Nature Genet 5: 386–391

Segawa M, Ohmi K, Itoh S, Aoyama M, Hayakawa H (1971) Childhood basal ganglia disease with remarkable response to L-dopa, hereditary basal ganglia disease with marked diurnal fluctuation. Shinryo (Tokyo) 24: 667–672

Segawa M, Nomura Y, Kase M (1986) Diurnally fluctuating hereditary progressive dystonia. In: Vinken PJ, Bruyn GW, Klawans HL (eds) Handbook of clinical neurology. Elsevier, New York, pp 529–539

Takahashi H, Lovine RA, Galloway MP, Snow BJ, Calne DB, Nygaard TG (1994) Biochemical and fluorodopa positron emission tomographic findings in an asymptomatic carrier of the gene for dopa-responsive dystonia. Ann Neurol 35: 354–356

Togari A, Ichinose H, Matsumoto S, Fujita K, Nagatsu T (1992) Multiple mRNA forms of human GTP cyclohydrolase I. Biochem Biophys Res Commun 187: 359–365

Yokochi M (1993) Clinical pathological identification of juvenile parkinsonism in reference to dopa-responsivedisorders. In: Segawa H (ed) Hereditary progressive dystonia. Parthenon, Carnforth, pp 37–48

Authors' address: Prof. T. Nagatsu, Institute for Comprehensive Medical Science, School of Medicine, Fujita Health University, Toyoake, Aichi 470-11, Japan.

MRI as a method to reveal in-vivo pathology in MS

D. W. Paty

Division of Neurology, The University of British Columbia, Vancouver, Canada

Summary. Multiple sclerosis is primarily a disease of myelin. The degeneration of neurons and axons in MS has not been considered to be a major factor until recently. Magnetic resonance (MR) techniques have helped a great deal in describing the evolution of MS lesions over time. Some pathological and MR studies have shown that in the chronic stage of MS, axonal loss occurs. Further studies using MR techniques should help to understand this phenomenon as it occurs in vivo.

Magnetic resonance imaging (MRI) has revolutionized the approach to diagnosis in multiple sclerosis (MS) and has also been established as an objective measure of following disease activity and natural history studies (Paty, 1987; Miller et al., 1988) and in clinical trials (Paty and Li, 1993; Miller et al., 1993). In addition to these major contributions, MR techniques now have the promise of revealing the actual tissue characteristics of MS in vivo pathology as it evolves. Table 1 lists some of the MR techniques that have the promise of identifying and measuring specific pathologies.

In the early 1980's, pathological correlation studies (Stewart et al., 1984; Stewart, 1989) showed that MRI could identify the chronic lesions of MS with precision. Additional MRI pathological correlations studies Nesbit and colleagues (1991) have studied pathological and MRI correlations and found that enhancement in life correlated with pathological inflammation. Barnes et al. (1991) showed that chronic plaques seen on MRI were pathologically heterogeneous. Most lesions (87%) had an expanded extracelluar space (referred to as "open"). They also concluded that axonal loss increased with the age of the lesions. Newcombe et al. (1991) felt that MRI over estimated the extent of the pathology due to the edema and other forms of increased water content without specific pathology.

The acute evolving pathology as revealed by carefully repositioned serial MRI studies begins with breakdown in the blood brain barrier (BBB). Disruption in the BBB is the first identifiable change to be detected (Kermode et al., 1990) (see Fig. 1). Enhancement usually lasts less than 4 weeks and is sensitive to steroids and other anti-inflammatory treatments. The BBB disruption allows leakage of immune competent cells, macrophages and other inflammatory components into brain substance to produce a reversible inflammatory

Table 1. Evolution of the MS lesion as detected by MRI

Sequence #	Pathological feature	Mechanisms	MRI techniques and applications	Therapeutic strategy
1	Breakdown in the BBB	IFNγ and inflammatory cells activate endothelial cells and cause loosening of tight junctions	Gadolinium enhancement	a) Steroids b) IFNβ or α to block IFNγ effects
2	Inflammation	Inflammatory cells & increased water content w/lymphokine release, IFNγ TNFα and others	T2 imaging; increasing water content causes increasing size lesion	Immunosuppresion IFNγ or α to block gamma effects. Blockade of TNF or other lymphokines
3	Demyelination and remyelination	Inflammation w/macrophage stripping of myelin lamellae, some remyelination occurs form preserved oligodendrocytes	a) loss of normal T2 myelin associated signal b) Appearance of myelin breakdown products (lipid) on MRS	Block macrophages by immunosuppression, possibly other mechanisms ? metabolic poisons
4	Gliosis	Astrocytepoliferation, active metabolism, Inositol, ? phosphate	a) T1 signal b) Changes on phosohate MRS c) Inositol changes on proton MRS as a maker of metabolism	Block astrocyte trophic effects
5	Axonal loss	a) Possible by-stander injury b) Gradual loss due to neuro-toxicity factors in an injured neuron	MRS-loss of NAA peak	Neuroprotection: NMDA blockade, anti oxidants block excitatory neuro-transmitter effects

IFN Interferon, *TNF* tumour necrosis factor, *NAA* N-acetyl aspartate

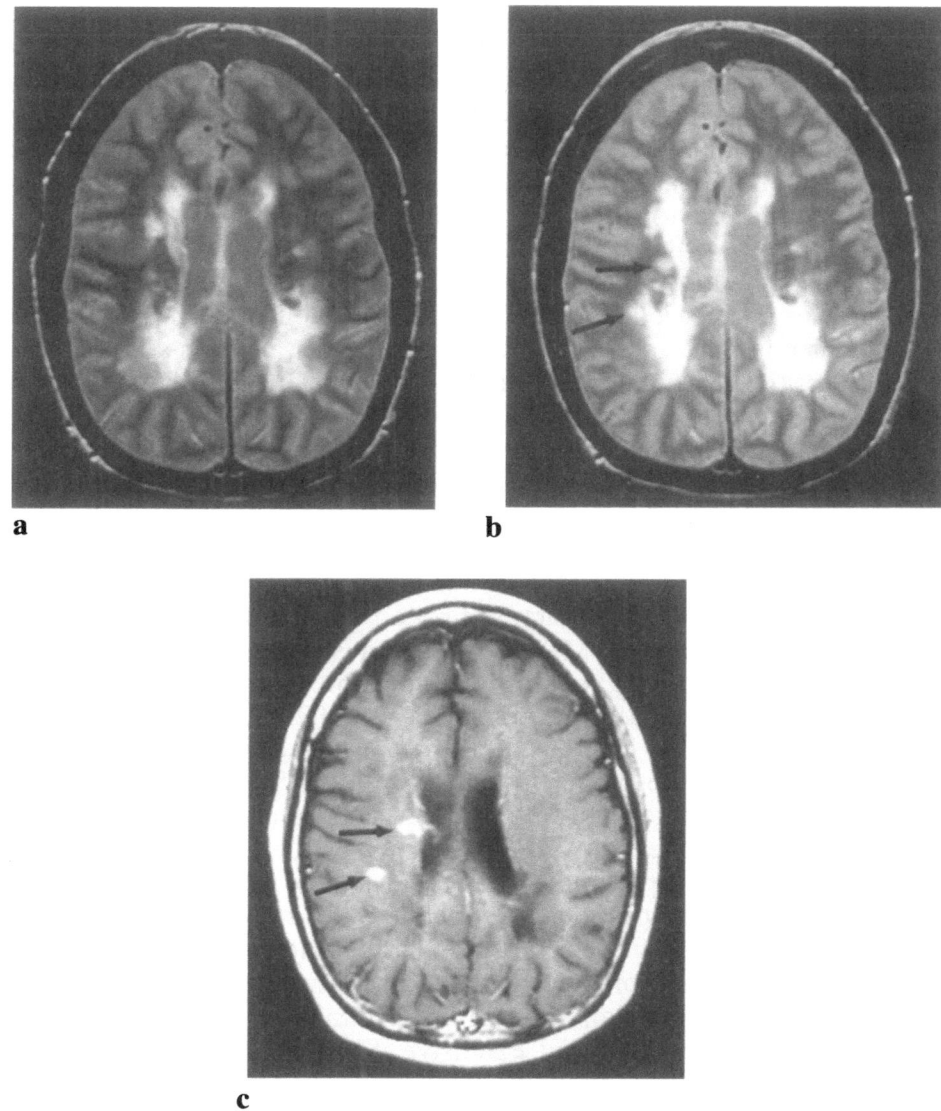

Fig. 1. These are MRI scans one month apart showing two new lesions (**b**) that also enhance (**c**). Note the changes (arrows) in the proton density scans from time (**a**) to time (**b**). C is the post gadolinium T1 scan taken at the same time as proton density scan 1b

lesion that can be seen on the proton density or T2 weighted MRI scan (Katz et al., 1993; Barnes et al., 1988). Many of these acute lesions are also seen on the unenhanced T1 image as well.

After the appearance of a bright lesion on the proton density or T2 image, the lesion will then enlarge to a variable extent. After a period of enlargement the lesion will become smaller (Willoughby, 1989; Koopmans et al., 1989a). At some point that same lesion will reactivate usually by re-enhancing and re-enlarging (Koopmans et al., 1989b; Miller et al., 1988). This cycle of breakdown in the BBB and inflammation is probably repeated several times before significant demyelination occurs (Prineas et al., 1993).

When the lesion is demyelinated it is probably seen as a permanent lesion on the MRI. Further activation of such a lesion can be as recurrent enhancement or enlargement or both. In many instances, the enhancement that occurs is at the centre of the margin of the lesion or as a round enlarging area budding off the side of the previously stable lesion (McDonald et al., 1992). As active demyelination occurs, there are two MRI features that may also change. First, with the loss of the normal lamellar structure of myelin, the water trapped in those lamellae is released into the surrounding neuropile. MacKay and his colleagues (1991) have found that by T2 relaxation analysis a minor component (13%) of the T2 signal seen only at short echo times (< 20 ms) in normal appearing white matter is lost. They propose that signal to be a myelin marker coming from the myelin associated water. Therefore, in demyelination one would expect to see loss of that particular signal. In fact, MacKay's experience is that some, but not all, MS lesions seen on MRI have lost that component of the T2 signal. In addition, during active demyelination, probably by macrophage action (Prineas et al., 1993), one would expect to see the tightly bound lipids of the normal lamellar structure broken down in to neutral lipid (Tourtellotte, 1992). As expected, MR spectroscopy studies have also shown that lipid-like resonances can be seen in the evolution of some newly formed MS lesions (Wolinsky et al., 1990; Naranya et al., 1991; Grossman et al., 1992; Koopmans et al., 1993). Minderhoud and his colleagues (1992) have used phosphorous magnetic resonance spectroscopy (MRS) to show that in certain MS lesions the phosphate spectrum changes suggest active metabolism going on in the most chronic lesions.

The next element of pathology that occurs in the MS lesion is gliosis. The Queen Square group (Barnes et al., 1988) and Stewart (1989) have both shown that the T1 signal may be the most sensitive MR measurement to detect gliosis. In fact, in one pathological correlation study, a case was found in which there was extensive gliosis and partial "remyelination" present in which the post-mortem MR image showed extensive T1 and T2 abnormality. At the same time there were no grossly visible demyelinated lesions seen (Stewart, 1989). Recent MRS studies have also revealed Inositol to be elevated in MS lesions (Zhu et al., 1992) and reversibly elevated in normal appearing white matter (Wolinsky, personal communication). Inositol may have an effect on astrocyte proliferation, and could possibly be used as a metabolic marker for active gliosis.

The final and most irreversible component to the MS lesion is axonal loss. The most characteristic pathological change in MS is demyelination with preservation of the axones (Lumsden, 1951). The early component of the pathology is clearly inflammatory (Prineas, 1993). Remyelination can occur (Ludwin, 1988) but does not likely play a large part in the recovery from most acute MS lesions (Moore et al., 1985). However, when and to what extent axonal loss occurs in MS lesions is poorly understood. Axonal stains in most MS lesions show relative preservation of axones. However, in some lesions the axonal loss is severe.

Axonal loss may be able to be detected using MR by measuring the loss of N-acetyl aspartate (NAA) peak on proton spectroscopy. NAA is a compound found in the CNS only in neurons (Birken and Oldendorf, 1989). It can be detected by MRS and quantitated by using new techniques (Ross et al., 1991). A number of investigators have found NAA to be low in some but not all MS lesions (Arnold et al., 1991; Matthews et al., 1991; Koopmans et al., 1993; Grossmann et al., 1992). It may be a reliable marker for axonal loss. In fact, Arnold and his colleagues (1994) have proposed a global measure of brain NAA as a measure of the total amount of neuronal damage. It may be that axonal loss occurs at the time of acute inflammation. However, it is equally likely that the axonal loss occurs due to gradually occurring death in injured and therefore vulnerable neurones. The mechanism of the neuronal death could be excitatory toxicity, free radical injury, etc. Axonal loss may also be responsible for the chronic irreversible neurological deficit in MS (Fog and Linnemann, 1970; Paty, 1987). At this time it is not known to what degree the chronic neurological impairment in MS is due to axonal loss. However, as a component of the pathology it is obviously the one most likely to be associated with a permanent deficit. For example, it is well understood that demyelination is compatible with preserved axonal conduction (Waxman, 1988) and preserved function. All neurologists have had the experience of seeing patients, who, following optic neuritis recover to normal or near normal visual function in spite of persisting severe optic atrophy and severely prolonged visual evoked potential latencies.

In summary, MR techniques hold the promise of revealing with some accuracy, the actual tissue changes that occur as the pathology of MS evolves. Such knowledge will help in planning strategies for future therapies in MS. For example, should the therapy of MS include a neuro-protection strategy (see Table 1), in order to slow down or prevent axonal loss?

References

Arnold DL, Matthews PM, Francis G, Antel J, et al (1991) Proton MR spectroscopic imaging (SI) for metabolic characterization of the brain plaques of demyelinating disease. Neurology 41: 144

Arnold DL, Riess GT, Matthews PM, Francis GS, Collins DL, Wolfson C, Antel JP (1994) Use of proton magnetic resonance spectroscopy for monitoring disease progression in multiple sclerosis. Ann Neurol 36: 76–82

Barnes D, McDonald WI, Tofts PS, Johnson G, Landon DN (1986) Magnetic resonance imaging of experimental cerebral oedema. J Neurol Neurosurg Psychiatry 49: 1341–1347

Barnes D, McDonald WI, Landon DN, Johnson G (1988) The characterization of experimental gliosis by quantitative nuclear magnetic resonance imaging. Brain 111: 83–94

Barnes D, Munro PMG, Youl BD, Prineas JW, McDonald WI (1991) The longstanding MS lesion. Brain 114: 1271–1280

Birken DL, Oldendorf WH (1989) N-acetyl-L-aspartic acid: a literature review of a compound prominent in H-NMR spectroscopic studies of brain. Neurosci Biobehav Rev 13: 23–31

Fog T, Linnemann F (1970) The course of multiple sclerosis in 73 cases with computer-designed curves. Acta Neurol Scand [Suppl] 46: 1–175

Grossman RI, Lenkinski RE, Ramer KN, Gonzalez-Scarano F, Cohen JA (1992) MR proton spectroscopy in multiple sclerosis. AJNR 13: 1535–1543

Katz D, Taubenberger JK, Cannella B, McFarlin DE, Raine CS, McFarland HF (1993) Correlation between magnetic resonance imaging and lesion development in chronic, active multiple sclerosis. Ann Neurol 34: 661–669

Kermode AG, Thompson AJ, Tofts P, MacManus DG, Kendall BE, Kingsley DPE, Moseley IF, Rudge P, McDonald WI (1990) Breakdown of the blood-brain barrier precedes symptoms and other MRI signs of new lesions in multiple sclerosis. Brain 113: 1477–1489

Koopmans RA, Li DKB, Oger JJ, Kastrukoff LF, Jardine C, Costley L, Hall S, Grochowski EW, Paty D W (1989a) Chronic progressive multiple sclerosis: serial magnetic resonance brain imaging over six months. Ann Neurol 26: 248–256

Koopmans RA, Li DKB, Oger JJF, Mayo J, Paty DW (1989b) The lesion of multiple sclerosis: imaging of acute and chronic stages. Neurology 39: 959–963

Koopmans RA, Li DK, Zhu G, Allen PS, Penn A, Paty DW (1993) Magnetic resonance spectroscopy of multiple sclerosis: in-vivo detection of myelin breakdown products [letter]. Lancet 341: 631–632

Ludwin SK (1988) Remyelination in the central nervous system and peripheral nervous system. Adv Neurol, Funct'l Rec In Neurol Dis 47: 215–254

Lumsden CE (1951) Fundamental problems in the pathology of multiple sclerosis and allied demyelinating diseases. BMJ 1: 1035–1043

MacKay AL, Whittall KP, Cover KS, Li DK, Paty DW, et al (1991) In vivo T2 relaxation measurements of brain may provide myelin concentration. Society of Magnetic Resonance in Medicine, 10th Annual Meeting, Aug 1991. Book of Abstracts 1: 917

Matthews PM, Francis G, Antel J, Arnold DL (1991) Proton magnetic resonance spectroscopy for metabolic characterization of plaques in multiple sclerosis. Neurology 41: 1251–1256

McDonald WI, Miller DH, Barnes D (1992) The pathological evolution of multiple sclerosis. Neuropathol Appl Neurobiol 18: 319–334

Miller DH, Rudge P, Johnson G, Kendall BE, Macmanus DG, Moseley IF, Barnes D, McDonald WI (1988) Brain 111: 927–939

Miller DH, Barkhof F, Nauta JJ (1993) Gadolinium enhancement increases the sensitivity of MRI in detecting disease activity in multiple sclerosis. Brain 116: 1077–1094

Minderhoud JM, Mooyaart EL, Kamman RL, et al (1992) In vivo phosphorus magnetic resonance spectroscopy in multiple sclerosis. Arch Neurol 49: 161–165

Moore GRW (1995) Pathology. In: Paty DW, Ebers G (eds) Multiple sclerosis. FA Davis, Philadelphia

Moore GRW, Neumann PE, Suzuki K, Lijtmaer HN, Traugott U, Raine CS (1985) Balo's concentric sclerosis: new observations on lesion development. Ann Neurol 17: 604–611

Narayana PA, Wolinsky JS, Jackson EF, McCarthy M (1991) Magnetic resonance of multiple sclerosis: correlation between proton spectroscopy and Gd-DTPA enhancement. Society of Magnetic Resonance in Medicine, 10th Annual Meeting, Aug 1991. Book of Abstracts 1: 82

Nesbit GM, Forbes GS, Scheithauer BW, Okazaki H, Rodriguez M (1991) Multiple sclerosis: histopathologic and MR and/or CT correlation in 37 cases at biopsy and three cases at autopsy. Radiology 180: 467–474

Newcombe J, Hawkins CP, Henderson CL, Patel HA, Woodroofe MN, Hayes GM, Cuzner ML, MacManus D, du Boulay EP, McDonald WI (1991) Histopathology of multiple sclerosis lesions detected by magnetic resonance imaging in unfixed post-mortem central nervous system tissue. Brain 114: 1013–1023

Paty DW (1987) Multiple sclerosis: assessment of disease progression and effects of treatment. Can J Neurol Sci 14: 518–520

Paty DW, Li DKB (1993) Interferon beta-1b is effective in relapsing-remitting multiple sclerosis. II. MRI analysis results of a multicenter, randomized, double-blind, placebo-controlled trial. UBC MS/MRI Study Group and the IFNB Multiple Sclerosis Study Group. Neurology 43: 662–667

Prineas JW, Barnard RO, Ruesz T, Kwon EE, Sharer L, Cho ES (1993) Multiple sclerosis. Pathology of recurrent lesions. Brain 116: 681–693

Ross BD (1991) Biochemical considerations in H spectroscopy. Glutamate and glutamine; myo-inositol and related metabolites. NMR in Biomedicine 4: 59–63

Stewart WA (1989) Proton spin relaxation as a means of characterizing the pathology of multiple sclerosis. Thesis, The University of British Columbia, Vancouver, Canada

Stewart WA, Hall LD, Berry K, Paty DW (1984) Correlation between NMR scan and brain slice: data in multiple sclerosis. Lancet ii: 412

Tourtellotte WW (1992) Issues in multiple sclerosis. A focused disease oriented research program. Ital J Neurol Sci 13: 47–53

Waxman SG (1988) Advances in neurology. Functional recovery in neurological disease. Adv Neurol 47: 157–184

Willoughby EW, Grochowski E, Li DKB, Oger J, Kastrukoff LF, Paty DW (1989) Serial magnetic resonance scanning in multiple sclerosis: a second prospective study in relapsing patients. Ann Neurol 25: 43–49

Wolinsky JS, Narayana PA, Fenstermacher MJ, et al (1990) Proton magnetic resonance spectroscopy in multiple sclerosis. Neurology 40: 1764–1769

Zhu G, Allen PS, Koopmans R, Li DK, Paty DW (1992) A marked elevation of inositol in MS lesions. SMRM. Book of Abstracts 1: 1948

Author's address: D. W. Paty, MD, FRCPC, Division of Neurology, The University of British Columbia, 222-2775 Heather Street, Vancouver, BC V5Z 3J5, Canada.

Familial amyotrophic lateral sclerosis

T. Siddique[1,2,3], **D. Nijhawan**[1], and **A. Hentati**[1]

[1]Department of Neurology, [2]Department of Cell and Molecular Biology, and
[3]Northwestern Institute of Neuroscience, Northwestern University Medical School,
Chicago, IL, U.S.A.

Summary. Amyotrophic lateral sclerosis is sporadic in ninety percent of cases and familial (FALS) in ten percent. Both forms of FALS whether transmitted as an autosomal dominant (DFALS) or as an autosomal recessive (RFALS) trait is genetically heterogeneous. The locus for one form of RFALS maps to chromosome 2q33. Fifteen percent of DFALS families have mutations in the gene for Cu, Zn superoxide dismutase (SOD1) gene which is coded on chromosome 21. These mutations result in decreased SOD1 activity and shortened half-life of the protein in most instances. Transgenic mice overexpressing mutated SOD1 protein develop an ALS-like disease which suggests that the degeneration of motor neurons in DFALS is caused by the gain of a novel toxic function by mutated SOD1 rather than by the decrease of SOD1 activity. Possible mechanisms of the novel neurotoxic function of mutated SOD1 are discussed.

Introduction

Amyotrophic lateral sclerosis (ALS) is also called motor neuron disease (MND), Lou Gehrig disease, or Charcot disease the latter two after the N.Y. Yankees baseball player and the French neurologist J.L. Charcot. ALS is caused by the degeneration of the large motor neurons of the motor cortex, brain stem, and spinal cord thus involving both the upper and lower motor neuron.

Amyotrophic lateral sclerosis is a progressive neurodegenerative disease in adults that occurs in familial and sporadic forms. Sporadic ALS (SALS) comprises approximately 90% of reported cases (Emery and Holloway, 1982; Siddique et al., 1989). The mean age of onset of symptoms is 56 years and the mean duration of disease is 3 years (Tandan and Bradley, 1985). The cause of SALS is unknown and current treatments are marginally useful in reducing the progression of disease.

Five to ten percent of reported ALS cases are familial (FALS). In the majority of cases, the pattern of inheritance is autosomal dominant (DFALS); although, a few families with an autosomal recessive form have been reported (RFALS).

Autosomal dominant familial amyotrophic lateral sclerosisv (DFALS)

DFALS is clinically indistinguishable from SALS. However, the mean age of onset for DFALS is 46 years, approximately ten years earlier than sporadic patients (Tandan and Bradley, 1985; Siddique, 1991a). The penetrance is age related where 50% of patients develop the disease by age 46 and 90% by age 70 (Siddique, 1991a). There is no evidence for genetic anticipation, that is the age of onset does not decrease in consecutive generations (Applebaum et al., 1992). The mean age of onset is somewhat later in DFALS families not associated with mutations in the Cu, Zn superoxide dismutase gene (SOD1).

Fifteen percent of DFALS families are linked to chromosome 21q21 (Siddique et al., 1991b) (genetic nomenclature: ALS1) and are associated with mutations in Cu, Zn superoxide dismutase (SOD1) (Rosen et al., 1993; Deng et al., 1993). The genetic locus (or loci) for the remaining 85% of DFALS families has not been identified.

Autosomal recessive familial amyotrophic lateral sclerosis (RFALS)

RFALS is rare but has been reported in settings of high consanguinity such as Tunisia (Hentati, 1989; Ben Hamida et al., 1990). RFALS is completely penetrant. The mean age at onset of symptoms in RFALS is 12 years (range 3 to 23 years) and the duration of disease ranges from 15 to 20 years.

Three clinical variants of RFALS are known (Hentati, 1989; Ben Hamida et al., 1990). RFALS type I is characterized by the relative predominance of the lower motor neuron involvement beginning in the early stages of the disease. Atrophy and weakness predominate in the hand and feet muscles whereas the tongue might be affected in later stages and bulbar involvement progresses slowly. The involvement of upper motor neurons is moderate in the early stages and becomes more apparent as the disease progresses. On the other hand, in RFALS type 3 the symptoms of upper motor neuron degeneration predominate. RFALS type 3 patients present with spasticity of limb and facial muscles, spastic and slurred speech which can lead to anarthria and uncontrolled weeping and laughter. Lower motor neuron involvement occurs in later stages of the disease and affect particularly the hands and feet. In RFALS type 2 the symptoms are confined to the lower limbs.

A large inbred Tunisian family with RFALS type 3 (genetic nomenclature: ALS2) has been linked to chromosome 2q33–35 (Hentati et al., 1994). The ALS2 locus was subsequently refined to 2q33 (Hentati A, Deng H-X, Siddique T, unpublished). The genetic analysis of families with RFALS type 1, the more common form of RFALS, did not show linkage to markers on either chromosome 2q33 or chromosome 21q21 demonstrating genetic locus heterogeneity in RFALS (Nijhawan et al., 1995). The locus for RFALS type I has not been identified.

DFALS and cytosolic Cu, Zn superoxide dismutase (SOD1)

Fifteen percent of DFALS families in Northern America have mutations in the gene for the cytosolic Cu, Zn superoxide dismutase.

What is cytosolic Cu, Zn superoxide dismutase (SOD1)?

Superoxide dismutases (SOD) are a group of enzymes that catalyze the conversion of the superoxide anion ($O_2^{\cdot-}$) to hydrogen peroxide and oxygen. SOD provides cellular defense against $O_2^{\cdot-}$ and its toxic derivatives. There are three isoforms of SOD in humans: cytosolic Cu, Zn superoxide dismutase (SOD1, Cu, Zn SOD), mitochondrial Mn superoxide dismutase (SOD2, MnSOD) and extracellular superoxide dismutase (SOD3, ECSOD). Each isozyme is encoded by a different gene on chromosomes 21q21, 6q27, and 4p respectively (Groner et al., 1986). No abnormality in SOD2 (Parboosingh et al., 1995) or SOD3 has been reported in FALS patients.

SOD1 is present in the cytoplasm of most cells including red blood cells and is particularly abundant in neurons (Pardo et al., 1995). Each SOD1 monomer is 153 amino acids and 16 kilodaltons. Each monomer contains one atom each of Cu and Zn. Stable dimers are formed by strong hydrophobic interactions between individual monomers at the interface. $O_2^{\cdot-}$ is guided to the Cu containing active site through a positively charged electrostatic guidance channel (Getzoff et al., 1989). The positive charges are provided by the amino acids Lys 122, Lys 134, and Arg 143. The latter stabilizes the position of $O_2^{\cdot-}$ in relation to the copper atom (Tainer et al., 1989). Twenty one highly conserved amino acids contribute to the electrostatic guidance channel. The channel narrows down in a stepwise fashion from a large shallow depression about 24 Å across to a deeper well about 10 Å wide and finally to an opening less than 4 Å wide just above the Cu atom (Tainer et al., 1989). Access to the active site is limited by size and charge which favors the small, negatively charged $O_2^{\cdot-}$ and excludes molecules with positive charge or larger size. The dismutase reaction proceeds at a rapid rate of $2 \times 10^9\ M^{-1}sec^{-1}$ (Klug et al., 1972). The two step dismutase reaction proceeds as follows:

$$O_2^{\cdot-} + Enz\text{-}Cu^{++} + H^+ \rightarrow O_2 + Enz\text{-}Cu^+ \tag{1}$$

$$O_2^{\cdot-} + Enz\text{-}Cu^+ + H^+ \rightarrow H_2O_2 + Enz\text{-}Cu^{++} \tag{2}$$

$$2\,O_2^{\cdot-} + 2H^+ \rightarrow SOD1 \rightarrow H_2O_2 + O_2 \tag{3}$$

Beside the dismutase activity, SOD1 has a marginal peroxidase activity (Hodgson and Fridovich, 1975). H_2O_2 which is the product of the dismutase reaction inactivates SOD1 (Symonyan and Nalbandyan, 1972) The inactivation of SOD1 by its own H_2O_2 product and peroxidase activity would proceed as follows:

$$H_2O_2 + Enz\text{-}Cu^{++} \leftrightarrow O_2^{\cdot-} + Enz\text{-}Cu^+ + 2H^+ \tag{4}$$

$$Enz\text{-}Cu^+ + H_2O_2 \leftrightarrow Enz\text{-}Cu^{++} - OH^{\cdot} + OH^- \tag{5}$$

$$Enz\text{-}Cu^{++} - OH^{\cdot} + ImHis63 \rightarrow Enz\text{-}Cu^{++} + ImHis63^{\cdot} + H_2O \tag{6}$$

Enz is the enzyme and ImH63 is the imidazole moiety of the histidine residue at position 63 of SOD1 polypeptide. In vivo, the peroxidase activity would be significant in the presence of an excess production of H_2O_2 which may result from a burst of $O_2^{\cdot-}$ molecules (Yim et al., 1993).

SOD1 mutations in DFALS

A single functional gene, 12 kb long, with five exons and four introns codes for SOD1 in humans and maps to chromosome 21q21 (Groner et al., 1986). There are at least four non-functional SOD1 pseudogenes on other chromosomes (Groner et al., 1986; Deng et al., unpublished).

Forty seven mutations in SOD1 have been identified in FALS patients (Table 1). All of the families previously reported to be linked to chromosome 21q markers showed a mutation in SOD1 and the mutation segregated with the disease in each of these families (Deng et al., 1993). Though mutations have been found in four of the five exons of SOD1 in FALS, none has been reported in exon 3. Mutations in exon 3 may be benign and rare, and thus not

Table 1. Roster of SOD1 mutations in FALS. Homozygotes with this mutation develop symptoms

Amino acid subtitution	Mutation symbol	Nucleotide substitution	Reference
Exon 1			
Ala[4]→Val	A4V[†]	GCC→GTC	Deng et al. (1993)
Ala[4]→Thr	A4T[†]	GCC→ACC	Nakano et al. (1994)
Val[7]→Glu	V7E	GTG→GAG	Hirano et al. (1994)
Leu[8]→Gln	L8Q	CTG→CAG	Bereznai et al. (1995)
Val[14]→Met	V14M[†]	GTG→ATG	Deng et al. (1995)
Gly[16]→Ser	G16S	GGC→AGC	Kawamata et al. (1995)
Glu[21]→Lys	E21K	GAG→AAG	Jones et al. (1994)
Glu[21]→Gly	E21G	GAG →GGG	Moulard et al. (1995)
Exon 2			
Gly[37]→Arg	G37R[†]	GGA→AGA	Rosen et al. (1993)
Leu[38]→Val	L38V	CTG→GTG	Rosen et al. (1993)
Gly[41]→Ser	G41S	GGC→AGC	Rosen et al. (1993)
Gly[41]→Asp	G41D[†]	GGC→GAC	Rosen et al. (1993)
His[43]→Arg	H43R[†]	CAT→CGT	Rosen et al. (1993)
His[46]→Arg	H46R[†]	CAT→CGT	Aoki et al. (1995)
His[48]→Gln	H48Q	CAT→CAG	De Belleroche et al. (1995)
Exon 3	NONE		
Exon 4			
Leu[84]→Val	L84V[†]	TTG→GTG	Aoki et al. (1994), Deng et al. (1995)
Gly[85]→Arg	G85R[†]	GGC→CGC	Rosen et al. (1993)
Asp[90]→Ala	D90A[†]	GAC→GCC	Andersen et al. (1995), Själander et al. (1995)

continued

Amino acid subtitution	Mutation symbol	Nucleotide substitution	Reference
Gly[93]→Ala	G93A[†]	GGT →GCT	Rosen et al. (1993)
Gly[93]→Cys	G93C	GGT→TGT	Rosen et al. (1993)
Gly[93]→Arg	G93R	GGT→CGT	Elshafey et al. (1994)
Gly[93]→Asp	G93D	GGT→GAT	Esteban et al. (1994)
Gly[93]→Ser	G93S	GGT→AGT	Kawamata et al. (1994)
Gly[93]→Val	G93V	GGT→GTT	De Belleroche et al. (1995)
Glu[100]→Gly	E100G[†]	GAA→GGA	Rosen et al. (1993)
Glu[100]→Lys	E100K[†]	GAA→AAA	Deng et al. (1995, unpublished)
Asp[101]→Asn	D101N	GAT→AAT	Jones et al. (1994)
Asp[101]→Gly	D101G	GAT→GGT	De Belleroche et al. (1995)
Ile[104]→Phe	I104F	ATC→TTC	Ikeda et al. (1995)
Leu[106]→Val	L106V[†]	CTC→GTC	Rosen et al. (1993)
Gly[108]→Val	G108V	GGA→GTA	De Belleroche et al. (1995)
Asp[109]→Asn	D109N	GAC→AAC	De Belleroche et al. (1995)
Ile[112]→Thr	I112T	ATC→ACC	Esteban et al. (1994)
Ile[113]→Thr	I113T[†]	ATT→ACT	Rosen et al. (1993)
Arg[115]→Gly	R115G	CGC→GGC	Kostrzewa et al. (1994)

Exon 5

Asp[124]→Val	D124V	GAT→GTT	Laing N. (1995, unpublished)
Asp[125]→His	D125H	GAC→CAC	De Belleroche et al. (1995)
Leu[126]→stop	L126stop[†]	TTG→TAG	Deng et al. (1995, unpublished)
Ser[134]→Asn	S134N	AGT→AAT	Watanabe et al. (1995)
Asn[139]→Lys	N139K	AAC→AAA	Pramatarova et al. (1995)
Leu[144]→Phe	L144F[†]	TTG→TTC	Deng et al. (1993)
Leu[144]→Ser	L144S	TTG→TCG	Sapp et al. (1994)
Ala[145]→Thr	A145T	GCT→ACT	Sapp et al. (1994)
Cys[146]→Arg	C146R	TGT→CGT	Kawamata et al. (1995)
Val[148]→Gly	V148G[†]	GTA→GGA	Deng et al. (1993)
Val[148]→Ile	V148I	GTA→ATA	Ikeda et al. (1995)
Ile[149]→Thr	I149T	ATT→ACT	Pramatarova et al. (1995)

Deletion

Leu	GACT*T*GGC[126] GAC GGC	
(with stop codon at Lys[130],23 aa deleted)		Nakashima et al. (1995)

Splice Junction

+Phe-Leu-Glu	T→G 10 bp before exon 5	Sapp et al. (1994)
+Phe-Phe-Thr-Gly-Pro-stop	A→G 11 bp betore exon 5	Deng et al. (1995, unpublished)
(all exon 5 deleted)		

[†] Identified at Northwestern University Medical School

ascertained, or lethal in utero and therefore not seen. Alternatively, mutations may not have been detected in exon 3 of the SOD1 of FALS patients because its integrity might be necessary for a toxic effect of mutant SOD1.

Except for two splice junction mutations, a dinucleotide deletion and a premature stop codon all other mutations reported in SOD1 in FALS are missense (Table 1). These exceptions affect the product of exon 5 and are: 1) Lys126Stop (Deng H-X, Kaplan J, Mitsumoto H, Caliendo J, Hung W-Y, Siddique T, unpublished); 2) The TT deletion in codon 126 which shifts the frame and also creates a stop codon, reducing the protein to 130 amino acids including five novel amino acids (Nakashima et al., 1995); 3) One of the splice juction mutation adds three novel codons to the 5' end of exon 5 (Sapp et al., 1994); 4) The other splice junction mutation adds 5 novel codons to the S' end of exon 4 and deletes all of exon 5 (Deng H-X, Caliendo J, Hung W-Y, Siddique T, unpublished).

The SOD1 protein with FALS mutations is deficient

When the initial mutations observed in FALS patients were mapped onto the crystallographic structure of SOD1 (Deng et al., 1993, 1995), they clustered in loops joining the β barrel. The mutated side chains cluster near the two Greek key connections closing off the ends of the β barrel, in the dimeric interface, and at the base of the active side loop. These side chains, which are structurally conserved in the wild type SOD1 structure between species appears to be critical for the structural integrity of the dimeric enzyme. Some of the side chain mutations such as Ala4Val, Leu38Val, Leu106Val and Val148Ala would also be expected to destabilize the subunit fold or the dimer contact. His46Arg, Leu84Val and Asp90Ala occur outside the β strand and the loop forming the β barrel (Deng et al., 1995; Själander et al., 1995). The mutation His46Arg directly affects the ligation of the Cu atom in the protein, since His46 is one of four histidines that bind to Cu in SOD1 (Carri et al., 1994). The mutation Leu84Val occurs at the end the Zn binding loop where it joins the β strand, while Asp90Ala is expected to cause little changes in the structure of the protein (Själander et al., 1995). Most of the mutations in SOD1 were predicted to destabilize the protein and result in a less active rather than an inactive SOD1 protein. Asp90Ala is expected to be relatively stable without appreciable decrease in dismutase activity.

The analysis of the SOD1 enzyme activity supports the structural prediction of loss of activity. The cytosolic superoxide dismutase activity assayed in red blood cells (RBC) (Deng et al., 1993; Tsuda et al., 1994; Bowling et al., 1995), in transformed lymphoblastoid cells (Borchelt et al., 1994), and in brain tissues (Bowling et al., 1993) from FALS patients with mutations in SOD1 gene, was found to be between 30% to 70% of normal controls (Fig. 1). As predicted, FALS patients homozygous with the D9OA mutation had near normal SOD1 activity (Själander et al., 1995). The specific activity of the mutated SOD1 polypeptide expressed in COS-1 cells (Borchelt et al., 1994) and in yeast (Rabizadeh et al., 1995), was between 0% to 70% of the normal

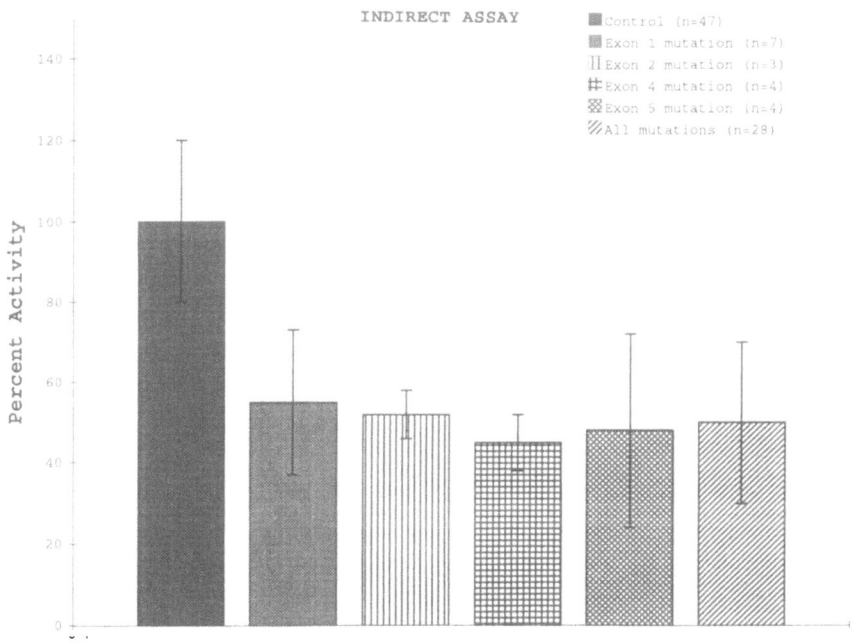

Fig. 1. Cytosolic Cu/Zn superoxide dismutase activity in red blood cell from FALS patients and controls (with permission of Emery and Rimoin)

polypeptide. The random dimerization of mutated and wild type monomers (Deng et al., 1993) would predict that 25% of the SOD1 dimers have both normal monomers, 25% of the dimers have both mutated monomers and 50% of the dimers have one normal and one mutated monomer. The decrease in SOD1 activity in FALS patients may result from instability of mutant-wild type heterodimers and mutant homodimers.

The reduction in SOD1 activity in RBCs of FALS patients is partly due to a reduction in the amount of SOD1 protein. SOD1 in RBCs of FALS patients is 35% to 75% that of controls (Siddique and Hentati, 1996a, b) (Fig. 2). No decrease in SOD1 protein was observed in non chromosome 21 FALS patients. The turnover of SOD1 carrying FALS mutations is significantly increased in all cases. The half life of mutant SOD1 protein in a COS-1 cell culture system is between 7.5 and 20 hours while the half life of the normal SOD1 protein is about 30 hours (Borchelt et al., 1994).

Mutation Asp90Ala which maintains a near normal dimutase activity, has the following peculiarities. First, about 2.5% of the population in the Tornadelan Valley of Sweden and the adjoining region of Finland are heterozygous carriers of the Asp90Ala mutation (Själander et al., 1995). Second, the homozygous carriers of this mutation have a significantly increased risk of developing ALS compared to Asp90Ala heterozygotes and wildtype homozygous individuals (Andersen et al., 1995). This makes the FALS phenotype caused by Asp90Ala mutation an autosomal recessive rather than autosomal dominant trait. Third, the penetrance of FALS in individuals homozygous for the Asp90Ala mutation was estimated at 38% (Andersen et al., 1995). Fourth, the

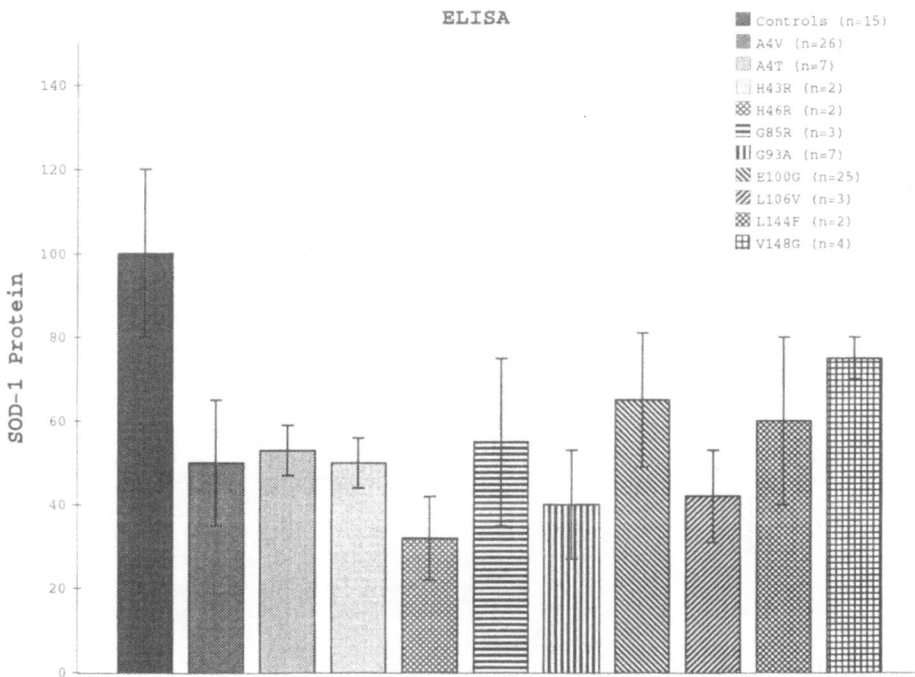

Fig. 2. SOD1 protein levels in red blood cells from FALS patients with SOD1 mutations, non chromosome 21 FALS patients and controls. The amount is determined by ELISA using polyclonal antibody against human SOD1 (with permission of Emery and Rimoin)

predicted effect of Asp90Ala on the structure would be minor which is consistent with the near normal cytosolic dismutase activity in both homozygous and heterozygous individuals with this mutation (Själander et al., 1995). Thus, Asp90Ala is an enigmatic mutation and does not conform to theories suggesting the loss of function of SOD1 results in motor neuron degeneration in FALS patients. However, it provides a model of recessive disease with reduced penetrance, which could fit the presentation of sporadic ALS.

Relationship of SOD1 mutants to the FALS phenotype

We analyzed the effect of SOD1 mutations on the age at onset of symptoms and duration of the disease in FALS patients. None of the mutations specifically influenced the age at onset of symptoms and 50% of FALS patient developed symptoms by the age of 46 years. There was a correlation between some mutations and the progression of the disease (Juneja et al., 1996). The mean duration of the disease in FALS patients with Ala4Val mutation was 1.2 years which was significantly shorter than the mean duration of 4.7 years observed in patients with Glu100Gly mutation (Juneja et al., 1996). The mutation Ala4Val was associated with the most rapid progression of FALS and the mutation His46Arg was associated with the slowest progression of FALS, with a duration of the disease of about 20 years (Aoki et al., 1994; Juneja et al., 1996). No correlation between the decrease of SOD1 activity assayed in RBC from FALS patients and the duration of the disease was observed.

FALS patients with mutations in SOD1 are about 15% of all FALS patients in North America. Mutations in SOD1 observed in a very small proportion of apparently SALS patients are due to low penetrance of some of SOD mutations especially, Asp90Ala and Ile113The. No mutation in SOD1 or decrease of the activity of the protein was observed in red cells of FALS patients without mutations in SOD1 or in SALS patients (Robbertecht et al., 1994; Puymirat et al., 1994; O'Reilly et al., 1995; Iqbal Z, Qureshi I, Hung W-Y, Ahmed A, Siddique T, unpublished). Bergeron et al. (1994) found a 42% increase of SOD1 mRNA levels in motor neurons in SALS patients, but this finding has not been confirmed.

A transgenic mice model for FALS

Transgenic mice overexpressing mutated human SOD1 protein is the most proximate model for familial ALS. The cloned human gene for wildtype and mutant SOD1 were seperately introduced into pronuclei of fertilized embryos and implanted in pseudopregnant female mice (Gurncy et al., 1994). Several of the offspring of these mice overexpress the human SOD1 gene. The human SOD1 gene carrying the mutations Ala4Val, Gly93Ala (Gurney et al., 1994), Gly37Arg (Wong et al., 1995), and the mouse SOD1 gene carrying Gly86Arg mutation (Ripps et al., 1995) as well as the wildtype human SOD1 gene have been analyzed in this fashion. The transgenic mice overexpressing the human SOD1 gene carrying mutation Gly93Ala, Gly37Arg and the mouse SOD1 mutation Gly86Arg develop weakness and paralysis. These symptoms which are similar to those of ALS patients progress rapidly to cause death at variable rates in different transgenic mice lines.

The transgenic mice bearing 18 or 36 copies of mutated human Gly93Ala SOD1 gene and a 3 to 5 fold increase of the SOD1 activity relative to the non-transgenic mice, develop the disease earlier and die faster than mice overexpressing 5 to 10 copies of the mutated Gly93Ala gene (Dal Canto and Gurney, 1995). Mice with 2 copies of mutated Gly93Ala SOD1 develop symptoms much later in life and the penetrance is reduced. The transgenic mice expressing the normal human SOD1 with seven copies of human wildtype SOD1 gene and a 3 fold increase of SOD1 activity, do not develop clinical symptoms at over two years of age. Recently, transgenic mice with the human Ala4Val mutation were noted to develop paralysis and exhibit pathological changes of motor neuron degeneration (Siddique T, Dal Canto M, unpublished) at age greater than 700 days. Thus, the burden of mutant SOD1 protein inversely correlates with the age at onset and directly correlates with the severity of pathology.

The pathological study of affected mice confirmed that the paralysis is due to degeneration of motor neurons in the spinal cord (Gurney et al., 1994; Dal Canto and Gurney, 1994). Light and electron microscopic analysis show a vacuolar degeneration of motor neurons. Vacuoles arising from the rough endoplasmic reticulum and the mitochondria were observed in the early stages of the disease in mice bearing large copy number of mutant SOD1 (Dal Canto and Gurney, 1995). The vacuoles displace the normal structures of the motor neurons and the Golgi apparatus is fragmented. At a later stage in the disease, the mitochondria degenerate and a fibrillary deposit is observed in axons of degenerating neurons. At this stage a substantial reduction in the number of motor neurons occurs. The mice overexpressing lower levels of mutated SOD1 show a fibrillary degeneration of the anterior horn with fibrillary deposits into the motor neurons resembling the pathological features of human ALS. Structures such as the olfactory bulb, thalamus, cerebellum, are affected in transgenic mice expressing high levels of mutated SOD1 protein in the late stages of the disease (Wong et al., 1995; Dal Canto and Gurney, 1995). Some minor vacuolar changes are observed in the very proximal portion of axons of motor neurons in mice overexpressing the wildtype human SOD1 gene (Dal Canto and Gurney, 1995).

Pathogenesis of DFALS with mutations in SOD1

When mutations in SOD1 were first discovered in FALS families, it was asked how may a mutation in an enzyme produce a dominant trait? Generally mutations in enzyme result in recessive disorders because most enzyme function is redundant. Whether a decrease of SOD1 dismutase activity contributes to the degeneration of motor neurons or if motor neurons degenerate solely by the acquisition of a new and toxic function by mutated SOD1 were then considered (Roos et al., 1995). The loss of normal dismutase function of SOD1 due to haploinsufficiency or a dominant negative effect as the cause of neurodegeneration was one hypothesis. In vitro the mutated SOD1 monomer was not found to have a dominant negative effect on the normal monomer (Borchelt et al., 1995). However a contrary result was shown in mutated SOD1 from Drosophilia (Phillips et al., 1995), as such a dominant negative effect in all SOD1 mutations has not been ruled out. The loss of dismutase function hypothesis assumes that motor neurons degenerate by the direct affect of the free radical O_2^- not properly scavenged by mutated SOD1.

Transgenic mice overexpressing mutant human SOD1 provide direct evidence of the involvement of mutant SOD1 in the pathogenesis of FALS and challenge the hypothesis that FALS is caused solely by loss of SOD1 activity postulated (Borchelt et al., 1994; Gurney et al., 1994; Roos et al., 1995). Loss of function would not explain the development of disease in transgenic mice expressing high levels of mutated SOD1 enzyme and not in mice expressing comparable levels of normal SOD1 protein. Also, the identification of mutations in SOD1 retaining normal dismutase activity, such as Asp90Ala supported the possibility that the adverse effect of mutated SOD1 protein was due to a gain of a new toxic function. The novel neurotoxic function of SOD1 has not been identified yet and the following hypothesis are being investigated:

1. An increase of the normal fuctions of dismutase or peroxidase. An increse of dismutase function is unlikely due to the relatively rapid rate of the dismutase reaction. An early report (Borchelt et al., 1994) suggesting such a gain in one of the mutants was probably an artifact of measurement; artifact of SOD1 activity measurement is not uncommon. An in vitro gain of peroxidase activity has been recently reported for the mutants Ala4Val and Gly93Ala (Wiedau-Pazos et al., 1996). An increase of the peroxidase activity of mutated SOD1 in vivo need to be investigated.

2. A gain of a novel enzymatic function due to an exposure of the active site of mutated SOD1. Large molecules that are not accessible to the active site of wildtype SOD1 would be accessible to the mutant SOD1 active site. The possible interaction of peroxynitrite (ONOO$^-$) with the Cu^{++} at the SOD1 active site resulting in the production of the nitronium species, NO$_2$, that may cause a nitration of tyrosine residues in protein was proposed as the novel neurotoxic function of mutated SOD1 (Beckman et al., 1993). Reduced dismutase activity in mutant SOD1 might result in increased levels of O$_2^-$ which reacts with nitric oxide (NO) at twice the rate it reacts with Cu^{++} to produce ONOO$^-$ (Koppenol et al., 1992). But nitration of protein tyrosine residues in FALS has not been demonstrated. Preliminary data from transgenic mice brain tissue using monoclonal antityrosine antibody does not suggest increased nitration, but additional experiments with increased sensitivity are in progress.

3. A toxin-like function specifically deleterious to motor neurons. This hypothesis was supported by (i) the conservation of exon 3 in SOD1 as no mutation were reported in this exon, and (ii) the possible exposure of the exon 3 motif in mutated SOD1.

Conclusion

Genetic linkage analysis has proved to be a powerful tool in understanding the causation of ALS and identification of mutations in SOD1 in FALS has opened a new field of inquiry into the pathogenesis of ALS. Transgenic mouse models overexpressing mutant SOD1 can now be used to rapidly screen therapeutic agents that may affect neurodegeneration. As the mechanism of degeneration of motor neurons that is caused by mutations in SOD1 still unknown, a more coherent picture of pathogenesis and rational therapy may emerge as additional genes associated with other forms of ALS are identified.

References

Andersen PM, Nilsson P, Ala-Hurula V, et al (1995) Amyotrophic lateral sclerosis associated with homozygosity for an Asp90 Ala mutation in Cu, Zn-superoxide dismutase. Nature Genet 10: 61–66

Aoki M, Ogasawara M, Matsubara Y, et al (1994) Familial amyotrophic lateral sclerosis (ALS) in Japan associated with H46R mutation in Cu/Zn superoxide dismutase gene: a possible new subtype of familial ALS. J Neurol Sci 126: 77–83

Applebaum JS, Roos RP, Salazar-Grueso EF, et al (1992) Intrafamilial heterogeneity in hereditary motor neuron disease. Neurology 42: 1488–1492

Beckman JS, Carson M, Smith CD, Koppenol WH (1993) ALS, SOD and peroxynitrite. Nature 364: 584

Ben Hamida M, Hentati F, Ben Hamida C (1990) Hereditary motor system diseases (chronic juvenile amyotrophic lateral sclerosis): conditions combining a bilateral pyramidal syndrome with limb and bulbar amyotrophy. Brain 113: 347-363

Bereznai B, Borasio GD, Winkler A, et al (1995) SOD1 Punktmutation in einer Familie mit ALS (abstract). Annual Meeting of the German Society of Neurogenetics, München

Bergeron C, Muntasser S, Somerville MJ, Weyer L, Percy ME (1994) Copper/zinc superoxide dismutase mRNA levels are increased in sporadic amyotrophic lateral sclerosis motor neurons. Brain Res 659: 272–276

Borchelt DR, Lee MK, Slunt HS, et al (1994) Superoxide dismutase I with mutations linked to familial amyotrophic lateral sclerosis possesses significant activity. Proc Natl Acad Sci USA 91: 8292–8296

Borchelt DR, Guarnieri M, Wong PC, et al (1995) Superoxide dismutase I subunits with mutations linked to familial amyotrophic lateral sclerosis do not affect wild type subunit function. J Biol Chem 270: 3234–3238

Bowling AC, Schulz JB, Brown RH Jr, Beal MF (1993) Superoxide dismutase activity, oxidative damage, and mitochondrial energy metabolism in familial and sporadic amyotrophic lateral sclerosis. J Neurochem 61: 2322–2325

Bowling AC, Barkowski EE, McKenna-Yasek D, et al (1995) Superoxide dismutase concentration and activity in familial amytrophic lateral sclerosis. J Neurochem 64: 2366–9

Carri MT, Battistoni A, Polizio F, Desideri A, Rotilio G (1994) Impaired copper binding by the H46R mutant of human Cu, Zn superoxide dismutase, involved in amyotrophic lateral sclerosis. FEBS Lett 356 (2–3): 314–316

Dal Canto MC, Gurney ME (1994) Development of central nervous system pathology in a murine transgenic model of human amyotrophic lateral sclerosis. Am J Pathol 145: 1271–1279

Dal Canto MC, Gurney ME (1995) Neuropathological changes in two lines of mice carrying a transgene for mutant human Cu, Zn SOD, and in mice overexpressing wild type human SOD: a model of familial amyotrophic lateral sclerosis (FALS). Brain Res 676: 25–40

De Belleroche J, Orrell R, Marklund S, et al (1995) Functional and structural correlates of 12 superoxide dismutase-1 mutations in UK families with amyotrophic lateral sclerosis (abstract). 6th International Symposium on ALS/MND, Dublin

Deng HX, Hentati A, Tainer JA, et al (1993) Amyotrophic lateral sclerosis and structural defects in Cu, Zn superoxide dismutase. Science 261: 1047–1051

Deng HX, Tainer JA, Mitsumoto H, et al (1995) Two novel SOD1 mutations in patients with familial amyotrophic lateral sclerosis. Hum Mol Genet 4: 1113–1116

Elshafey A, Lanyon WG, Connor JM (1994) Identification of a new missense point mutation in exon 4 of the Cu/Zn superoxide dismutase (SOD-1) gene in a family with amyotrophic lateral sclerosis. Hum Mol Genet 3: 363–364

Emery AEH, Holloway S (1982) Familial motor neuron disease. In: Rowland LP (ed) Human motor neuron diseases. Raven Press, New York, pp 139–147

Esteban J, Rosen DR, Bowling AC, et al (1994) Identification of two novel mutations and a new polymorphism in the gene for Cu/Zn superoxide dismutase in patients with amyotrophic lateral sclerosis. Hum Mol Genet 3: 997–998

Getzoff ED, Tainer JA, Stempien MM, Bell GI, Hallewell RA (1989) Evolution of Cu, Zn superoxide dismutase and the Greek key 8-Barrel structural motif. Proteins 5: 322–336

Groner Y, Gieman-Hurwitz J, Dafri N, et al (1986) The human Cu/Zn superoxide dismutase gene family: architecture and expression of the chromosome 21-encoded functional gene and its processed pseudogenes. In: Rotilis G (ed) Superoxide and superoxide dismutase in chemistry, biology and medicine. Elsevier Science Publishers, Biochemical Division, Amsterdam

Gurney ME, Pu H, Chiu AY, et al (1994) Motor neuron degeneration in mice that express a human Cu, Zn superoxide dismutase mutation. Science 264: 1772–1775

Hentati A (1989) Contribution a l'etude des paraplegies spasmodiques et familiales pure (strumpelllorrain) et associees en Tunisie. Thesis, Faculte de Medecine de Sfax, Tunisia

Hentati A, Bejaoui K, Pericak-Vance MA, et al (1994) Linkage of recessive familial amyotrophic lateral sclerosis to chromosome 2q33-q35. Nature Genet 7: 425–428

Hirano M, Fujii J, Nagai Y, et al (1994) A new variant Cu/Zn superoxide dismutase (Val7Glu) deduced from lymphocyte mRNA sequences from Japanese patients with familial amyotrphic lateral sclerosis. Biochem Biophys Res Commun 204: 572–577

Hodgson EK, Fridovich I (1975) The interaction of bovine erythrocyte superoxide dismutase with hydrogen peroxide: inactivation of the enzyme. Biochemistry 14: 5299–5303

Ikeda M, Abe K, Aoki M, et al (1995a) A novel point mutation in the Cu/Zn superoxide dismutase gene in a patient with familial amyotrophic lateral sclerosis. Hum Mol Genet 4: 491–492

Jones CT, Swinger RJ, Brock DJH (1994a) Identification of a novel SOD1 mutaiton in an apparently sporadic amyotrophic lateral sclerosis patient and the detection of I1e113Thr in three others. Hum Mol Genet 3: 649–650

Jones CT, Shaw PJ, Chari G, Brock DJ (1994b) Identification of a novel exon 4 SOD1 mutation in a sporadic amyotrophic lateral sclerosis patient. Mol Cell Probes 8: 329–330

Juneja T, Pericak-Vance M, Laing NG, Dave S, Siddique T (1997) Prognosis in familial ALS: progression and survival in patients with E100G and A4V mutations in Cu, Zn superoxide dismutase. Neurology 48: 55–57

Kawamata J, Hasegawa H, Shimohama S, Kimura J, Tanaka S, Ueda K (1994) Leu106 → Val (CTC → GTC) mutation of superoxide dismutase-1 gene in patient with familial amyotrophic lateral sclerosis in Japan (letter). Lancet 343: 1501

Kawamata J, Shimohama S, Hasegawa H, Imura T, Kimura J, Ueda K (1995) Deletion and point mutations in superoxide dismutase-1 gene in amyotrophic lateral sclerosis (abstract). XIth TMIN International Symposium, Tokyo

Klug D, Rabani J, Fridovich I (1972) A direct demonstration of the catalytic action of superoxide dismutase through the use of pulse radiolysis. J Biol Chem 247: 4839

Koppenol WH, Moreno JJ, Pryor WA, Ischiropoulos H, Beckman JS (1992) Peroxynitrite, a cloaked oxidant formed by nitric oxide and superoxide. Chem Res Toxicol 5: 834–842

Kostrzewa M, Burch-Lehmann U, Muller U (1994) Autosomal dominant amyotrophic lateral sclerosis: a novel mutation in the Cu/Zn superoxide dismutase-1 gene. Hum Mol Genet 3: 2261–2262

Moulard B, Camu W, Brice A, et al (1995) A previously undescribed mutation in the SOD1 gene in a French family with atypical ALS (abstract). 6th International Symposium on ALS/MND, Dublin

Nakano R, Sato S, Inuzuka T, et al (1994) A novel mutation in Cu/Zn superoxide dismutase gene in Japanese familial amyotrophic lateral sclerosis. Biochem Biophys Res Commun 200: 695–703

Nakashima K, Watanabe Y, Kuno N, Nanba E, Takahashi K (1995) Abnormality of Cu/Zn superoxide dismutase (SOD1) activity in Japanese familial amyotrophic lateral sclerosis with two base pair deletion in the SOD1 gene. Neurology 45: 1019–1020

Nijhawan D, Hentati A, Hentati E, et al (1995) Genetic locus heterogeneity in autosomal recessive familial amyotrophic lateral sclerosis (abstract). Am J Hum Genet 57 [Suppl]: A199

O'Reilly SA, Roedica J, Nagy D, et al (1995) Motor neuron-astrocyte interactions and levels of Cu, Zn superoxide dismutase in sporadic amyotrophic lateral sclerosis. Exp Neurol 131: 203–210

Parboosingh JS, Rouleau GA, Meninger V, McKenna-Yasek D, Brown RH Jr, Figlewicz DA (1995) Absence of mutations in the Mn superoxide dismutase or catalase genes in familial amyotrophic lateral sclerosis. Neuromuscul Disord S: 7–10

Pardo CA, Xu Z, Borchelt DR, Price DL, Sisodia SS, Cleveland DW (1995) Superoxide dismutase is an abundant component in cell bodies, dendrites, and axons of motor neurons and in a subset of other neurons. Proc Natl Acad Sci USA 92: 954–958

Phillips JP, Tainer JA, Getzoff ED, Boulianne G, Kirby K, Hilliker AJ (1995) Subunit-destabilizing mutations in Drosophila copper/zinc superoxide dismutase: neuro-pathology and a model of dimer dysequilibrium. Proc Natl Acad Sci USA 92: 8533–8534

Pramatarova A, Figlewicz DA, Krizus A, et al (1995) Identification of new mutations in the Cu/Zn superoxide dismutase gene of patients with familial amyotrophic lateral sclerosis. Am J Hum Genet 56: 592–596

Puymirat J, Cossette L, Gosselin F, Bouchard JP (1994) Red blood cell Cu/Zn superoxide dismutase activity in sporadic amyotrophic lateral sclerosis. J Neurol Sci 127: 121–123

Rabizadeh S, Gralla EB, Borchelt DR, et al (1995) Mutations associated with amyotrophic lateral sclerosis convert superoxide dismutase from an antiapoptotic gene to a proapoptotic gene: studies in yeast and neural cells. Proc Natl Acad Sci USA 92: 3024–3028

Ripps ME, Huntley GW, Hof PR, Morrison JH, Gordon JW (1995) Transgenic mice expressing an altered murine superoxide dismutase gene provide and animal model of amyotrophic lateral sclerosis. Proc Natl Acad Sci USA 92: 689–693

Robbertecht W, Sapp P, Viaene MK, et al (1994) Cu/Zn superoxide dismutase activity in familial and sporadic amyotrophic lateral sclerosis. J Neurochem 62: 384–387

Roos RR, Siddique T, Tainer JA (1995) Summary of Superoxide dismutase (SOD) and free radicals in amyotrophic lateral sclerosis and neurodegeneration. Neurology 45: 1779–1780 (conference)

Rosen DR, Siddique T, Patterson D, et al (1993) Mutations in Cu/Zn superoxide dis-mutase gene are associated with familial amyotrophic lateral sclerosis [published erratum appears in Nature (1993) 364: 362]. Nature 362: 59–62

Sapp PC, Rosen DR, Hosler BA, et al (1994) Identification of three novel mutations in the gene for Cu/Zn superoxide dismutase in patients with familial amyotrophic lateral sclerosis. Neuromuscul Disord 5: 353–357

Siddique T (1991a) Molecular genetics of familial amyotrophic lateral sclerosis. Adv Neurol 56: 227–231

Siddique T, Hentati A (1996a) Motor neuron disease. In: Rimoin DL, Connor JM, Pyeritz RE, Emery AE (eds) Emery & Rimoins's principle and practice of medical genetics. Churchill Livingston, New York, pp 2457–2472

Siddique T, Hentati A (1996b) Familial motor neuron disease. In: Appel SH (ed) Current neurology. Mosby-Year Book, Chicago, pp 281–301

Siddique T, Pericak-Vance MA, Brooks BR, et al (1989) Linkage analysis in familial amyotrophic lateral sclerosis. Neurology 39: 919–925

Siddique T, Figlewicz DA, Pericak-Vance MA, et al (1991b) Linkage of a gene causing familial amyotrophic lateral sclerosis to chromosome 21 and evidence of genetic-locus heterogeneity. N Engl J Med 324: 1381–1384

Själander A, Beckman G, Deng HX, Iqbal Z, Tainer JA, Siddique T (1995) The D9OA mutation results in a polymorphism of Cu, Zn superoxide dismutase that is prevalent in northern Sweden and Finland. Hum Mol Genet 4: 1105–1108

Symonyan MA, Nalbandyan RM (1972) Interaction of hydrogen peroxide with superox-ide dismutase from erythrocytes. FEBS Lett 28: 22–24

Tainer JA, Hallewell RA, Roberts VR, et al (1989) Probing enzyme-substrate recognition and catalytic mechanism. In: Simic MG, Taylor KA, Ward JF, Sontag CV (eds) Oxygen radicals in biology and medicine. Plenum Press, New York

Tandan R, Bradley WG (1985) Amyotrophic lateral sclerosis, part 1. Clinical features, pathology, and ethical issues in management. Ann Neurol 18: 271–280

Tsuda T, Munthasser S, Fraser PE, et al (1994) Analysis of the functional effects of a mutation in SOD1 associated with familial amyotrophic lateral sclerosis. Neuron 13: 727–736

Watanabe M, Aoki M, Abe K, et al (1995) A novel missense mutation (S134N) of the SOD1 gene in a patient with familial motor neuron disease (abstract). XIth TMIN International Symposium, Tokyo

Wiedau-Pazos M, Goto JJ, Rabizadeh S, et al (1996) Altered reactivity of superoxide dismutase in familial amytrophic lateral sclerosis. Science 271: 515–518

Wong PC, Pardo CA, Borchelt DR, et al (1995) An adverse property of familial ALS-linked SOD1 mutation causes motor neuron disease characterized by vacuolar degeneration of mitochondria. Neuron 14: 1105–1116

Yim MB, Chock PB, Stadtman ER (1993) Enzyme function of Copper, Zinc superoxide dismutase as a free radical generator. J Biol Chem 286: 4099–4105

Authors' address: Dr. T. Siddique, Department of Neurology, Northwestern University Medical School, Tarry Building 13-175, 300 East Chicago Avenue, Chicago, IL 60611, U.S.A.

Chronic administration of a partial agonist at strychnine-insensitive glycine receptors: a novel experimental approach to the treatment of ischemias

L. H. Fossom and **P. Skolnick**

Laboratory of Neuroscience, NIDDK, National Institutes of Health, Bethesda, MD,
U.S.A.

Summary. During the past decade, converging lines of evidence have linked the abnormal release or leak of excitatory amino acids to the neurodegeneration associated with a wide range of pathologies including cerebral ischemias, Huntington's disease, and AIDS dementia (Coyle and Robinson, 1987; Lipton, 1994; Meldrum, 1994). Pharmacological studies indicate that activation of both ionotropic and metabotropic glutamate receptors can substantially contribute to excitotoxic cell damage (Choi, 1992; Pizzi et al., 1993; Sheardown et al., 1993; Xue et al., 1994). Based on these findings, therapeutic strategies based on blunting or blocking glutamatergic transmission may be useful in treating a variety of neurodegenerative disorders.

Among ionotropic glutamate receptors, the family of N-methyl-D-aspartate (NMDA) receptors has been a prominent target for therapeutic intervention in the treatment of neurodegenerative disorders. Like other ligand-gated ion channels, NMDA receptors are likely to be constituted as heterooligomers (Kutswada et al., 1992; Meguro et al., 1992; Sheng et al., 1994). While neither the composition nor stoichiometry of wild type NMDA receptors has been determined, studies using both recombinant receptors expressed in a variety of systems and subunit specific antibodies indicate these receptors are constituted as a supramolecular complex containing one or more NMDAR-1 (ζ) subunit and one or NMDAR-2 (ε) subunit.

NMDA receptors possess multiple, allosterically coupled recognition sites that represent potential targets for modulating glutamatergic transmission. The apparent *requirement* for coordinate occupation of two recognition sites (the glutamate and strychnine-insensitive glycine sites) for operation of NMDA receptors (Kleckner and Dingledine, 1988) is a property that appears to be unique across structurally diverse families of ligand-gated ion channels. We reasoned that this feature might prove particularly vulnerable to therapeutic intervention, and embarked on in vivo studies with 1-aminocyclopropane-carboxylic acid (ACPC), a high affinity, partial agonist ligand at strychnine-insensitive glycine receptors (Marvizon et al., 1989; Watson and Lanthorn, 1990; Fossom et al., 1995a). Thus, if the in vitro studies demonstrating a requirement for coordinate occupation of glutamate and glycine sites also obtained in vivo, then a high affinity glycine ligand with partial agonist properties would function as an NMDA antagonist (Skolnick et al., 1989). Consistent with this hypothesis, ACPC has been shown to attenuate glutamate-induced neurotoxicity in primary neuronal cultures (Boje et al., 1993; Fossom et al., 1995a) and to reduce neuronal cell loss in vulnerable brain regions when administered at the time of ischemic insult (Long and Skolnick, 1994; Fossom et al., 1995a). The ability of glycine to block the neuroprotective actions of ACPC both in vivo (Long and Skolnick, 1994) and in vitro (Boje et al., 1993) is fully consistent with this proposed mode of action. While these actions of ACPC were predictable, its neuroprotective actions following chronic treatment [in the *absence* of significant amounts of ACPC in brain at the time of ischemic insult (von Lubitz et al., 1992)] would not be readily predictable, and offer some interesting therapeutic possibilities. This paper will review the data obtained in three animals models: global ischemia in gerbils; dynorphin A-induced spinal ischemia in rats, and middle carotid artery occlusion in mice which demonstrate that chronic administration of ACPC is neuroprotective despite the absence of measurable brain or plasma concentrations of drug (Miller et al., 1992; von Lubitz et al., 1992; Boje, 1994). Evidence obtained in primary neuronal cultures suggesting that sustained exposure to glycinergic ligands can affect NMDA receptor composition will also be reviewed.

In vivo studies

The notion that chronic administration of ACPC would be neuroprotective stems from the observation that the CD_{90} (that dose producing convulsions in 90% of the animals) of NMDA (125 mg/kg) in mice was significantly reduced (to a CD_{50}) following a seven day regimen of ACPC (Skolnick et al., 1992). NMDA challenge was performed 24 h after the last dose of ACPC, and both brain and blood levels of ACPC were below the limits of detection (Miller et al., 1992; Skolnick et al., 1992). These observations led us to hypothesize that chronic treatment with ACPC produced some adaptive change in NMDA receptors (Skolnick et al., 1992).

The first study examining the effects chronic treatment with ACPC in ischemic brain damage employed bilateral carotid artery occlusion (20 min.) in gerbils (von Lubitz et al., 1992). In addition to the hippocampal damage common to other models of global ischemia using a brief (typically 5–10 min.) occlusion of the carotid arteries, this paradigm results in a high mortality (the 7 day survival is typically ~25%) and profound neurological deficits in the surviving animals. The ischemic insult was applied 24 h after six daily injections of ACPC (300 mg/kg) or vehicle. Chronic administration of ACPC resulted in a threefold increase in 7 day survival (to > 80%), and significant improvements in both neurological status and neuronal survival in vulnerable brain regions compared with vehicle treated animals (Fig. 1). While a single

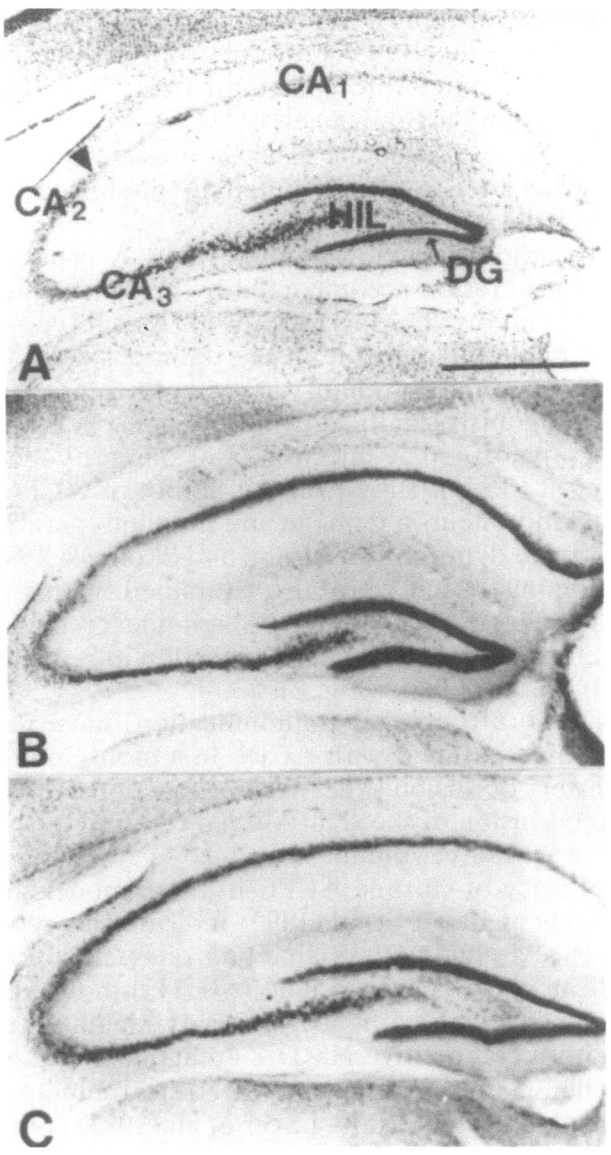

Fig. 1. Effects of chronic ACPC administration on ischemia-induced neurodegeneration in the hippocampal formation. These are representative sections from gerbils obtained 1 week after receiving: **A** 7 days of vehicle administration followed by 20 min. of forebrain ischemia; **B** 7 days of vehicle administration (control); and **C** 6 daily injections of ACPC (300 mg/kg) followed by a single injection-free day and 20 min. of forebrain ischemia. This is a Nissl stained section; bar: 500 μm. Ischemia remarkably reduces neurons in CA₁ and CA₂ (Panel A) compared to controls (Panel B). Chronic ACPC produces a statistically significant increase in neuron survival in hippocampus (Panel C). Administration of an additional bolus of ACPC 30 min. prior to ischemia did not significantly improve the neuronal sparing in hippocampus produced by the chronic regimen (data not presented).
This figure was reproduced from von Lubitz et al. (1992)

injection of ACPC administered 5 min. after ischemic insult produced comparable effects (Fossom et al., 1995a), the neuroprotective action of chronic
ACPC is likely to be mediated through a fundamentally different mechanism.
Thus, in the chronic regimen, brain levels of ACPC were at or below the limits
of detection at the time of ischemic insult and administration of a bolus of
ACPC 30 min. prior to ischemia resulted in no further improvement in
outcome (von Lubitz et al., 1992).

Lumbar subarachnoid injection of dynorphin A causes an ischemia-induced neuronal degeneration accompanied by persistent hindlimb paralysis.
Activation of the N-methyl-D-aspartate (NMDA) subtype of glutamate receptor appears to play a pivotal role in the pathophysiology associated with
dynorphin A-induced spinal cord injury, since both competitive and noncompetitive NMDA antagonists significantly improve recovery of hindlimb motor
function following dynorphin-A injection (Bakshi and Faden, 1990; Isaac et
al., 1990; Long et al., 1994). Based on the ability of ACPC to significantly
reduce both spinal motoneuron damage and hindlimb paralysis when administered 30 min. prior to dynorphin-A (Long and Skolnick, 1994), the effects of
chronic ACPC administration were also examined in this model. Six daily
injections of ACPC (200 mg/kg; with the last administered 24 h prior to insult)
proved to be as efficacious as acute administration in reducing dynorphin A-
induced hindlimb motor deficits (Fig. 2).

Lopez and Lanthorn (personal communication) have recently examined
the effects of chronic treatment with ACPC in a mouse model of permanent
middle cerebral artery occlusion (MCAo). Six injections of ACPC (300 mg/kg,
i.p.) 24 h prior to occlusion of the right MCAo resulted in a significant (18%;
$p < 0.05$) reduction in infarct volume.

Despite the efficacy of chronic ACPC in these models of ischemic brain
damage, the findings of Layer et al. (1993) indicate this regimen will not be
palliative in all cases of glutamate-related neurodegeneration. Pharmacological evidence indicates methamphetamine (METH)-induced degeneration of
nigrostriatal nerve terminals is at least in part mediated through NMDA
receptors. Thus, both competitive NMDA antagonists (e.g. CGS-19755) and
use-dependent channel blockers (e.g. MK-801) attenuate METH-induced
damage (Sonsalla et al., 1991). While Layer et al. (1993) confirmed the ability
of MK-801 to significantly reduce METH-induced depletion of striatal
dopamine and DOPAC, administration of ACPC (200 mg/kg) for 7 days prior
to METH was ineffective. Moreover, acute treatment (using a regimen identical to that employed for MK-801, drug administration 15 min prior to each
METH injection) with ACPC, the low efficacy glycine partial agonist HA-966,
and the competitive glycine antagonist 7-chlorokynurenic acid (7-ClKYN)
were also ineffective against METH-induced neurotoxicity. The failure of
these glycinergic ligands to block METH-induced neurotoxicity is particularly
striking in view of their ability to block glutamate-induced neurotoxicity both
in a variety of in vitro cell cultures (Hartley et al., 1990; Patel et al., 1990; Boje
et al., 1993; Fossom et al., 1995a) and in several models of ischemic brain
damage when administered at the time of insult (Beaughard et al., 1990; Long
and Skolnick, 1994; Fossom et al., 1995a). Nonetheless, Lombard et al. (1994)

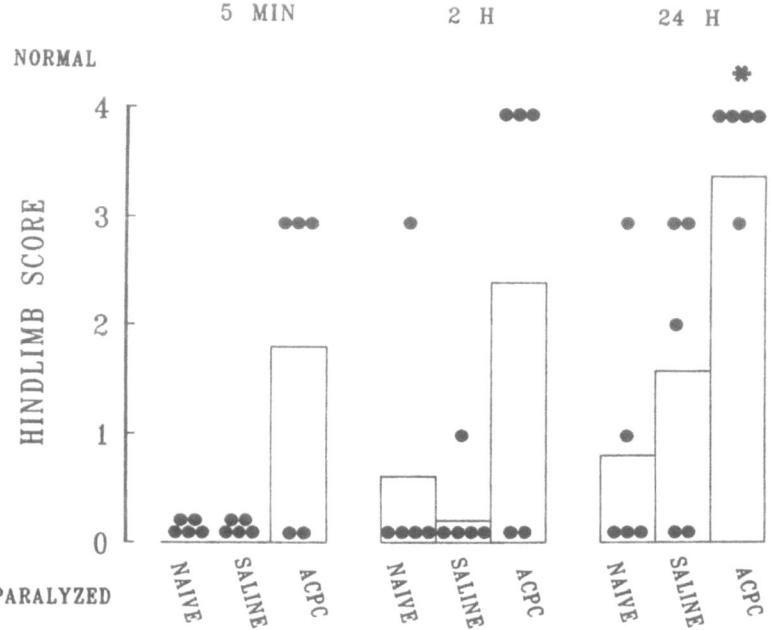

Fig. 2. Effects of chronic ACPC administration on recovery of hindlimb motor function after spinal subarachnoid injection of dynorphin A. ACPC (200 mg/kg, i.p.) or saline was injected daily for 6 days. After one injection-free day, all rats were treated with 20 nmol of dynorphin A. Rats in the naive control group were untreated preceding dynorphin A administration. Dynorphin A was injected between L4-L5 of male Sprague-Dawley rats using a 30 gauge needle as described in Long and Skolnick (1994). Symbol: * p < 0.05 when compared to saline or naive control groups. This figure was reproduced from Long and Skolnick (1994)

have recently reported that while competitive NMDA antagonists and use-dependent blockers protect against ischemia-induced retinal damage, the glycine antagonist 7-chlorothiokynurenate acid proved ineffective. In view of the heterogeneous nature of NMDA receptors (Kutswada et al., 1992), we hypothesize that the failure of acute or chronic application of glycinergic ligands to protect against these insults may be related to regional differences in the composition of NMDA receptors. Consistent with this hypothesis, several reports have documented that differences in subunit composition can effect remarkable changes in the potencies of several NMDA receptor ligands (Wafford et al., 1993; Williams, 1993; Laurie and Seeburg, 1994).

In vitro studies

In order to study the molecular mechanisms responsible for effects of chronic exposure to ACPC, we examined the effects of sustained (20–24 h) exposure to this and other glycinergic ligands in cerebellar granule neurons in culture. Consistent with data obtained in cortical cultures (Hartley et al., 1990; Patel et al., 1990) glycinergic ligands including HA-966, 7-ClKYN, and ACPC reduced glutamate-induced neurotoxicity in cerebellar granule cell cultures (Boje et al., 1993; Fig. 3). However, following sustained exposure of these cultures to glycinergic ligands (glycine, D-cycloserine, ACPC and HA-966), ACPC no longer protected neurons against glutamate (25 µM)-induced neurotoxicity, and the neuroprotective actions of HA-966 and 7-ClKYN were significantly reduced (Fig. 3 and Boje et al., 1993). While this in vitro culture system clearly does not model the in vivo situation, these findings are consistent with the hypothesis (Skolnick et al., 1992; von Lubitz et al., 1992) that chronic exposure to glycinergic ligands can affect NMDA receptor function. Subsequent experiments were directed at determining the mechanisms responsible for the

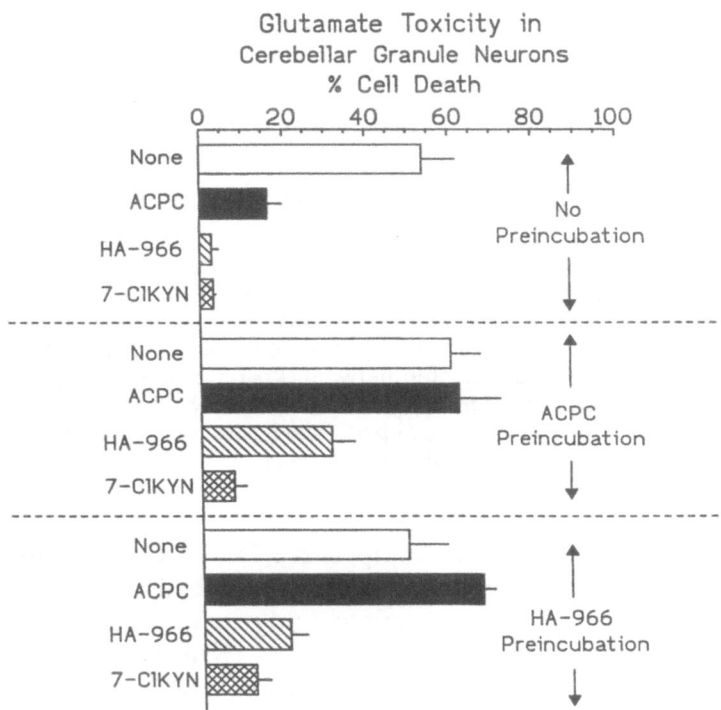

Fig. 3. The neuroprotective actions of glycinergic ligands in neuronal cell cultures are modified by sustained exposure. Cerebellar granule cell neurons were prepared from 6–8 day old rat pups as described (Boje et al., 1994). Following 8 days in culture, cells were exposed to glutamate (25 µM) in the presence or absence of ACPC (1 mM), HA-966 (0.1 mM) and 7-chlorokynurenic acid (7-CIKYN) (0.1 mM) and the neurotoxicity assessed 18–24 hours later by trypan blue exclusion. Concentration-effect curves were performed to determine optimally effective concentrations of these compounds. Under these conditions these glycinergic ligands provided a significant protection against glutamate-induced neurotoxicity (top panel, "No Preincubation"). The middle and lower panels demonstrate that a 24 h exposure of granule cells to ACPC (1 mM) or HA-966 (1 mM) followed by washout and glutamate challenge in the presence of ACPC, HA-966, or 7-CIKYN abrogates the neuroprotective actions of ACPC and significantly reduces the neuroprotection produced by HA-966. The reduction in the neuroprotective actions of 7-CIKYN produced by these pretreatments did not achieve statistical significance. This figure was reproduced from Boje (1994)

apparent loss in neuroprotective actions of these glycinergic ligands. These studies revealed that chronic exposure to glycinergic ligands *increases* the potency of glutamate to kill granule cells in culture (Fossom et al., 1995b). Examining a more proximal event in the excitotoxic cascade, NMDA-mediated increases in $[Ca^{+2}]_i$, we observed that sustained exposure to glycinergic ligands (in this case, ACPC; Fig. 4) resulted in a remarkable increase in $[Ca^{+2}]_i$ in response to NMDA (5 µM) compared to control neurons. Thus, sustained exposure to glycinergic ligands creates a population of neurons which are "supersensitive" to subsaturating concentrations of NMDA or glutamate. Since NMDA receptor subunit composition can have a profound influence on ligand potency (Wafford et al., 1993; Williams, 1993; Laurie and Seeburg, 1994), we examined the effects of sustained exposure to ACPC on NMDA receptor mRNA levels by slot-blot analysis. These studies revealed a selective, ~2.5-fold increase in NMDA-R2C (ε3) mRNA levels compared to control cultures (Fossom et al., 1995b). Since studies with recombinant NMDA receptors have demonstrated a significantly higher potency of glutamate and glycine in cells expressing expressing NMDA-R1/2C compared to cells expressing NMDA-R1/2A (Wafford et al., 1993) these findings suggest that a change in subunit composition, with a concommittant increase in receptors

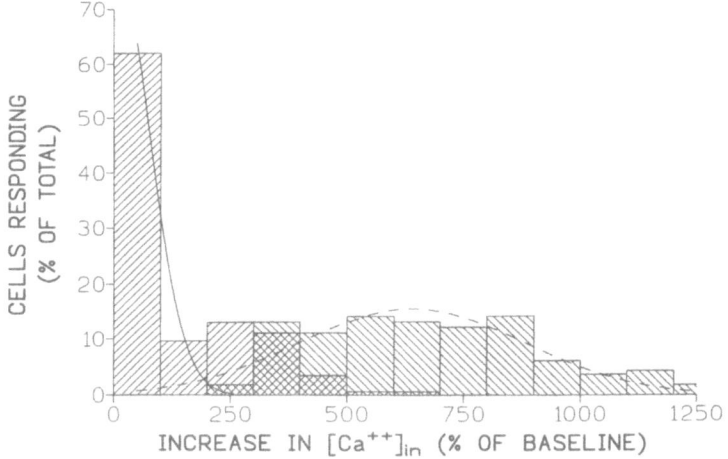

Fig. 4. Sustained exposure of cerebellar granule cells to ACPC amplifies NMDA-stimulated increases in $[Ca^{+2}]_i$: Cerebellar granule cells were plated and grown on glass coverslips. After 8 days in culture, cells were exposed to ACPC (1 mM) for 24 h. Media were removed and cells were loaded with the Ca^{+2} sensitive dye Fura-2 (as the acetoxymethyl ester). The coverslips were washed and fluoresence was measured until the baseline was stable (~ 10 min.). NMDA (5 µM) was added, and data was acquired for an additional 20 min. Note that in >60% of control neurons (left diagonals), this concentration of NMDA increased $[Ca^{+2}]_i$ by ≤ 100% . In contrast, NMDA increased $[Ca^{+2}]_i$ by > 300% in > 90% of cells subjected to a sustained exposure to ACPC (right diagonals). The crosshatched bars represent areas of overlap between control and ACPC treated neurons. The solid and dashed lines represent a Gaussian fit of the increases in $[Ca^{+2}]_i$ in control (solid line) and ACPC pretreated (dashed line) neurons. Resting $[Ca^{+2}]_i$ in control and ACPC pretreated neurons was 27 ± 4.2 and 21 ± 3.8 nM, respectively (n = 3 coverslips/group). This is a representative experiment consisting of the data collected from > 175 cells/group

containing NMDA-R2C subunits, could account for the diminished effects of ACPC, HA-966 and 7-ClKYN following sustained exposure to glycinergic ligands. Perhaps more important, these findings provide a "proof of concept" that sustained exposure to glycinergic ligands can exert transcriptional control over NMDA receptor subunits, which may explain the neuroprotective effects of chronic ACPC in vivo. Further studies are underway to determine whether a similar effect on transcriptional or post-transcriptional control of NMDA receptor mRNA is also apparent in vivo, and to find other in vitro systems which may model the in vivo situation with greater fidelity.

Conclusions

Chronic treatment with ACPC attenuates ischemic brain damage in three different animal models. Preclinical toxicology studies required for Phase I trials have demonstrated chronic treatment with ACPC is well tolerated in rodents and primates. For example, despite a high bioavailability, no mortality was observed in monkeys following oral ACPC doses of ≤ 2.25 g/kg/day for 14 days or intravenous doses of 0.6 g/kg/day for 28 days. Moreover, preliminary Phase I data using a single intravenous dose of ACPC produced no overt behavioral effects or remarkable changes in cardiovascular and respiratory dynamics (R. Meibach and M. Maccecchini, personal communication). Based on preclinical efficacy, both competitive NMDA antagonists and use dependent blockers are generally considered candidates for acute treatment of (e.g.) stroke, despite psychotomimetic side effects which would clearly limit their use to all but life threatening situations (Sveinbjornsdottr et al., 1993; Muir et al., 1994). No evidence of psychotomimetic behavior (that is, mimickry of the behavioral profile of compounds such as phenyclidine and MK-801) has been noted following ACPC (Skolnick et al., 1989; Witkin and Steele, 1991; Evoniuk et al., 1992). In view of the preclinical and clinical profile of ACPC, clinical trials using a chronic treatment regimen prior to coronary artery bypass grafting (CABG) are planned (R. Meibach and M. Maccecchini, personal communication). While no animal model of ischemic brain damage has been pharmacologically validated in humans, chronic treatment with ACPC or a related compound may offer a novel, highly efficient means of treating ischemias, and perhaps other neurodegenerative disorders associated with excitotoxicity.

Acknowledgements

The authors thank Drs. T. Lanthorn (Astra Arcus USA), M. Maccecchini, and R. Meibach (Symphony Pharmaceuticals) for permitting us to cite their findings prior to publication.

References

Bakshi R, Faden AI (1990) Competitive and non-competitive NMDA antagonists limit dynorphin A-induced rat hindlimb paralysis. Brain Res 507: 1–5

Beaughard M, Michelin M, Massingham R (1990) Effects of the putative glycine antagonist HA-966 on the neurological and histological changes induced by transient global ischemia in rats and gerbils. In: Krieglstein J, Oberpichler H (eds) Pharmacology of cerebral ischemia. Wissenschaftliche Verlagsgesellschaft, Stuttgart, pp 275–280

Boje K (1994) In vitro and in vivo studies with glycine partial agonists: a novel strategy for preventing NMDA receptor-mediated tissue damage. In: Palfreyman MG, Reynolds IJ, Skolnick P (eds) Direct and allosteric control of glutamate receptors. CRC Press, Boca Raton, pp 119–126

Boje KM, Wong G, Skolnick P (1993) Desensitization of the NMDA receptor complex by glycinergic ligands in cerebellar cell cultures. Brain Res 603: 207–214

Choi DW (1992) Excitotoxic cell death. J Neurobiol 23: 1261–1276

Evoniuk GE, Hertzman RP, Skolnick P (1991) A rapid method for evaluating the behavioral effects of dissociative anesthetics in mice. Psychopharmacol 105: 125–128

Fossom LH, Von Lubitz DKJE, Lin RC-S, Skolnick P (1995a) Neuroprotective actions of 1-aminocyclopropanecarboxylic acid, a partial agonist at strychnine-insensitive glycine sites. Neurol Res 17: 265–269

Fossom LH, Basile AS, Skolnick P (1995b) Sustained exposure to 1-aminocyclopropanecarboxylic acid, a glycine partial agonist, alters NMDA receptor function and subunit composition. Mol Pharmacol 48: 981–987

Hartley DM, Moyner H, Colamarino SA, Choi DW (1990) 7-Chlorokynurenate blocks NMDA receptor mediated neurotoxity in murine cortical culture. Eur J Neurosci 2: 291–295

Isaac L, van Zandt O'Malley T, Ristic H, Stewart P (1990) MK-801 blocks dynorphin A (1-13)-induced loss of the tail-flick reflex in the rat. Brain Res 531: 83–87

Kleckner NW, Dingledine R (1988) Requirement for glycine in activation of NMDA-receptors expressed in Xenopus oocytes. Science 241: 835–837

Kutsuwada T, Kashiwabuchi N, Mori H, Sakimura K, Kushiya E, Araki K, Meguro H, Masaki H, Kumanishi T, Arakawa M, Mishina M (1992) Molecular diversity of the NMDA receptor channel. Nature 358: 36–41

Laurie DJ, Seeburg PH (1994) Ligand affinities at recombinant N-methyl-D-aspartate receptors depend on subunit composition. Eur J Pharmacol 268: 335–345

Layer RT, Bland LR, Skolnick P (1993) MK-801, but not drugs acting at strychnine-insensitive glycine receptors, attenuate methamphetamine nigrostriatal toxicity. Brain Res 625: 38–44

Lipton SA (1994) Ca2+, N-methyl-D-aspartate receptors, and AIDS-related neuronal injury. Int Rev Neurobiol 36: 1–27

Lombardi G, Moroni F, Moroni F (1994) Glutamate receptor antagonists protect against ischemia-induced retinal damage. Eur J Pharmacol 271: 489–495

Long JB, Skolnick P (1994) 1-Aminocyclopropanecarboxylic acid protects against dynorphin A-induced spinal injury. Eur J Pharmacol 261: 295–301

Long JB, Rigamonti DD, Oleshansky MA, Wingfield CP, Martinez-Arizala A (1994) Dynorphin A-induced rat spinal cord injury: evidence for excitatory amino acid involvement in a pharmacological model of ischemic spinal cord injury. J Pharmacol Exp Ther 269: 358–366

Marvizon JC, Lewin AH, Skolnick P (1989) 1-Aminocyclopropane carboxylic acid: a potent and selective ligand for the glycine modulatory site of the N-methyl-D-aspartate receptor complex. J Neurochem 52: 992–994

Meguro H, Mori H, Araki K, Kushiya E, Kutsuwada T, Yamazaki M, Kumanishi T, Arakawa M, Sakimura K, Mishina M (1992) Functional characterization of a heteromeric NMDA receptor channel expressed from cloned cDNAs. Nature 357: 70–74

Meldrum B (1994) Neuroprotection by NMDA and non-NMDA glutamate antagonists. In: Palfreyman MG, Reynolds IJ, Skolnick P (eds) Direct and allosteric control of glutamate receptors. CRC Press, Boca Raton, pp 127–138

Miller R, La Grone J, Skolnick P, Boje KM (1992) High-performance liquid chromatographic assay for 1-aminocyclopropanecarboxylic acid from plasma and brain. J Chromatogr 578: 103–108

Muir KW, Grosset DG, Gamzu E, Lees KR (1994) Pharmacological effects of the noncompetitive NMDA antagonist CNS 1102 in normal volunteers. Br J Clin Pharmacol 38: 33–38

Patel J, Zinkand WC, Thompson C, Keith R, Salama A (1990) Role of glycine in the N-methyl-D-aspartate-mediated neuronal cytotoxicity. J Neurochem 54: 849–854

Pizzi M, Fallacara C, Arrighi V, Memo M, Spano PF (1993) Attenuation of excitatory amino acid toxicity by metabotropic glutamate receptor agonists and aniracetam in primary cultures of cerebellar granule cells. J Neurochem 61: 683–689

Robinson MB, Coyle JT (1987) Glutamate and related acidic excitatory neurotransmitters: from basic science to clinical application. FASEB J 1: 446–455

Sheardown MJ, Suzdak PD, Nordholm L (1993) AMPA, but not NMDA, receptor antagonism is neuroprotective in gerbil global ischaemia, even when delayed 24 h. Eur J Pharmacol 236: 347–353

Sheng M, Cummings J, Roldan LA, Jan YN, Jan LY (1994) Changing subunit composition of heteromeric NMDA receptors during development of rat cortex. Nature 368: 144–147

Skolnick P, Marvizon J, Jackson B, Monn J, Rice K, Lewin A (1989) Blockade of N-methyl-D-aspartate induced convulsions by 1-aminocyclopropane-carboxylates. Life Sci 45: 1647–1655

Skolnick P, Miller R, Young A, Boje K, Trullas R (1992) Chronic treatment with 1-aminocyclopropane-carboxylic acid desensitizes behavioral responses to agents acting at the N-methyl-D-aspartate receptor complex. Psychopharmacol 107: 489–496

Sonsalla PK, Riordan DE, Heikkila RE (1991) Competitive and noncompetitive antagonists at N-methyl-D-aspartate receptors protect against methamphetamine-induced dopaminergic damage in mice. J Pharmacol Exp Ther 256: 506–512

Sveinbjornsdottir S, Sander JWAS, Upton D, Thompson PJ, Patsalos PN, Hirt D, Emre M, Lowe D, Duncan DS (1993) The excitatory amino acid antagonist D-CPP-ene (SDZ EAA-494) in patients with epilepsy. Epilepsy Res 16: 165–174

von Lubitz D, Lin R, McKenzie R, Devlin T, McCabe RT, Skolnick P (1992) A novel treatment of global cerebral ischemia with a glycine partial agonist. Eur J Pharmacol 219: 153–158

Wafford KA, Bain CJ, Le Bourdelles B, Whiting PJ, Kemp JA (1993) Preferential co-assembly of recombinant NMDA receptors composed of three different subunits. Neuro Report 4: 1347–1349

Watson G, Lanthorn TH (1990) Pharmacological characteristics of cyclic homologues of glycine at the N-methyl-D-aspartate receptor associated glycine site. Neuropharmacol 29: 727–730

Williams K (1993) Ifenprodil discriminates subtypes of N-methyl-D-aspartate receptor: selectivity and mechanisms at recombinant heteromeric receptors. Mol Pharmacol 44: 851–859

Witkin J, Steele T (1992) Effects of strychnine-insensitive glycine receptor ligands on discriminative stimulus effects of N-methyl-D-aspartate (NMDA) channel antagonists. Abstr Soc Neurosci 18: 447, #192.16

Xue D, Huang Z-G, Barnes K, Lesiuk HJ, Smith KE, Buchan AM (1994) Delayed treatment with AMPA, but not NMDA, antagonists reduce neocortical infarction. J Cereb Blood Flow Metab 14: 251–261

Authors' address: Dr. P. Skolnick, Laboratory of Neuroscience, NIH, NIDDK/LN, Building 8, Room 111, Bethesda, MD 20892, U.S.A.

Apoptosis in neurodegenerative disorders: potential for therapy by modifying gene transcription

W. G. Tatton[1,2,4], **R. M. E. Chalmers-Redman**[1,4], **W. Y. H. Ju**[1], **J. Wadia**[1], and **N. A. Tatton**[3,4]

Departments of [1]Physiology/Biophysics, [2]Psychology, [3]Pharmacology and
[4]The Institute for Neuroscience, Dalhousie University, Halifax, Nova Scotia, Canada

Summary. Apoptotic, rather than necrotic, nerve cell death now appears as likely to underlie a number of common neurological conditions including stroke, Alzheimer's disease, Parkinson's disease, hereditary retinal dystrophies and Amyotrophic Lateral Sclerosis. Apoptotic neuronal death is a delayed, multistep process and therefore offers a therapeutic opportunity if one or more of these steps can be interrupted or reversed. Research is beginning to show how specific macromolecules play a role in determining the apoptotic death process. We are particularly interested in the critical nature of gradual mitochondrial failure in the apoptotic process and propose that a maintenance of mitochondrial function through the pharmacological modulation of gene expression offers an opportunity for the effective treatment of some types of neurological dysfunction.

Our research into the development of small diffusible molecules that reduce apoptosis has grown from studies of the irreversible MAO-B inhibitor (–)-deprenyl. (–)-Deprenyl can reduce neuronal death independently of MAO-B inhibition even after neurons have sustained seemingly lethal damage. (–)-Deprenyl can also influence the process outgrowth of some glial and neuronal populations and can reduce the concentrations of oxidative radicals in damaged cells at concentrations too small to inhibit MAO. In accord with earlier work of others, we showed that (–)-deprenyl alters the expression of a number of mRNAs or of proteins in nerve and glial cells and that the alterations in gene expression/protein synthesis are the result of a selective action on transcription. The alterations in gene expression/protein synthesis are accompanied by a decrease in DNA fragmentation characteristic of apoptosis and the death of responsive cells. The onco-proteins Bcl-2 and Bax and the scavenger proteins Cu/Zn superoxide dismutase (SOD1) and Mn superoxide dismutase (SOD-2) are among the 40–50 proteins whose synthesis is altered by (–)-deprenyl. Since mitochondrial membrane potential correlates with mitochondrial ATP production, we have used confocal laser imaging techniques in living cells to show that the transcriptional changes induced by (–)-deprenyl result in a maintenance of mitochondrial membrane potential, a decrease in intramitochondrial calcium and a decrease in cytoplasmic oxidative radical levels. We therefore propose that (–)-deprenyl acts on gene expression to maintain mitochondrial function and decrease cytoplasmic oxidative radical levels and thereby reduces apoptosis. An understanding of the molecular steps by which (–)-deprenyl selectively alters transcription may lead to the development of new therapies for neurodegenerative diseases.

Overview

Nervous system disorders most often disable rather than kill their victims. Often the deficits caused by neurological or psychiatric diseases continue over a number of years, even decades. Consequently, diseases involving neuronal dysfunction are responsible for more than half of the economic burden of ill health in the Western World.

Nervous system operation depends on precisely organized interconnections between specific assemblies of functionally-differentiated neurons. Mammalian neurons are post mitotic; and, if lost, are not normally replaced. Furthermore, if neuronal interconnections degenerate or are severed there is a limited capacity for their regrowth. The loss of functionally specific neuronal assemblies due to death or impaired interconnections underlies almost all neurological and psychiatric diseases. Accordingly, the two major quests of neurological research development are to discover therapies that reduce neuronal death and restore working neuronal interconnections.

Three different cellular processes have been proposed to underlie neuronal loss causing neurological dysfunction. Those processes have been termed necrosis, apoptosis and atrophy. The cardinal features of necrosis are cell swelling and the rapid disruption of external cellular membranes. Until recently, necrosis was believed to underlie most of the neuronal death causing neurological disease. The cardinal feature of apoptosis was thought to be the destruction of neuronal DNA by endonucleases while external membranes remained relatively intact and was believed to occur only as part of normal nervous system development. Finally, atrophy entails the marked shrinkage of neuronal cell bodies and processes so that they may not be recognizable as neuronal on standard pathological examination. Atrophy has been found after some forms of damage to neuronal processes in animal models, but it may also be part of brain aging.

Neuronal necrosis, apoptosis and atrophy usually differ in their time course and therefore might also be termed as immediate, delayed and chronic neuronal loss respectively. Necrosis usually is completed in minutes or hours after an insult, while neuronal apoptosis may require a number of hours or days. Atrophy is protracted and can proceed over weeks or months. The delay between insult and irreversible cellular loss in apoptosis and the possibility of functionally recovering atrophic neurons offers therapeutic windows not available with necrosis. That is, therapy may be effective in apoptosis and atrophy after the neuron has been damaged while therapy must be undertaken at the same time or prior to the onset of the damage in necrosis.

During the last several years, evidence has accumulated indicating that apoptotic neuronal death may be an important component of a number of common neurological and psychiatric disorders (Altman, 1992; Margolis et al., 1994; Thompson, 1995; Cotman and Anderson, 1995). Some of those disorders include stroke (Linnik et al., 1993; Rosenbaum et al., 1994), Alzheimer's disease (AD) (Cotman, 1994; Johnson, 1994; Su et al., 1994; Cotman and Anderson, 1995; Dragunow et al., 1995; Lassmann et al., 1995; Smale et al., 1995), Parkinson's disease (PD) (Langston, 1994; Anglade et al., 1995), Amyotrophic Lateral Sclerosis (ALS) (Eisen and Krieger, 1993; Yoshiyama et al., 1994;), Huntington's disease (Price et al., 1992), hereditary retinal degenerations (Shahinfar et al., 1991; Chang et al., 1993; Lolley et al., 1994; Portera et al., 1994; Steinberg, 1994; Tso et al., 1994), glaucoma (Buchi, 1992; Berkelaar et al., 1994; Garcia et al., 1994; Rabacchi et al., 1994; Silveira et al., 1994; Buys et al., 1995), and spinal muscular atrophy (Lefebvre et al., 1995; Roy et al., 1995). Apoptosis of glial cells (Pender et al., 1991 et al., 1993) may also contribute to disorders like multiple sclerosis and diabetic peripheral neuropathy. The role and prevalence of neuronal atrophy in human neurological disorders is not well understood [see (Finch, 1993) for a review] and may just be a prolonged form of apoptosis.

The recognition that apoptosis may be a fundamental component of many common neurological disorders [see (Thompson, 1995; Cotman and Anderson, 1995)] has shifted some of the emphasis of research aimed at preventing neuronal loss. Previously, the research endeavored to develop molecules that could protect neurons from sustaining damage that might lead to necrosis. The therapeutic approach was termed neuronal protection and the agents as neuronal protectants. The 21 aminosteroids (lazeroids) are typical neuronal protective agents since they are believed to decrease the production of potentially damaging oxidative radicals (Hall and McCall, 1994).

Currently, efforts are focused on the development of agents that maintain the survival of neurons which have already sustained sufficient damage to cause their death. Agents with that capacity are said to rescue neurons. Neurotrophic factors were the first agents shown to have a capacity for neuronal rescue. They are members of a number of macromolecular families including the neurotrophin, mitogen and cytokine/neurokine families. Each neurotrophic factor acts on one or more specific receptors which usually induce a number of diverse cellular processes, only one of which is a reduction of neuronal death.

A great deal of effort has been expended in exploring the potential of different neurotrophic factors in the treatment of neurological disorders. Since most neurotrophic factors are proteins, they do not readily cross the blood brain barrier. It was anticipated that systematically delivered neurotrophic factors could circumvent the blood brain barrier via retrograde transport along neuronal axons projecting to the periphery, or by diffusion across local areas of the blood brain barrier made permeable by disease. Other promising approaches were intrathecal delivery into the cerebral spinal fluid or the development of small diffusible molecules that could activate specific neurotrophic factor receptors. However, recent animal studies (Henderson et al., 1994; Zhang et al., 1995) and human clinical trials (Baringaga, 1994), using either systemic or intrathecal delivery of neurotrophins and neurotrophic cytokines, have shown that systemic or intrathecal doses of neurotrophic factors sufficient to increase neuronal survival also cause disabling weight loss. The finding of serious side effects with systemic or intrathecal administration appears to suggest that the intracellular mechanisms activated by neurotrophic factor receptors may be too extensive to allow for a reduction of neuronal death without disturbing other important cellular functions. Neurotrophic factors will therefore have a role in neurological therapy if methods can be found for specifically targeting their actions to cells at risk. Another approach is to develop small, easily diffusable molecules that have a range of actions that are more limited than those of neurotrophic factors, that is, molecules that would specifically reduce neuronal death or promote neuronal regrowth.

Neuronal death by apoptosis versus necrosis

Classically, neuronal death caused by necrosis was considered to include three major features in varying proportions: 1) failure of mitochondrial respiration with a resultant loss of ATP-dependent cellular functions, 2) increased cytosolic levels of oxidative radicals, in part due to a reduction in mitochondrial oxidative respiration, causing lipid peroxidation and lysis of plasma membranes together with oxidative damage to proteins and DNA, and 3) increased intracellular Ca^{2+} levels with the activation of calcium-dependent proteolytic enzymes. Histologically, necrosis featured plasma membrane fracture and cytoplasmic organelle dissolution with a relative maintenance of nuclear integrity. An inflammatory cell reaction due to the extruded cytoplasmic contents accompanied these neuronal changes.

A typical example of necrotic neuronal death was believed to be brain ischemia/hypoxia causing oxygen deprivation and acidic tissue pH. Oxygen deprivation and intracellular acidosis caused mitochondrial failure and a loss of electrical potential across the inner mitochondrial membrane which induced massive release of synaptic glutamate. The glutamate activated receptors on nearby neurons caused an immense influx of Ca^{2+} into their cytoplasm. Necrotic neuronal death has been modeled by the application of excitotoxic glutaminergic agonists (Choi and Rotham, 1990), MPTP (Tipton and Singer, 1993), 6-hydroxydopamine (6-OHDA) and iron infusion (Sengstock et al., 1992) or by direct mechanical trauma (Hall and McCall, 1994).

Apoptotic neuronal death was initially only considered to involve developing neurons which failed to compete adequately for trophic support from their target cells (other neurons or muscle) or from nearby non-neuronal cells (Hamburger and Oppenheim, 1982). Developmental apoptosis was frequently called programmed cell death and was responsible for the reduction of neuronal numbers found during embryogenesis or early postnatal life. Apoptosis was believed to have five characteristics not found in necrosis: 1) cleavage of DNA, principally by Ca^{2+} dependent endonucleases, initially into large pieces (50–300 kilobases) and then into oligonucleosomal sized pieces with the formation of

185 kilobase "ladders" visible on agarose gels, 2) nuclear condensation and fractionation with the formation of nuclear bodies, 3) a requirement for new protein synthesis, 4) a maintenance of organelle and plasma membrane integrity with the formation of membrane wrapped cytoplasmic bodies or vacuoles, and 5) little or no inflammatory reaction (Wyllie et al., 1984).

Over the last 4 years, it has been shown that apoptosis can be initiated by a number of insults other than just trophic withdrawal or deprivation. Insults thought only to cause neuronal necrosis now have been shown to induce apoptosis, particularly when delivered at low levels or at a slow rate (Cotman and Anderson, 1995). To illustrate that point, Fig. 2, drawn from recent work in our laboratory, shows how a pro-oxidant, H_2O_2, can induce the death of partially-differentiated PC12 cells which fulfills the criteria for either apoptosis or necrosis, depending on the concentration of the H_2O_2.

For the left side of Fig. 1, PC12 cells were passaged in minimal essential media (MEM) and serum. They were then partially-differentiated by placing them in MEM with serum and NGF (M/S+N) for 6 days [see (Tatton et al., 1994a) for details, abbreviations and further data]. In order to trophically deprive the cells, they were washed repeatedly and replaced into MEM only (M/O). Cell death became significant at 12 hours and the cells showed DNA "laddering" on agarose gels, endonuclease digestion by in situ labeling of free 3' DNA ends (ApopTag, Oncor Ltd) and transcriptional changes characteristic of apoptosis (see details of similar transcriptional changes below), all considered typical of apoptosis. Importantly, the cells showed ApopTag positive nuclei when their external plasma membranes could be seen to be still intact when viewed with interference contrast or confocal microscopy. Control cells which were treated identically but replaced into M/S+N after washing did not demonstrate any of these apoptotic changes.

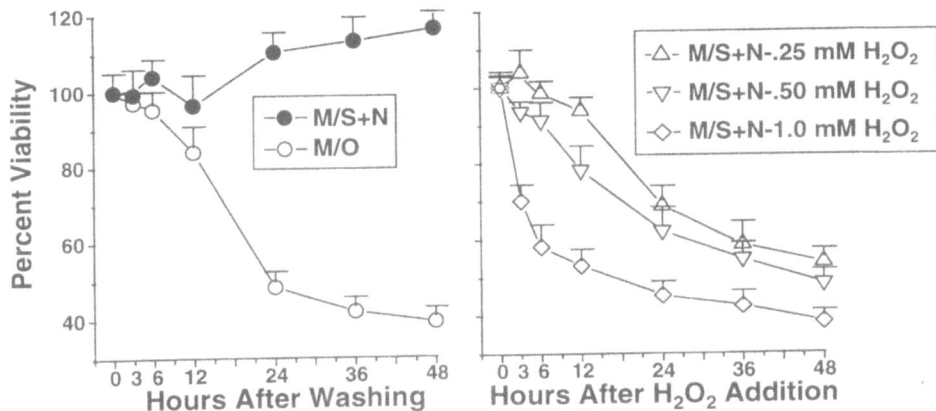

Fig. 1. H_2O_2 can induce either "necrosis" or apoptosis in partially differentiated PC 12 cells – Experimental methods are described in text. Low concentrations of H_2O_2 induced "delayed" cell death that displayed a time course and nuclear DNA stigmata that are similar to apoptosis induced by trophic withdrawal caused by serum and NGF removal. Higher concentrations of H_2O_2 induced "immediate" cell death with a time course and stigmata that were typical of necrosis

The right side of Fig. 1 shows that similarly prepared, partially differentiated PC 12 cells maintained in M/S+N (i.e. they were not trophically-withdrawn), died after exposure to varying concentrations of H_2O_2. The concentrations of H_2O_2 are the same as those previously used to induce either necrosis or apoptosis in a line of hemopoetic cells (Hockenbery et al., 1993). Addition of 0.25 mM H_2O_2 induced delayed cell death with a time course similar to that found after trophic withdrawal. It was accompanied by positive ApopTag staining of nuclear DNA with preserved external membranes suggesting apoptosis. Addition of 1.0 mM H_2O_2 caused almost 40% of the cells to die in less than 3 hours without evidence of ApopTag staining of nuclear DNA. A major proportion of those cells showed fractured external membranes. An intermediate H_2O_2 concentration of 0.5 mM H_2O_2 caused cell loss with an initial rapid phase followed by slower delayed death. Similar to the work of others in different cell systems, the above studies indicate that apoptosis and necrosis can be induced by the same insults, depending on the magnitude or time course of the insult to the cells.

Apoptosis has been shown to accompany nervous tissue damage caused by hypoxia-ischemia (Dragunow et al., 1993; Linnik et al., 1993; Rosenbaum et al., 1994), exposure to excitotoxins (Behl et al., 1993; Joseph et al., 1993; Kaku et al., 1993; Montpied et al., 1993; Samples and Dubinsky, 1993; Zhong et al., 1993; Csernansky et al., 1994; Dessi et al., 1994; Mitchell et al., 1994; Yan et al., 1994; Bonfoco et al., 1995a; Copani et al., 1995; Potera-Cailliau et al., 1995), MPTP/MPP+ (Dipasquale et al., 1991; Hartley et al., 1994; Mochizuki et al., 1994; Mutoh et al., 1994), 6-OHDA or dopamine (Walkinshaw and Waters, 1994; Ziv et al., 1994), pro-oxidants such as H_2O_2 (Slater et al., 1995), calcium channel blockers (Koh and Cotman, 1992), methamphetamine (Finnegan and Karler, 1992), excessive iron levels (Farinelli and Greene, 1996; Zsnagy et al., 1995), colchicine (Bonfoco et al., 1995b), ceramide (Brugg et al., 1996), agents found in cycad flour (Gobe, 1994), some sialoglycoproteins (Kobayashi et al., 1994), DNA synthesis blockers (Dessi et al., 1995) and mitochondrial respiratory chain inhibitors (Hartley et al., 1994), all of which were previously thought to cause only necrosis. Importantly, apoptosis of cultured neurons has been shown to result after exposure to two agents with possible direct clinical relevance, beta amyloid protein (Forloni, 1993; Forloni et al., 1993; Loo et al., 1993; Rabizadeh et al., 1994) and gp120, the AIDs protein (Muller et al., 1992). In situ evidence for nuclear DNA cleavage has been found in the nervous systems of humans suffering with Alzheimer's disease (Cotman, 1994; Johnson, 1994; Su et al., 1994; Cotman and Anderson, 1995; Dragunow et al., 1995; Lassmann et al., 1995; Smale et al., 1995), Parkinson's disease (Anglade et al., 1995), Huntington's disease (Price et al., 1992), AIDS encephalitis (Petito and Roberts, 1995), spinal muscular atrophy (Lefebvre et al., 1995; Roy et al., 1995) and ALS (Eisen and Krieger, 1993; Yoshiyama et al., 1994; Thomas et al., 1995). Although cautions about the reliabilty of in situ DNA labeling as a conclusive means of showing apoptosis have been raised (Charriaut-Mariangue and Ben-Ari, 1995), the balance of data indicates that apoptosis is an important component of some neurodegenerative diseases.

What are the essential differences between apoptosis and necrosis?

The realization that apoptosis may be an important contributor to many conditions causing neurological dysfunction has sparked a controversy as to how apoptosis and necrosis can be reliably recognized. Initially it was thought that the findings of oligonucleosomal DNA digestion, as evidenced by "laddering" of DNA gels at 185 bp intervals, and a requirement for new protein synthesis (the synthesis of so-called "death proteins") were necessary to define apoptosis (see Deckwerth and Johnson, 1993). In contrast, mitochondrial failure and high cytoplasmic levels of oxidative radicals and calcium were believed to define necrosis. Work in our and other laboratories have shown that mitochondrial failure and oxidative radical damage are involved in apoptosis caused by cytokine/neurotrophin withdrawal and appear to contribute to apoptosis caused by a variety of insults (see below). Other studies have shown that DNA degradation and new protein synthesis are not essential to the progression of apoptosis (Rukenstein et al., 1991; Batistatou et al., 1993).

We believe that one key difference between apoptosis and necrosis depends on whether there is adequate time for cells to respond to an insult with "compensatory" changes in transcriptional activity. Damage that is insufficient to cause almost immediate cell body lysis may initiate transcriptional events aimed at limiting further damage or at initiating repair mechanisms. Neurons are particularly vulnerable to that kind of damage because of their extensive processes which allow damage to occur at relatively great distances from the synthetic machinery. Damage frequently occurs to dendrites or axons at distances that are hundreds or thousands of microns from the cell body. It has been proposed that a "decisional" process for choosing between cellular "suicide" and repair (Oltvai and Korsmeyer, 1994) occurs in parallel to or as part of the transcriptional events.

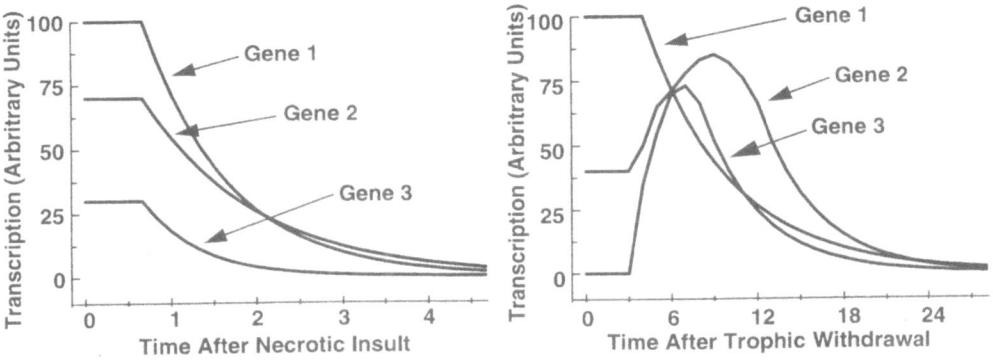

Fig. 2. Changes in gene expression with necrosis or apoptosis in neurons. Schematic of transcriptional changes in necrosis versus those in apoptosis. Time is in arbitrary units. The figure emphasized the cardinal feature of apoptosis. That is, that some genes undergo transient increases in expression in the interval between the insult initiating apoptosis and the death of the cells. In necrosis, the neurons' external membranes are rapidly fractured and there is insufficient time for increases in gene expression

Johnson and coworkers (Deckwerth and Johnson, 1993) have begun to define the timing of progression of selective changes in gene expression that occur in sympathetoblasts as they enter into apoptosis. We have found similar selective changes in gene expression in PC 12 cells entering apoptosis caused by trophic withdrawal (Ju and Tatton, unpublished data). Importantly, the selective changes in apoptosis differ from the rapid decline in the expression of all genes that occurs with necrosis in the same cells (schematized in Fig. 2). The absence of transcriptional increases in necrosis may be the most reliable feature to differentiate necrosis from apoptosis.

Apoptosis and Bcl-2/Bax

Increased expression of the oncogene bcl-2 reduces the apoptotic threshold in a variety of cell types [see (Korsmeyer, 1992; Reed, 1994) for reviews] including PC12 cells, facial motoneurons (Dubois-Dauphin et al., 1994) and neurons in C. elegans (Hengartner and Horvitz, 1994). Bcl-2 can suppress apoptosis after a wide variety of insults (Zhong et al., 1993) including trophic withdrawal and pro-oxidant challenges and may regulate a distal step in a final common pathway for apoptotic cell death. There are a number of other members of the Bcl-2 family including Bcl-x_L, Bcl-x_S, Ccl-1, A1, Bad and several open reading frames in DNA viruses. All of the members have highly conserved BH1 and BH2 domains. Bcl-2 oncoprotein is 26 kd in size and forms a homodimer. It also forms a heterodimer with other family members and Bax (Oltvai et al., 1993; Yin et al., 1994). Increased expression of bax or bcl-x_S increases the susceptibility of cells to apoptotic death. Hence Bcl-2 (and/or Bcl-x_L) and Bax (and/or Bcl-x_S) function in a "push-pull" manner to influence cell survival. The Bcl-2/Bax ratio has been proposed as one determinant of a cells' susceptibility to apoptosis (Oltvai and Korsmeyer, 1994). Bad dimerizes with Bcl-2 (and Bcl-x) and displaces Bax thereby freeing Bax to promote apoptosis (Yang et al., 1995). Bad may be one of a number of molecules that modulate apoptosis by interacting with Bcl-2. BAG-1, an acidic protein, shares no homology with Bcl-2 but binds with the protein and potentiates the anti-apoptotic action of Bcl-2 (Takayama et al., 1995). By itself, BAG-1 does not reduce apoptosis but its presence may be a requirement if Bcl-2 is to be effective in reducing apoptosis. Bcl-2 appears to reduce apoptosis through cellular pathways and mechanisms that do not require neurotrophic factor support.

Although the bcl-2 and bax oncogene systems are clearly effective in decreasing or promoting apoptosis, the cellular mechanisms underlying their actions are unclear. Molecular biology and protein chemistry have provided a great deal of insight into the structure of these proteins relative to apoptosis but relatively little is known about the cellular mechanisms mediated by the proteins. Understanding those cellular mechanisms will be important to the rationale for the therapeutic use of Bcl-2/Bax systems, particularly the development of molecules that can exploit these systems. Recent work has been interpreted to show that increased Bcl-2 expression can reduce necrotic death in some cell systems (Kane et al., 1995). This may indicate that if necrosis and apoptosis share one or more mechanistic steps, then Bcl-2 acts at a common step (see below).

Apoptosis and SOD1/SOD2

Some cases of familial ALS (FALS) have been shown to be associated with multiple CAG repeats in the exons coding for Cu/Zn superoxide dismutase (SOD1) (Deng et al., 1993; Rosen et al., 1994). We (Tsuda et al., 1994) and others have shown that the coding abnormalities reduce the scavenger capacity of SOD1 by 40–50%. We have provided evidence that neurons which suffer high oxidative loads may be preferentially vulnerable to decreased scavenging by SOD1 (Tsuda et al., 1994). SOD1 deficiency causes apoptosis in PC12 cells (Troy and Shelanski, 1994) and increased SOD1 levels delay apoptosis in sympathetoblasts (Greenlund et al., 1995).

A deficiency of SOD1 activity may not be the only basis for neuronal death in FALS. Expression of a FALS SOD1 mutation in mice has produced ALS-like deficits (Dal-Canto and Gurney, 1994; Gurney et al., 1994), even though overall SOD1 activity is within or beyond the normal range (Siddique, this volume). These findings suggest that SOD1 may have another function that prevents neuronal death other than just the scavenging of superoxide radicals. The structure of the protein may be important to the mediation of that unknown function.

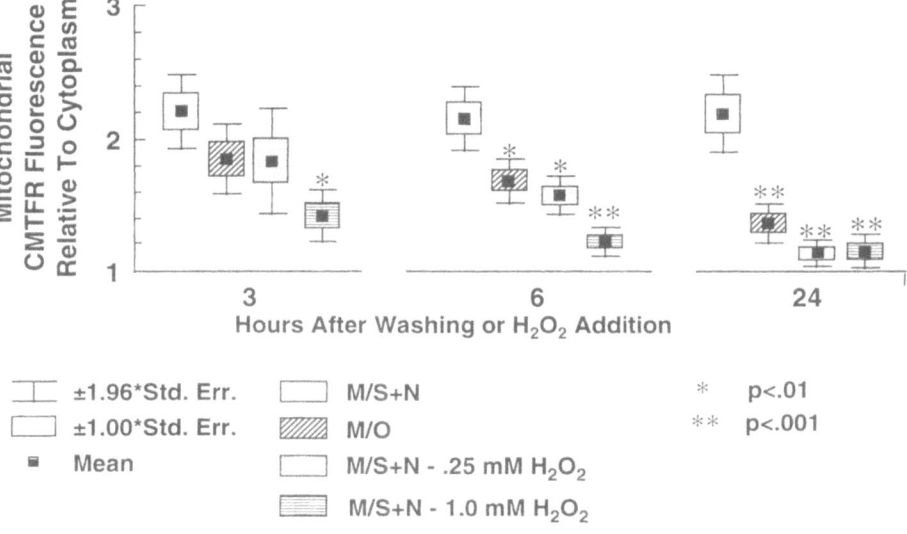

Fig. 3. Loss of mitochondrial membrane potential ($\delta\psi$), as shown by the decrease in mitochondrial/cytosolic fluorescence ratio, occurs in partially differentiated PC12 cells that appear to be dying by either apoptosis or necrosis. See Fig. 1 for the time course of cell death associated with the trophic withdrawal (M/O) or the application of different concentrations of H_2O_2. Note that a loss of mitochondrial membrane potential precedes the cell death by a similar time interval whether the cells die by necrosis or apoptosis

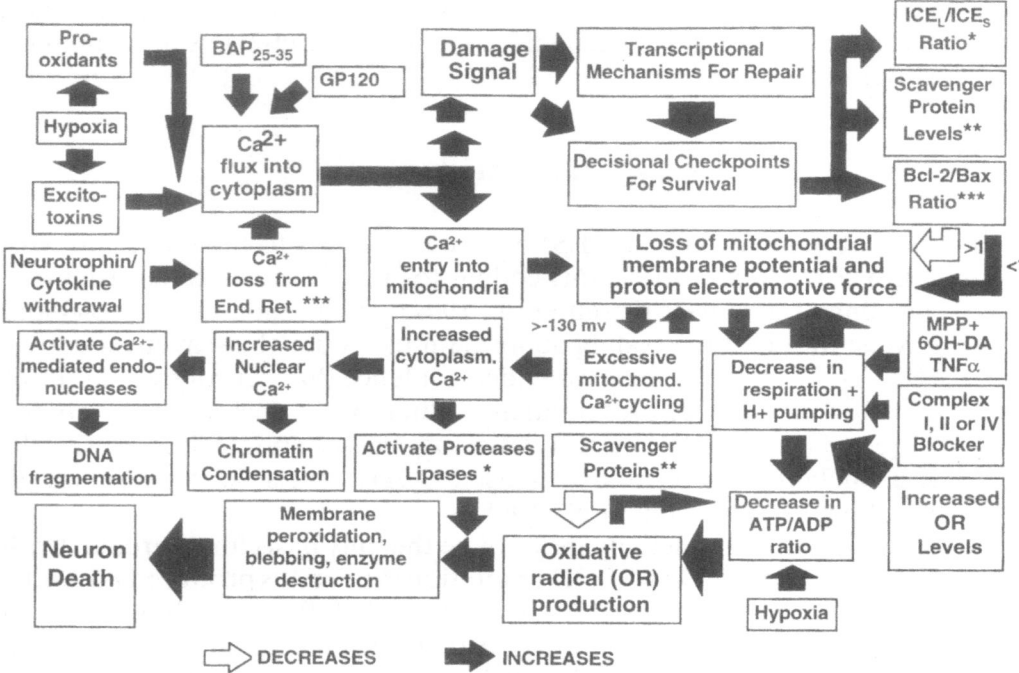

Fig. 4. A proposed schematic model for the transcriptional changes and the mitochondrial failure that accompany either apoptosis or necrosis. The schematic emphasizes the role of Ca^{2+} fluxes, mitochondrial failure and mitochondrially-derived oxidative radicals in apoptosis and necrosis and attempts to show the relationships between those factors for a number of factors that cause neuronal death. The major difference between apoptosis and necrosis is proposed as the time necessary for the cell to respond to a damage signal with alterations in protein synthesis. See Oltvai and Korsmeyer (1994) and the text for further details

A number of other proteins have now been proposed as being important to the progression of apoptosis, for example the interleukin converting enzymes (ICE). Details on some of those proteins and a model as to how they might govern the progress of apoptosis are provided in (Oltvai and Korsmeyer 1994). Also see Fig. 4.

Apoptosis and mitochondrial ATP production

Mitochondria are critically important to the normal functioning of most cells as sites for oxidative phosphorylation, the reduction of reactive oxygen species, and the sequestration of intracellular free Ca^{2+} (Werth and Thayer, 1994). Mitochondria normally maintain a high mitochondrial membrane potential ($\delta\psi$). $\delta\psi$ is linearly related to proton electromotive force (δp) in mitochondria ($\delta p = \delta\psi - 60\,\delta pH$), where δpH= mitochondrial pH-cytosol pH) (Chacon et al., 1994). The maintenance of $\delta\psi$ is essential to mitochondrial respiration so that the ATP/ADP ratio in a mitochondrion is proportional to δp and $\delta\psi$. $\delta\psi$ is a

Nernstian potential determined by the free intramitochondrial concentrations of Na+, H+, Ca^{2+} etc. relative to the cytosolic concentrations of the ions. A uniporter electrogenically transports Ca^{2+} into the mitochondria while anti-porters exchange Ca^{2+}/Na^+ and Ca^{2+}/H^+. Normally, mitochondria contain small amounts of free Ca^{2+} and maintain $\delta\psi$ values of about -150 mv.

The high negative intramitochondrial potential largely reflects the electro-motive force necessary to maintain a high internal proton concentration and if cytosolic Ca^{2+} levels are elevated, the uptake of Ca^{2+} is driven by the $\delta\psi$ – allowing mitochondria to store relatively large amounts of Ca^{2+} (Richter and Kass, 1991). The $\delta\psi$ also supplies the electrical component of the proton-motive force required for oxidative phosphorylation (Richter and Kass, 1991), driving the intramitochondrial accumulation of many cations, and affecting the stability of nascent mitochondrial proteins (Cote et al., 1990). Factors which cause a marked increase in mitochondrial membrane permeability and the free distribution of mitochondrial ions and small solutes result in a loss of $\delta\psi$ and failure of energy production (Bernardes et al., 1994). Situations in which Ca^{2+} and other ions are freely lost from mitochondria or in which high levels of Ca^{2+} accumulate in mitochondria due to high cytosolic levels (van de Water et al., 1994) will result in a compromise of mitochondrial energy production.

Mitochondrial impairment appears to contribute to the progress of apop-totic neuronal death in neurodegenerative diseases (Beal, 1992; Beal et al., 1993; Frim et al., 1993a, b; Mattson et al., 1993; Tipton and Singer, 1993; Mutisya et al., 1994). Furthermore, experimental models of apoptosis includ-ing apoptosis induced by tumor necrosis factor-α (Schulze-Osthoff et al., 1992), 1-methyl-4-phenylpyridinium ions (Tipton and Singer, 1993), manga-nese (Brouillet et al., 1993), 3-nitropropionic acid (Brouillet et al., 1993), and H_2O_2 (Richter and Frei, 1988) have all been shown to involve mitochondrial impairment.

It has been shown that apoptosis can involve a reduction in $\delta\psi$ and the uncoupling of electron transport from ATP production (Vayssiere et al., 1994). There has been disagreement whether the fall in $\delta\psi$ occurs early in the apoptotic process, before DNA fragmentation, or after the onset of DNA fragmentation. For example, Raff and coworkers have argued that cells without mitochondrial DNA can undergo apoptosis and therefore mitochon-dria are unlikely to be essential to the process (Jacobson et al., 1993a, b). Alternately, experiments in cell free systems have shown that mitochondrial factors are essential to apoptosis (Newmeyer et al., 1994). Furthermore, a reduction in $\delta\psi$ has been shown to be an early irreversible step of the apoptotic death of lymphocytes in vivo (Zamami et al., 1995a, b).

We have used the potentiometric dyes, chloromethyl tetramethylrosamine (CMTFR), a rhodamine derivative, and 5,5',6,6'- tetrachloro 1,1',3,3'-tetra-ethylbenzimidazolo carbocyanine (JC-1), a carbocyanine derivative, with fluo-rescence and confocal microscopy to examine the relationship between cell death and mitochondrial failure. Studies were carried out in the partially-differ-entiated PC12 model as illustrated in Fig. 1 above. As shown in Fig. 3, we found that $\delta\psi$ was not significantly decreased in the cells at 3 hours after trophic deprivation or exposure to 0.25 mM H_2O_2. There was a gradual decline in $\delta\psi$ which became significant at 6 hours and continued to 24 hours after trophic withdrawal or exposure to 0.25 mM H_2O_2. In contrast, exposure to 1.0 mM H_2O_2 caused a rapid and progressive reduction of $\delta\psi$ that was significant by 3 hours. Hence the "low" concentration of H_2O_2 caused mitochondrial failure that was similar in time course to that found after trophic withdrawal. That is, significant reductions in $\delta\psi$ were delayed to 6 hours and progressed gradually thereafter. Both insults induced the DNA changes in the cells that are linked to apoptosis with nuclear DNA degradation (see figure 1 in Tatton et al., 1994) becoming significant after 8 hours. Therefore, $\delta\psi$ was decreased before most cells showed evidence of nuclear DNA fragmentation (Tatton et al., 1994).

In the cells undergoing necrosis due to a "high" concentration of H_2O_2, as in the right side of Fig. 2, loss of $\delta\psi$ occurred rapidly, largely within 3 hours and there was no evidence of DNA fragmentation commensurate with apoptosis suggesting that the cells presumably died due to necrosis.

Bcl-2-Bax, $\delta\psi$ and oxidative radical levels

Both Bcl-2 and Bax have been localized to mitochondrial membranes, although there has been disagreement as to whether they are located in the outer or the inner mitochondrial membrane (Chen-Levy and Cleary, 1990; Hockenbery et al., 1990; Monaghan et al., 1992; Janiak et al., 1994; Lithgow et al., 1994). Bcl-2, however, has also been localized to the endoplasmic reticular and nuclear membranes (Hockenbery et al., 1990; Akao et al., 1994; Janiak et al., 1994). It was thought that the truncated form of Bcl-2 which does not dock in mitochondrial membranes and remains in the cytosol was less effective in reducing apoptosis than the similar protein which is located in the mitochondrial membrane (Hockenbery et al., 1993). Recent research has shown that the critical portions of Bcl-2 are located just outside the portion that attaches to the mitochondrial membrane, near the membrane surface (Givol et al., 1994; Nguyen et al., 1994). Expression of human bcl-2 in C. Elegans blocks developmental nerve cell death in the worms in a manner similar to the endogenous gene, ced-9, which shares sequence homology with bcl-2. Ced-9 is an element of a polycistronic locus that contains the cyt-1 gene which encodes a protein similar to cytochrome b560 of complex II of the mitochondrial respiratory chain in mammals (Hengartner and Horvitz, 1994).

Bcl-2 overexpression in a fibrosarcoid cell line was shown to prevent a decrease in $\delta\psi$ that was associated with apoptosis caused by tumor necrosis factor (Hennet et al., 1993). Conversely, we have shown that a decrease in Bcl-2 levels in trophically-deprived PC12 cells is associated with a progressive decrease in $\delta\psi$ in the cells (see Fig. 3 and Tatton et al., 1994b). The capacity of bcl-2 overexpression to block apoptosis is overridden by mitochondrial dysfunction caused by inhibitors of the mitochondrial respiratory complexes (Smets et al., 1994; Wolvetang et al., 1994) and those inhibitors can induce apoptosis in cells that express normal levels of Bcl-2 (Wolvetang et al., 1994). It has also been shown that the reduction in apoptosis induced by Bcl-2 overexpression is associated with a decrease in oxidative radical levels and reduced peroxidation of membrane lipids (Hockenbery et al., 1993; Reed, 1994).

We have used confocal microscopy to show that a decrease in Bcl-2 levels in trophically deprived PC12 cells entering apoptosis is associated with high cytosolic levels of oxidative radicals (Wadia and Tatton, unpublished findings). Bcl-2 has been shown to alter the subcellular partitioning of Ca^{2+} with an increase in the cytosolic fraction which would be expected to compromise mitochondrial membrane potential and therefore mitochondrial respiration (Baffy et al., 1993). Therefore, the increases in cytosolic oxidative radical levels are likely secondary to a failure of mitochondrial function due to a loss of mitochondrial Bcl-2 since a reduction of ATP production has been shown to induce high levels of mitochondrially-derived superoxide radicals (Richter, 1993). Other authors oppose that view and have proposed that Bcl-2 does not decrease oxidative radical production but rather increases the action of an antioxidant pathway which eliminates radicals (Korsmeyer et al., 1993).

Based on our data and the findings of others (Chacon et al., 1994), we have argued that mitochondrial failure is an early and determining event in both necrosis and apoptosis and that a number of the events that lead to neuronal death are similar in the two processes (see Fig. 4). Bredesen and his coworkers have now presented data showing that increased Bcl-2 levels may reduce either apoptosis or necrosis (Kane et al., 1995). That finding would appear consistent with the view that a number of steps are the same in the two processes and that Bcl-2 influences a common step. Similarly, some workers have found evidence that necrosis can be associated with DNA "laddering" in some cells. We would argue that if necrosis proceeds slowly enough, and there is sufficient time for increased nuclear Ca^{2+} to activate endonucleases, then oligonucleosomal DNA digestion could occur.

Therefore, the fundamental difference may lie in the rate and extent of the cell damage. In necrosis, the somal membranes are rapidly lysed due to a combination of a failure of ATP production, massive Ca^{2+} influx and oxidative radical damage (see Fig. 4). The rapidity of the events may leave insufficient time for the activation of transcriptionally-based "repair" mechanisms and "decisional" processes relative to survival [see Oltvai and Korsmeyer (1994) and Fig. 4] or for the slower increases in nuclear Ca^{2+} that cause chromatin condensation and endonuclease activation associated with DNA digestion.

In contrast, we propose that neuronal damage leading to apoptosis is less extensive and usually involves, for example, the cells' processes rather than the somata directly. The relative slowness of the process degeneration allows time for the transcriptional and nuclear events that are characteristic of apoptosis (see Fig. 4). We believe that mitochondrial failure is the critical event in apoptosis and is responsible for the increased oxidative radical levels and consequent lipid peroxidation found in apoptosis (Hockenbery et al., 1993). Based on our findings (see below), an increase in mitochondrial Bcl-2 levels (or a decrease in Bax levels) may maintain mitochondrial membrane potential and therefore proton electromotive force and the ATP/ADP ratio. This would prevent the increases in oxidative radical levels and nuclear Ca^{2+} levels that are responsible for cell death and inducing the nuclear stigmata of apoptosis.

(–)-Deprenyl, a MAO-B inhibitor, reduces neuronal death, selectively alters protein synthesis, increases glial and neuronal process growth and reduces mitochondrially-derived oxidative radical levels without inhibiting MAO-B

Between 1991 and 1993, our group was the first to establish that (–)-deprenyl (selegiline) could reduce neuronal death by an apparently novel mechanism. Three aspects of the reduction in neuronal death by (–)-deprenyl were of particular note: 1) that (–)-deprenyl was effective if delivered within a short period of time after the neuronal damage was completed indicating that the drug could mediate neuronal rescue rather than neuronal protection (Tatton and Greenwood, 1991; Tatton et al., 1993), 2) that the doses of (–)-deprenyl which increased neuronal survival were inadequate to inhibit MAO-B, even after prolonged administration (Ansari et al., 1993), and 3) that in contrast to the stereoselectivity of MAO-B inhibition, the action on neuronal survival was stereospecific which indicated that a site other than the FAD portion of MAO-B mediated the action on neuronal survival (Ansari et al., 1993). A number of neuronal phenotypes subjected to a variety of insults, both in vivo and in vitro have now been shown to be rescued by (–)-deprenyl (Finnegan et al., 1990; Tatton and Greenwood, 1991; Salo and Tatton, 1992; Ansari et al., 1993; Tatton, 1993, 1994a; Roy and Bedard, 1993; Barber et al., 1993; Oh et al., 1994; Ju et al., 1994; Koutsiliere et al., 1994; Iwasaki et al., 1994; Zhang et al., 1995; Buys et al., 1995).

Over the last 10 years, a number of studies have shown that (–)-deprenyl can alter the expression of a variety of genes or the synthesis of a variety of proteins in either neurons or glia and that for most of the genes/proteins the changes in gene expression/protein synthesis were independent of MAO-B inhibition. The changes in gene expression/protein synthesis involved glial fibrillary acidic protein (GFAP) (Biagini et al., 1993; Li et al., 1993; Ju et al., 1994) ciliary neurotrophic factor (CNTF) (Seniuk et al., 1994), basic fibroblast growth factor (bFGF) (Biagini et al., 1994), trk C (Ekblom et al., 1993), superoxide dismutase (SOD) (Carrillo et al., 1992; Thiffault et al., 1994), and aromatic amino acid decarboxylase (AAAD) (Li et al., 1992). Relatively little attention was paid to the reported changes in protein synthesis/gene expression since each was reported independently without recognition that each finding was part of a broader transcriptional action of (–)-deprenyl.

Microdialysis showed that (–)-deprenyl (at concentrations as small as 10^{-11} to 10^{-12} M) markedly reduced the concentrations of hydroxyl radicals in the rat striatum after exposure to MPP+ (Wu et al., 1993). The increased production of oxidative radicals after MPP+ poisoning is presumeably due to complex I damage. It was not known whether (–)-deprenyl scavenged the radicals itself, whether it interfered with their production (Schapira, 1994) or whether increases in scavenging proteins like SOD1 served to reduce radical levels. More recently, it has been shown that (–)-deprenyl increases in nigral neuronal survival is coupled with the decreased levels of oxidative radicals found with microdialysis (Wu et al., 1995). Since increases in survival were found for (–)-deprenyl concentrations as small as 10^{-11} M, it seems that scavenging by (–)-deprenyl is unlikely to account for the decreased radical levels and the increased neuronal survival.

Recent studies have shown that (–)-deprenyl markedly increased the process length of process bearing astrocytes in "wounded" cultures (Seniuk et al., 1994) and the growth of neurites in explanted rat spinal motoneurons (Iwasaki et al., 1994). Both actions occur at concentrations as small as 10^{-11} to 10^{-12} M, concentrations that are insufficient to inhibit MAO.

Increases in neuronal survival by (–)-deprenyl require new protein synthesis

In order to determine whether the protein synthesis/gene expression changes reported for (–)-deprenyl were in any way related to the reduction in neuronal death, particularly apoptosis, we developed an in vitro model of apoptosis using trophic withdrawal in partially-differentiated PC12 cells. (–)-Deprenyl reduced the death of the trophically-withdrawn PC12 cells and also decreased the percentage of the cells with evidence of nuclear DNA digestion [see figures 1 and 2 in Tatton et al. (1994)]. Studies with transcriptional and translational blocking agents showed that the anti-apoptotic action of (–)-deprenyl in the partially-differentiated PC12 cells required new protein synthesis [see figure 3A3 in Tatton et al. (1994)]. Kinetic studies of trancriptional or translational blockade showed that (–)-deprenyl induced the new protein synthesis [see figures 3B-3E in Tatton et al. (1994)]. In the PC12 cells, 4 hours of unblocked transcription or 6 hours of unblocked translation was sufficient to reduce the apoptosis.

What are the proteins induced by (–)-deprenyl and why do they reduce neuronal death?

Work is in progress in our laboratory using differential display polymerase chain reaction (PCR) and two dimensional protein gels as a means of confirming whether (–)-deprenyl did in fact alter gene transcription and new protein synthesis in the partially differentiated PC 12 cells entering into apoptosis. The differential display PCR has revealed a reduction in most PCR products in PC12 cells by 3–6 hours after trophic deprivation caused by placement of the washed cells into minimum essential media (MEM) compared with those which retained trophic support by replacement into serum and NGF (M/S+N). Addition of (–)-deprenyl (10^{-9} M) to MEM maintained the presence and apparent levels of many of the PCR products. It also increased the levels of some of the products and induced some new products that were not detectable when compared with cells that retained trophic support by replacement into serum and NGF.

Autoradiograms for two dimensional gels for proteins extracted from PC12 cells treated with ^{35}S labeled methionine just prior to washing to remove trophic proteins, showed: a) that more than 100 newly synthesized proteins could be detected in PC12 cells replaced into M/S+N at 6 hours after washing; b) that at 6 hours after washing the number of detectable newly synthesized proteins were markedly reduced to fewer than 20 in cells placed into M/O. Despite the marked overall decrease, several proteins showed increased levels and several proteins that were not apparent in cells replaced into M/S+N were evident in M/O; and c) that placement into M/-d maintained or increased 30–40 proteins that were decreased or lost in M/O by 6 hours. Furthermore, several proteins not detectable in either M/S+N or M/O were evident in M/-d and some proteins that were detectable in M/O seemed to be decreased in M/-d.

The differential display PCR and two dimensional protein gels showed: 1) that partially-differentiated PC12 cells entering into apoptosis markedly reduce overall gene expression and protein synthesis by 3–6 hours after trophic withdrawal. A small number of gene products/proteins were increased or first became evident by 6 hours after trophic withdrawal in accord with the concept of "death proteins". Like earlier studies (Rukenstein et al., 1991), our research with transcriptional and translational blockers showed that those newly synthesized proteins could not be essential to the progression of the apoptosis and that the reduction of a constitutive protein was more likely responsible [see figures 3A1 and 3A2 in Tatton et al. (1994)]; and 2) the fact that (–)-deprenyl (10^{-9} M) caused approximately 40 genes/proteins that were lost at 6 hours after trophic withdrawal to be retained or increased. Furthermore, a smaller number of proteins/genes, that were not expressed after trophic withdrawal, were found to be expressed with (–)-deprenyl treatment.

We have begun to use RT PCR and western blots complemented by in situ hybridization, immunocytochemistry and pulse-chase methods to identify the genes and their proteins whose transcription is altered in the trophically-deprived PC12 cells. These studies show that the expression/synthesis of most genes/proteins in the PC12 cells entering into apoptosis are not effected (neurofilament light protein, for example). To date, more than a dozen proteins have been shown to be altered by (–)-deprenyl in the cells including c-Jun, c-Fos, Bcl-2, Bcl-X_L, Bax, NADH dehydrogenase subunit 1, SOD1, SOD2, glutathione peroxidase and inducible heat shock protein 70. Most other proteins levels, like tubulin or MAP-2, are unaltered so that the changes in gene expression are selective. The expression of c-Fos, Bcl-2, SOD1 and SOD2 were increased in the cells entering apoptosis to levels similar to those found in cells supported by NGF and serum while the expression of Bax was decreased compared to that of cells entering apoptosis or those maintained in NGF and serum.

These results raise the possibility that an increase in Bcl-2 and a decrease in Bax in partially-differentiated PC12 cells combine to reduce the vulnerability of PC12 cells to trophic withdrawal and that an increase in SOD1 and SOD2 reduce the concentrations of superoxide radicals which would contribute to the progression of apoptosis in the cells (see Fig. 4).

We have carried out confocal microscopy experiments similar to those shown in Fig. 3 using (–)-deprenyl in the trophically-withdrawn PC12 cells and have found that (–)-deprenyl prevents a reduction in mitochondrial membrane potential after trophic-withdrawal (Tatton et al., 1994b). That finding would seem to be consistent with other work suggesting that a maintenance of Bcl-2 levels is accompanied by a maintenance of mitochondrial membrane potential (Hennet et al., 1993). Other work in progress in our laboratory using confocal microscopy imaging of living PC12 cells after trophic withdrawal have shown that (–)-deprenyl reduces mitochondrial free Ca^{2+} levels and cytoplasmic oxidative radical levels in the cells (Wadia and Tatton, unpublished findings).

Taken together these findings may explain all of the MAO-B independent actions of (–)-deprenyl described above including the capacity to decrease neuronal apoptosis in vivo and in vitro, the capacity to alter the expression/ synthesis of selected genes/proteins, the capacity to reduce the concentration of mitochondrially derived oxidative radicals and the capacity to increase neuronal or glial process growth. Most importantly, they indicate that at least part of the reduction in neuronal death caused by (–)-deprenyl results from selective alterations in gene transcription that maintain mitochondrial function and decrease levels of oxidative radicals. A full understanding of the molecular events underlying the MAO-B independent transcriptional actions of (–)-deprenyl, particularly identification of the binding site(s) for the drug and those events linking the non-MAO-B binding to selective alterations in gene expression may provide for the design of systemic agents that will effectively reduce apoptosis in neurodegenerative diseases without the side effects found so far for neurotrophic factor therapy.

References

Akao Y, Otsuki U, Kataoka S, Ito Y, Tsujimoto Y (1994) Multiple subcellular localization of bcl-2: detection in nuclear outer membrane, endoplasmic reticulum membrane, and mitochondrial membranes. Cancer Res 54: 2468–2471

Altman J (1992) Programmed cell death: the paths to suicide. TINS 15: 278–280

Anglade P, Michel P, Marquez J, Mouatt-Prient A, Ruberg M, Hirsch EC, et al (1995) Apoptotic degeneration of nigral dopaminergic neurons in Parkinson's disease. Proc Natl Acad Sci 21: 489–493

Ansari KS, Yu PH, Kruck TX, Tatton WG (1993) Rescue of axotomized immature rat-facial motoneurons by R(–)-deprenyl: stereospecificity and independence from monoamine oxidase inhibition. J Neurosci 13: 4042–4053

Baffy G, Miyashita T, Williamson JR, Reed JC (1993) Apoptosis induced by withdrawal of Interleukin 3 (IL-3) from an IL-3-dependent hematopoietic cell line is associated with partitioning of intracellular calcium and is blocked by enforced Bcl-2 oncoprotein production. J Biol Chem 268: 6511–6519

Barber AJ, Paterson IA, Gelowitz DL, Voll CL (1993) Deprenyl protects rat hippocampal pyramidal cells from ischemic insult. Soc Neurosci Abstr 19: 1646

Baringaga M (1994) Neurotrophic factors enter the clinic. Science 264: 772–774

Batistatou A, Merry DW, Korsmeyer SJ, Greene LA (1993) Bcl-2 affects survival but not neuronal differentiation of PC12 cells. J Neurosci 13: 4422–4428

Beal MF (1992) Does impairment of energy metabolism result in excitotoxic neuronal death in neurodegenerative illnesses? Ann Neurol 31: 119–130

Beal MF, Hyman T, Koroshetz W (1993) Do defects in mitochondrial energy metabolism underlie the pathology of neurodegenerative diseases? TINS 16: 125–131

Behl C, Hovey L, Krajewski S, Schubert D, Reed JC (1993) Bcl-2 prevents killing of neuronal cells by glutamate but not by amyloid beta protein. Biochem Biophys Res Commun 197: 949–956

Berkelaar M, Clarke DB, Wang YC, Bray GM, Aguayo AJ (1994) Axotomy results in delayed death and apoptosis of retinal ganglion cells in adult rats. J Neurosci 14: 4368–4374

Bernardes CF, Meyer-Fernandes JR, Basseres DS, Castilho RF, Vercesi AE (1994) Ca(2+)-dependent permeabilization of the inner mitochondrial membrane by 4,4'-diisothicyana-tostilbene-2,2'-disulfonic acid (DIDS). Biochim Biophys Acta 1188: 93–100

Biagini G, Zoli M, Fuxe K, Agnati LF (1993) L-Deprenyl increases GFAP immunoreactivity selectively in activated astrocytes in rat brain. Neuroreport 4: 955–958

Biagini G, Frasoldati A, Fuxe K, Agnati LF (1994) The concept of astrocyte-kinetic drug in the treatment of neurodegenerative diseases: evidence for L-Deprenyl-induced activation of reactive astrocytes. Neurochem 25: 17–22

Bonfoco E, Krainc D, Ankarcrona M, Nicotera P, Lipton SA (1995a) Apoptosis and necrosis: two distinct events induced, respectively, by mild and intense insults with N-methyl-D-aspartate or nitric oxide/superoxide in cortical cell cultures. Proc Natl Acad Sci USA 92: 7162–7166

Bonfoco E, Ceccatelli S, Manzo L, Nicotera P (1995b) Colchicine induces apoptosis in cerebellar granule cells. Exp Cell Res 218: 189–200

Brouillet E, Jenkins BG, Hyman BT, Ferrante RJ, Kowall NW, Srivastave R, Roy DS, BRR, Beal MF (1993) Age-dependent vulnerability of the striatum to the mitochondrial toxin 3-nitropropionic acid. J Neurochem 60: 356–359

Brugg B, Michel PP, Agid Y, Ruberg M (1996) Ceramide induces apoptosis in cultured mesencephalic neurons. J Neurochem 66: 733–739

Buchi ER (1992) Cell death in the rat retina after a pressure-induced ischaemia-reperfusion insult: an electron microscopic study. I. Ganglion cell layer and inner nuclear. Exp Eye Res 55: 605–613

Buys YM, Trope GE, Tatton WG (1995) (-)-Deprenyl increases the survival of retinal ganglion cells after optic nerve crush. Curr Eye Res 14: 119–126

Carrillo MC, Kanai S, Nokubo M, Ivy GO, Sato Y, Kitani K (1992) (–)-Deprenyl increases activities of superoxide dismutase and catalase in striatum but not in hippocampus. The sex and age-related differences in the optimal dose in the rat. Exp Neurol 116: 286–294

Chacon E, Reece JM, Nieminen AL, Zahrebelski G, Herman B, Lemasters JJ (1994) Distribution of electrical potential, pH, free Ca2+, and volume inside cultured adult rabbit cardiac myocytes during chemical hypoxia: a multiparameter digitized confocal microscopic study. Biophys 66: 942–952

Chang G-Q, Hao Y, Wong F (1993) Apoptosis: final common pathway of photoreceptor death in rd, rds, and rhodopsin mutant mice. Neuron 11: 595–605

Chen-Levy S, Cleary ML (1990) Membrane topology of the Bcl-2 proto-oncogenic protein demonstrated in vitro. J Biol Chem 265: 4929–4933

Choi DW, Rotham SM (1990) The role of glutamate neurotoxicity in hypoxic-ischemic neuronal death. Ann Rev Neurosci 13: 171–182

Copani A, Bruno VM, Barresi V, Battaglia G, Condorelli DF, Nicoletti F (1995) Activation of metabotropic glutamate receptors prevents neuronal apoptosis in culture. J Neurochem 64: 101–108

Cote C, Boulet D, Poirier J (1990) Expression of the mammalian mitochondrial genome-role for membrane potential in the production of mature translational products. J Biol Chem 265: 7532–7538

Cotman CW, Anderson AJ (1995) A potential role for apoptosis in neurodegeneration and Alzheimer's disease. Mol Neurobiol 10: 19–45

Cotman CW, Whittemore ER, Watt JA, Anderson AJ, Loo DT (1994) Possible role of apoptosis in Alzheimer's disease. Ann NY Acad Sci 747: 36–49

Csernansky CA, Canzoniero LM, Sensi SL, Yu SP, Choi DW (1994) Delayed application of aurintricarboxylic acid reduces glutamate-induced. J Neurosci Res 38: 101–108

Dal-Canto RC, Gurney ME (1994) Development of central nervous system pathology in a murine transgenic model of human amyotrophic lateral sclerosis. Am J Pathol 145: 1271–1279

Deckwerth TL, Johnson EM (1993) Temporal analysis of events associated with programmed cell death (apoptosis) of sympathetic neurons deprived of nerve growth factor. J Cell Biol 123: 1207–1222

Deng HX, Hentati A, Tainer JA, Iqbal Z, Cayabyab A, Hung WY, Getzoff ED, Hu P, Herzgeldt B, Roos RP (1993) Amyotrophic lateral sclerosis and structural defects in Cu, Zn superoxide dismutase. Science 261: 1047–1051

Dessi F, Ben-Ari Y, Charriaut-Marlangue C (1994) Increased synthesis of specific proteins during glutamate-induced neuronal death in cerebellar culture. Brain Res 654: 27–33

Dessi F, Pollard H, Moreau J, Ben Ari Y, Charriaut-Marlangue C (1995) Cytosine arabinoside induces apoptosis in cerebellar neurons in culture. J Neurochem 64: 1980–1987

Dipasquale B, Marini AM, Youle RJ (1991) Apoptosis and DNA degradation induced by 1-methyl-4-phenyl-tetrahydropyridine. Biochem Biophys Res Commun 181: 1442–1448

Dragunow M, Young D, Hughes P, MacGibbon G, Lawlor P, Singleton K (1993) Is c-Jun involved in nerve cell death following status epilepticus and hypoxic-ischaemic brain injury? Mol Brain Res 18: 347–352

Dragunow M, Faull RL, Lawlor P, Beilharz EJ, Singleton K, Walker EB, et al (1995) In situ evidence for DNA fragmentation in Huntington's disease striatum and Alzheimer's disease temporal lobes. Neuroreport 6: 1053–1057

Dubois-Dauphin MFH, Tsujimoto Y, Huatte J, Martinou JC (1994) Neonatal motoneurons overexpressing the bcl-2 protooncogene in transgenic mice are protected from axotomy-induced cell death. Proc Natl Acad Sci USA 91: 3309–3313

Eisen A, Krieger C (1993) Pathogenic mechanisms in sporadic amyotrophic lateral sclerosis. Can J Neurol Sci 20: 286–296

Ekblom J, Jossan SS, Ebendal T, Soderstrom S, Oreland L, Aquilonius S-M (1993) Expression of mRNAs for neurotrophins and members of the Trk family in the rat brain after treatment with L-deprenyl. Acta Neurol Scand 84 [Suppl]: 79–86

Farinelli SE, Greene LA (1996) Cell cycle blockers mimosine, ciclopirox and deferoxamine prevent the death of PC12 cells and postmitotic sympathetic neurons after removal of trophic support. J Neurosci 16: 1150–1162

Finch CE (1993) Neuron atrophy during aging: programmed or sporadic? TINS 16: 104–110

Finnegan KT, Karler R (1992) Role for protein synthesis in the neurotoxic effects of methamphetamine in mice and rats. Br Res 591: 160–164

Finnegan KT, Skratt JJ, Irwin I, DeLanney LE, Langston JW (1990) Protection against DSP-4 induced neurotoxicity by deprenyl is not related to its inhibition of MAO B. Eur J Pharmacol 184:

Forloni G (1993) beta-Amyloid neurotoxicity. Funct Neurol 8: 211–225

Forloni G, Chiesa R, Smiroldo S, Verga L, Salmona M, Tagliavini F (1993) Apoptosis mediated neurotoxicity induced by chronic application of beta amyloid fragment 25–35. Neuroreport 4: 523–526

Frim DM, Simpson J, Uhler TA, Short MP, Bossi SR, Breakfield XO, Isacson O (1993a) Striatal degeneration induced by mitochondrial blockade is prevented by biologically delivered NGF. J Neurosci Res 35: 452–458

Frim DM, Uhler TA, Short MP, Ezzedine ZD, Klagsbrun M, Breakefield XO (1993b) Effects of biologically delivered NGF, BDNF and bFGF on striatal excitotoxicity. Neuroreport 4: 367–370

Garcia VE, Gorczyca W, Darzynkiewicz Z, Sharma SC (1994) Apoptosis in adult retinal ganglion cells after axotomy. J Neurobiol 25: 431–438

Givol I, Tsarfaty I, Resau J, Rulong S, da Silva PP, Nasioulas G, DuHadaway J, Hughes SH, Ewert DL (1994) Bcl-2 expressed using a retroviral vector is localized primarily in the nuclear membrane and the endoplasmic reticulum of chicken embryo fibroblasts. Cell Growth Differ 5: 419–429

Gobe GC (1994) Apoptosis in brain and gut tissue of mice fed a seed preparation of the cycad Lepidozamia peroffshyana. Biochem Biophys Res Commun 205: 327–333

Greenlund LJS, Deckwerth TL, Johnson EM (1995) Superoxide dismutase delays neuronal apoptosis: a role for reactive oxygen species in programmed neuronal death. Neuron 14: 303–315

264   W. G. Tatton et al.

Gurney ME, Pu H, Chiu AY, Dal-Canto MC, Polchow CY, Alexander DD (1994) Motor
    neuron degeneration in mice that express a human Cu, Zn superoxide dismutase
    mutation. Science 264: 1771–1775
Hall ED, McCall JM (1994) Therapeutic potential of the lazeroids (21 aminosteroids) in
    acute CNS trauma, ischemia and subarachnoid hemorrhage. Adv Pharmacol 28: 221–
    268
Hamburger V, Oppenheim RW (1982) Naturally-occurring neuronal death in vertebrates.
    Neurosci Comment 1: 38–55
Hartley A, Stone JM, Heron C, Cooper JM, Schapira AH (1994) Complex I inhibitors
    induce dose-dependent apoptosis in PC12 cells. J Neurochem 63: 1987–1990
Henderson JT, Seniuk NA, Richardson PM, Gauldie J, Roder JC (1994) Systemic
    administration of ciliary neurotrophic factor induces cachexia in rodents. J Clin Invest
    93: 2632–2638
Hengartner MO, Horvitz HR (1994) C. elegans cell survival gene ced-9 encodes a
    functional homologue of the mammalian proto-oncogene bcl-2. Cell 76: 655–676
Hennet TG, Bertoni G, Richter C, Peterhans E (1993) Expression of Bcl-2 protein
    enhances the survival of mouse fibrosarcoid cells in tumor necrosis factor-mediated
    cytotoxicity. Cancer Res 53: 1456–1460
Hockenbery D, Nuez G, Milliman C, Schreiber RD, Korsmeyer SJ (1990) Bcl-2 is an inner
    mitochondrial membrane protein that blocks programmed cell death. Nature 348:
    334–336
Hockenbery DM, Ottval ZN, Xiao-Ming Y, Korsmeyer SJ (1993) Bcl-2 functions in an
    antioxidant pathway to prevent apoptosis. Cell 75: 241–251
Iwasaki Y, Ikeda K, Shoijima T, Kobayashi T, Tagaya N, Kinoshita M (1994) Deprenyl
    enhances neurite outgrowth in cultured rat spinal ventral horn neurons. J Neurol Sci
    125: 11–13
Jacobson MD, Burne JF, King MP, Miyashita T, Reed JC, Raff MC (1993) Bcl-2 blocks
    apoptosis in cells lacking mitochondrial DNA. Nature 361: 365–369
Janiak F, Leber B, Andrews DW (1994) Assembly of Bcl-2 into microsomal and outer
    mitochondrial membranes. Biol Chem 269: 9842–9849
Joseph R, Li W, Han E (1993) Neuronal death, cytoplasmic calcium and internucleosomal
    DNA fragmentation: evidence for DNA fragments being released from cells. Brain
    Res Mol Brain Res 17: 70–76
Johnson EM Jr (1994) Possible role of neuronal apoptosis in Alzheimer's disease.
    Neurobiol Aging 15(2): S187–189
Ju WYH, Holland DP, Tatton WG (1994) (–)-Deprenyl alters the time course of death of
    axotomized facial motoneurons and the hypertrophy of neighboring astrocytes in
    immature rats. Exp Neurol 126: 233–246
Kaku DA, Giffard RG, Choi DW (1993) Neuroprotective effects of glutamate antagonists
    and extracellular acidity. Science 260: 1516–1518
Kane DJ, Ord T, Anton R, Bredesen DE (1995) Rapid Communication: Expression of
    bcl-2 inhibits necrotic neural cell death. J Neurosci Res 40: 269–275
Kobayashi Y, Saheki T, Shinozawa T (1994) Induction of PC12 cell death, apoptosis, by
    a sialoglycopeptide from bovine brain. Biochem Biophys Res Commun 203: 1554–
    1559
Koh JY, Cotman CW (1992) Programmed cell death: its possible contribution to neuro-
    toxicity mediated by calcium channel antagonists. Brain Res 587: 233–240
Korsmeyer SJ (1992) Bcl-2: an antidote to programmed cell death. Cancer Surveys 15:
    105–118
Korsmeyer SJ, Shutter JR, Veis DJ, Merry DE, Oltvai ZN (1993) Bcl-2/Bax: a rheostat
    that regulates an anti-oxidant pathway and cell death. Semin Cancer Biol 4: 327–332
Koutsiliere E, O'Callaghan JFX, Chen T-S, Riederer P, Rausch W-D (1994) Selegiline
    enhances survival and neurite outgrowth of MPP-treated dopaminergic neurons. Eur
    J Pharmacol 269: R3–R4
Langston JW (1994) MPTP Parkinsonism. Mov Disord 9: 3

Lassmann H, Bancher C, Breitschopf H, Wegiel J, Bobinski M, Jellinger K, et al (1995) Cell death in Alzheimer's disease evaluated by DNA fragmentation in situ. Acta Neuropathol Berl 89: 35–41

Lefebvre S, Burglen L, Reboullet S, Clermont O, Burlet P, Viollet L, Benichou B, Cruaud C, Millasseau P, Zeviani M, Le Paslier D, Frezal J, Cohen D, Weissenbach J, Munnich A, Melki J (1995) Identification and characterization of a spinal muscular atrophy-determining gene. Cell 80: 155–165

Li XM, Juorio AV, Paterson IA, Zhu MY, Boulton AA (1992) Specific irreversible monoamine oxidase-B inhibitors stimulate gene expression of aromatic L-amino acid decarboxylase in PC12-cells. J Neurochem 59: 2324–2327

Li XM, Qi J, Juorio AV, Boulton AA (1993) Reduction in glial fibrillary acidic protein messenger RNA abundance induced by (–)-Deprenyl and other monoamine oxidase B-inhibitors in C6 glioma cells. J Neurochem 61: 1573–1576

Linnik MD, Zobrist RH, Hatfield MD (1993) Evidence supporting a role for programmed cell death in focal cerebral ischemia in rats. Stroke 24: 2002–2008

Lolley RN, Rong H, Craft CM (1994) Linkage of photoreceptor degeneration by apoptosis with inherited defect in phototransduction. Invest Ophthalmol Vis Sci 35: 358–362

Margolis RL, Chuang DM, Post RM (1994) Programmed cell death: implications for neuropsychiatric disorders. Biol Psychiatry 35: 946–956

Mitchell IJ, Lawson S, Moser B, Laidlaw SM, Cooper AJ, Walkinshaw G, et al (1994) Glutamate-induced apoptosis results in a loss of striatal neurons in the parkinsonian rat. Neurosci 63: 1–5

Mochizuki H, Nakamura N, Nishi K, Mizuno Y (1994) Apoptosis is induced by 1-methyl-4-phenylpyridinium ion (MPP+) in ventral mesencephalic-striatal co-culture in rat. Neurosci Lett 170: 191–194

Monaghan P, Robertson D, Amos AS, Dyer MJS, Mason DY, Greaves MF (1992) Ultrastructural localization of Bcl-2 protein. J Histochem Cytochem 40: 1819–1825

Montpied P, Weller M, Paul SM (1993) N-methyl-D-aspartate receptor agonists decrease protooncogene bcl-2 mRNA expression in cultured rat cerebellar granule neurons. Biochem Biophys Res Commun 195: 623–629

Mou L, Miller H, Li J, Wang E, Chalifour L (1994) Improvements to differential display method for gene analysis. Biochem Biophys Res Comm 199: 564–569

Muller WE, Schroder HC, Ushijima H, Dapper J, Bormann J (1992) gp120 of HIV-1 induces apoptosis in rat cortical cell cultures: prevention by memantine. Eur J Pharmacol 226: 209–214

Mutoh T, Tokuda A, Marini AM, Fujiki N (1994) 1-Methyl-4-phenylpyridinium kills differentiated PC12 cells with a concomitant change in protein phosphorylation. Br Res 661: 51–55

Newmeyer DD, Farschon DM, Reed JC (1994) Cell-free apoptosis in Xenopus egg extracts: inhibition by Bcl-2 and requirement for an organelle fraction enriched in mitochondria. Cell 79: 353–364

Nguyen M, Branton PE, Walton PA, Oltvai ZN, Korsmeyer SJ, Shore GC (1994) Role of membrane anchor domain of Bcl-2 in suppression of apoptosis caused by E1B-defective adenovirus. J Biol Chem 269: 16521–16524

Oh C, Murray B, Bhattacharya N, Holland D, Tatton WG (1994) (–)-Deprenyl alters the survival of adult facial motoneurons after axotomy: increases in vulnerable C57BL strain but decreases in Mnd mutants. J Neurosci Res 38:

Olanow CW (1993) A radical hypothesis for neurodegeneration. TINS 16: 439–444

Oltvai ZN, Korsmeyer SJ (1994) Checkpoints of dueling dimers foil death wishes. Cell 79: 189–192

Oltvai Z, Milliman C, Korsmeyer SJ (1993) Bcl-2 heterodimers in vivo with a conserved homologue, Bax, that accelerates programmed cell death. Cell 74: 609–619

Pender MP, Nguyen KB, McCombe PA, Kerr JFR (1991) Apoptosis in the nervous system in experimental allergic encephalomyelitis. J Neurol Sci 104: 81–87

Petito CK, Roberts B (1995) Evidence of apoptotic cell death in HIV encephalitis (see comments). Am J Pathol 146: 1121–1130

Portera CC, Sung CH, Nathans J, Adler R (1994) Apoptotic photoreceptor cell death in mouse models of retinitis pigmentosa. Proc Natl Acad Sci USA 91: 974–978

Portera-Cailliau C, Hedreen JC, Price DL, Koliatsos VE (1995) Evidence for apoptotic cell death in Huntington disease and excitotoxic animal models. J Neurosci 15: 3775–3787

Price DL, Martin LJ, Clatterbuck RE, Koliatsos VE, Sisodia SS, Walker LC, Cork LC (1992) Neuronal degeneration in human diseases and animal models. J Neurobiol 23: 1277–1294

Rabacchi SA, Bonfanti L, Liu XH, Maffei L (1994)Apoptotic cell death induced by optic nerve lesion in the neonatal rat. J Neurosci 14: 5292–5301

Rabizadeh S, Bitler CM, Butcher LL, Bredesen DE (1994) Expression of the low-affinity nerve growth factor receptor enhances beta-amyloid peptide toxicity. Proc Natl Acad Sci USA 91: 10703–10706

Reed JC (1994) Bcl-2 and the regulation of programmed cell death. J Cell Biol 124: 1–6

Richter C (1993) Pro-oxidants and mitochondrial Ca2+: their relationship to apoptosis and oncogenesis. FEBS Letts 325: 104–107

Richter C, Frei B (1988) Ca2+ release from mitochondria induced by prooxidants. Free Rad Biol Med 4: 365–375

Richter C, Kass GEN (1991) Oxidative stress in mitochondria: its relationship to cellular Ca2+ homeostasis, cell death, proliferation and differentiation. Chem Biol Interact 77: 1–23

Rosen DR, Siddique T, Patterson D, Figlewicz DA, Sapp P, Hentati P, Donaldson A, Goto J, O'Regan JP, Deng HK, Gusella JS, Horvitz HR, Brown RH (1994) Mutations in Cu/Zn superoxide dismutase gene are associated with familial amyotrophic lateral sclerosis. Nature 362: 59–62

Rosenbaum DM, Michaelson M, Batter DK, Doshi P, Kessler JA (1994) Evidence for hypoxia-induced, programmed cell death of cultured neurons. Ann Neurol 36: 864–870

Roy E, Bedard PJ (1993) Deprenyl increases survival of rat fetal nigral neurones in culture. Neuroreport 4: 1183–1186

Roy N, Mahadevan MS, McLean M, Shutler G, Yaraghi Z, Farahani R, Baird S (1995) The gene for neuronal apoptosis inhibitory protein is partially deleted in individuals with spinal muscular atrophy. Cell 80: 167–178

Rukenstein A, Rydel RE, Greene LA (1991) Multiple agents rescue PC12 cells from serumfree cell death by translation- and transcription-independent mechanisms. J Neurosci 11: 2552–2563

Salo PT, Tatton WG (1992) Deprenyl reduces the death of motoneurons caused by axotomy. J Neurosci Res 31: 394–400

Samples SD, Dubinsky JM (1993) Aurintricarboxylic acid protects hippocampal neurons from glutamate. J Neurochem 61: 382–385

Schapira AV (1994) Mechanisms of cell death in Parkinson's disease. Mov Disord 9: 2

Schulze-Osthoff K, Bakker AC, Vanhaesebroeck B, Beyaert R, Jacob WA, Fiers W (1992) Cytotoxic activity of tumor necrosis factor is mediated by early damage of mitochondrial functions. J Biol Chem 267: 5317–5323

Sengstock GJ, Olanow CW, Dunn AJ, Arendash GW (1992) Iron induces degeneration of nigrostriatal neurons. Brain Res Bull 28: 645–649

Seniuk NA, Henderson JT, Tatton WG, Roder JC (1994) Increased CNTF gene expression in process bearing astrocytes following injury is augmented by R(–)-deprenyl. J Neurosci Res 37:

Shahinfar S, Edward DP, Tso MO (1991) A pathologic study of photoreceptor cell death in retinal photic injury. Curr Eye Res 10: 47–59

Silveira LC, Russelakis CM, Perry VH (1994) The ganglion cell response to optic nerve injury in the cat: differential responses revealed by neurofibrillar staining. J Neurocytol 23: 75–86

Slater AF, Nobel CS, Orrenius S (1995) The role of intracellular oxidants in apoptosis. Biochim Biophys Acta 1271 : 59–62

Smale G, Nichols NR, Brady DR, Finch CE, Horton WE Jr (1995) Evidence for apoptotic cell death in Alzheimer's disease. Exp Neurol 133: 225–230

Smets LA, Van den Berg J, Acton D, Top B, Van Rooij H, Verwijs-Janssen M (1994) Bcl-2 expression and mitochondrial activity in leukemic cells with different sensitivity to glucocorticoid-induced apoptosis. Blood 84: 1613–1619

Steinberg RH (1994) Survival factors in retinal degenerations. Curr Opin Neurobiol 4: 515–524

Su JH, Anderson AJ, Cummings BJ, Cotman CW (1994) Immunohistochemical evidence for apoptosis in Alzheimer's disease. Clin Neurosci Neuropathol 5: 2529–2533

Takayama S, Takaaki S, Kajewski S, Kochel K, Irie S, Millan JA, Reed JC (1995) Cloning and functional analysis of BAG-1: a novel Bcl-2 binding protein with anti-cell death activity. Cell 80: 279–284

Tatton WG (1993) Selegiline ((–)-deprenyl) can mediate neuronal rescue rather than neuronal protection. Mov Disord 8

Tatton WG, Greenwood CE (1991) Rescue of dying neurons: a new action for Deprenyl in MPTP parkinsonism. J Neurosci Res 30: 666–672

Tatton WG, Seniuk NA, Ju WYH, Ansari KS (1993) Reduction of nerve cell death by deprenyl without monoamine oxidase inhibition. In: Leiberman A, Olanow W, Youdim MBH, Tipton K (eds) Monoamine oxidase inhibitors in neurological diseases. Raven Press, New York, pp 217–248

Tatton WG, Ju WYL, Holland DP, Tai CE, Kwan MM (1994) (–)-Deprenyl reduces PC12 cell apoptosis by inducing new protein synthesis. J Neurochem 63: 1572–1574

Thiffault C, Aumont N, Quirion R, Poirier J (1994) Antioxidant enzymes in an animal model of Parkinson's disease. Can J Physiol Pharmacol 72: 592

Thomas LB, Gates DJ, Richfield EK, Schweitzer JB, Steindler DA (1995) DNA end labeling (TUNEL) in Huntingtons disease and other neuropathological conditions. Exp Neurol 133: 265–272

Thompson CB (1995) Apoptosis in the pathogenesis and treatment of disease. Science 267: 1456–1462

Tipton KF, Singer TP (1993) Advances in our understanding of the mechanisms of the neurotoxicity of MPTP and related compounds. J Neurochem 61: 1191–1207

Troy CM, Shelanski ML (1994) Down-regulation of copper/zinc superoxide dismutase causes apoptotic death. Proc Natl Acad Sci USA 91: 6384–6387

Tso MO, Zhang C, Abler AS, Chang CJ, Wong F, Chang GQ, Lam TT (1994) Apoptosis leads to photoreceptor degeneration in inherited retinal dystrophy of RCS rats. Invest Ophthalmol Vis Sci 35: 2693–2699

Tsuda T, Munthasser S, Fraser PE, Percy ME, Rainero I, Vaula G, Pinessi L, Bergamini L, Vignocchi G, Crapper McLachlan DR, Tatton WG, St. George-Hyslop P (1994) Analysis of the functional effects of a mutation in SOD1 associated with familial Amyotrophic Lateral Sclerosis. Neuron 13: 727–736

van de Water B, Zoeteweij JP, deBont HJ, Mulder GJ, Nagelkerke JF (1994) Role of mitochondrial Ca2+ in the oxidative stress-induced dissipation of the mitochondrial membrane potential. Studies in isolated proximal tubular cells using the nephrotoxin 1,2-dichlorovinyl-L-cysteine. J Biol Chem 269: 14546–14552

Vayssiere JL, Petit PX, Risler Y, Mignotte B (1994) Commitment to apoptosis is associated with changes in mitochondrial biogenesis and activity in cell lines conditionally immortalized with simian virus 40. Proc Natl Acad Sci USA 91: 11752–11760

Walkinshaw G, Waters CM (1994) Neurotoxin-induced cell death in neuronal PC12 cells is mediated by induction of apoptosis. Neurosci 63: 975–987

Werth JL, Thayer SA (1994) Mitochondria buffer physiological Ca2+ loads in cultured rat dorsal root ganglion neurons. J Neurosci 14: 348–356

Wolvetang EF, Johnson KL, Krauer K, Ralph SJ, Linnane AW (1994) Mitochondrial respiratory chain inhibitors induce apoptosis. FEBS Lett 339: 40–44

Wu R-M, Chiueh CC, Pert A, Murphy DL (1993) Apparent antioxidant effect of 1-deprenyl on hydroxyl radical formation and nigral injury elicited by MPP+ in vivo. Eur J Pharmacol 243: 241–248

Wu RM, Murphy DL, Chiueh CC (1995) Neuronal protective and rescue effects of deprenyl against MPP+ dopaminergic toxicity. J Neural Transm 100: 53–61

Wyllie AH, Morris RG, Smith AL, Dunlop D (1984) Chromatin cleavage in apoptosis: association with condensed chromatin morphology and dependence on macromolecular synthesis. J Pathol 142: 67–77

Yan GM, Ni B, Weller M, Wood KA, Paul SM (1994) Depolarization or glutamate receptor activation blocks apoptotic cell death. Brain Res 656: 43–51

Yang E, Zha J, Jockel J, Boise LH, Thompson CB, Korsmeyer SJ (1995) Bad, a heterodimeric partner for Bcl-xL and Bcl-2 displaces Bax and promotes cell death. Cell 80: 285–291

Yin X-M, Oltvai ZN, Korsmeyer SJ (1994) BH1 and BH2 domains of Bcl-2 are required for inhibition of apoptosis and heterodimerization with Bax. Nature 369: 321–323

Yoshiyama Y, Yamada T, Asanuma K, Asahi T (1994) Apoptosis related antigen, Le(Y) and nick-end labeling are positive in spinal motor neurons in amyotrophic lateral sclerosis. Acta Neuropathol Berl 88: 207–211

Zamzami N, Marchetti P, Castedo M, Decaudin D, Macho A, Hirsch T, et al (1995a) Sequential reduction of mitochondrial transmembrane potential and generation of reactive oxygen species in early programmed cell death. J Exp Med 182: 367–377

Zamzami N, Marchetti P, Castedo M, Zanin C, Vayssiere J-L, Petit PX, Kroemer G (1995b) Reduction in mitochondrial potential constitutes an early irreversible step of programmed lymphocyte death in vivo. J Exp Med 181: 1661–1672

Zhang F, Richardson PM, Holland DP, Guo G, Tatton WG (1995) CNTF or (–)-deprenyl in immature rats: survival of axotomized facial motoneurons and weight loss. J Neurosci Res 40: 564–570

Zhong LT, Sarafian T, Kane DJ, Charles AC, Mah SP, Edwards RH, Bredesen DE (1993) bcl-2 inhibits death of central neural cells induced by multiple agents. PNAS 90: 4533–4537

Ziv I, Melamed E, Nardi N, Luria D, Achiron A, Offen D, Barzilai A (1994) Dopamine induces apoptosis-like cell death in cultured chick sympathetic. Neurosci Lett 170: 136–140

Zsnagy I, Steiber J, Jeney (1995) Induction of age pigment accumulation in the brain cells of young male rats through iron-injection into the cerebrospinal fluid. Gerontology 41: 145–156

Authors' address: Dr. W. G. Tatton, Institute for Neuroscience, Dalhousie University, Halifax, N.S., Canada B3H 4H7.

Disinhibition-Dementia-Parkinsonism-Amyotrophy Complex (DDPAC) is a non-Alzheimer's frontotemporal dementia

K. C. Wilhelmsen

Department of Neurology, University of California, San Francisco, and The Gallo Clinic and Research Center, San Francisco General Hospital, San Francisco, CA, U.S.A.

Summary.   DDPAC was defined based on the cardinal symptoms of the syndrome found in family Mo. Investigation of DDPAC cases in family Mo shows non-specific pathological changes in a distribution that is consistent with the cardinal features of the disease. Genetic analysis identified a locus on chromosome 17q21-22 that produces this syndrome which is inherited as an autosomal dominant trait. DDPAC overlaps clinically and pathologically with a variety of named neurodegenerative syndromes and falls within the spectrum called frontotemporal dementia. The localization of the disease locus for DDPAC allows the testing of whether other familial neurodegenerative conditions also map to the same chromosomal regions. It seems possible that ultimately these conditions that have thus far been difficult to categorize will be subject to a nosolgy based on genetic etiology.

Clinical features of DDPAC

DDPAC is a neurodegenerative disease that was defined based on the clinical features found in thirteen affected members of family Mo (Lynch et al., 1994; Wilhelmsen et al., 1994). The severe behavioral changes, and subsequent development of a frontal type dementia with parkinsonism, distinguish the disease in this family from more common neurodegenerative disorders (e.g. Alzheimer's and Parkinson's disease). In family Mo there was a characteristic prodromal period (Wszolek et al., 1995) with nonspecific behavioral abnormalities. This period typically begins in the second or third decade and lasts for up to twenty years. The unifying theme during this time is disinhibition. More typical examples of disinhibition include hyper-religiosity, inappropriate sexual behavior, excessive eating, and shoplifting. Five had major depression or alcoholism as young adults. Two of the affected individuals were institutionalized and were each given a provisional diagnosis of schizophrenia. Due to the insidious onset of DDPAC, it is difficult to demarcate when the disease begins. Patients could be clearly identified as affected at an average age of 45 with death following an average of 13 years later.

Dementia and parkinsonism eventually developed in all patients with DDPAC. The dementia was characterized by early memory loss; anomia; and poor construction with relative preservation of orientation, speech, and calculations (Lynch et al., 1994). Parkinsonism was manifest by rigidity, bradykinesia, and postural instability. Tremor was found less commonly.

Although amyotrophy is not a prominent clinical feature in most patients with DDPAC, it led to a respiratory arrest in one patient with DDPAC. Another recently developed proximal muscle group wasting.

DDPAC is linked to 17q21-22

The distribution of individuals with DDPAC in Family Mo (Fig. 1) strongly suggested that DDPAC is a highly penetrant autosomal dominant trait. The absence of significant contact between some affected individuals and the geographic distribution of affected individuals further supported the premise that a genetic, and not an environmental factor, determines susceptibility to DDPAC. To confirm that DDPAC is due to genetic factors, we mapped the trait to chromosome 17q21-22 using polymorphic microsatellite markers (Wilhelmsen et al., 1994). It was essential that diagnostic criteria be established prior to any attempt at localizing the disease gene. Based on observations of family Mo, an ad hoc model was constructed. To be considered affected, an individual had to have at least two of four cardinal features: a history of disinhibition, atypical dementia, parkinsonism, and amyotrophy. Individuals with only one cardinal feature were considered unknown for linkage analysis. Linkage was detected with a series of markers at 17q21-22 with a maximum multipoint Lod score of 4.2.

With a screen of the genome for linkage with sequential markers, a Lod score of 4 implies that there is less than a 0.5% probability of a false positive (Morton, 1955). This assumes that there has been no inadvertent source of bias. By defining the linkage model before the determination of any genotypes, the most likely source of bias was avoided. The eponym *wld* has been suggested for the disease gene locus that causes DDPAC (Fahn et al., 1995).

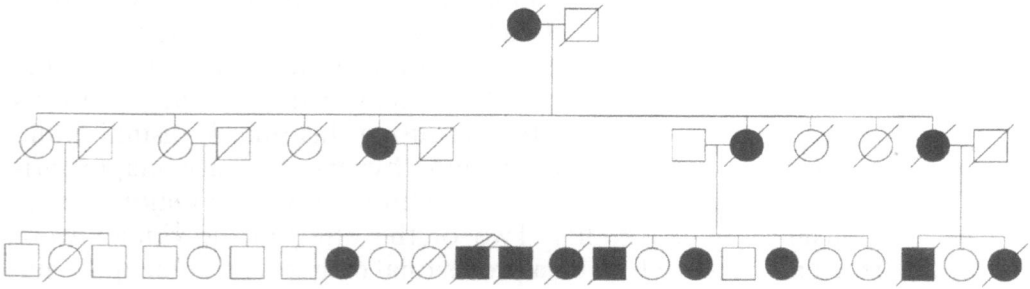

Fig. 1. Family Mo Patients with DDPAC are shaded in black. Sex and birth orders have been changed to protect confidentiality

The region that contains the locus responsible for DDPAC has many intriguing candidate genes (Genome Database) as well as a fragile site (Kormann-Bortolotto et al., 1992). The most intriguing candidate is the gene for t (tau), the major component in the paired helical filament found in Alzheimer's disease (Kosik, 1990). Other possible candidate genes include a nerve growth factor receptor and a homeobox gene cluster. Each of these genes could be considered a candidate for DDPAC, but it is at least as likely that an undiscovered gene is responsible.

The chromosomal region containing *wld*, as defined by meiotic segregation analysis of family Mo, is large and contains too many candidate genes to attempt a systematic positional cloning effort. To reduce the size of the genetic region, and thus the number of candidate genes, it is necessary to identify additional families with mutations in *wld*. The identification of the *wld* locus was based exclusively on clinical diagnosis. We were interested in knowing whether there is a characteristic pathology in DDPAC because its pleomorphic clinical features may obscure its relationship to other conditions.

Pathologic features of DDPAC

During this project, postmortem evaluations were done on six patients with DDPAC (Simma et al., 1995; Lynch et al., 1994). These patients had similar pathology, further supporting the linkage result. All had atrophy of the frontal and temporal lobes and depigmentation of the substantia nigra. All had a striking spongy rarefaction of the neuropil confined to cortical layers two and three, that was most extensive in the frontal and temporal cortices. The anterior cingulate cortex was severely affected. Rare ballooned neurons were seen in affected cortex that stained diffusely with tau and ubiquitin. No Lewy bodies, neurofibrillary tangles, or senile plaques were seen. Despite gliosis of the perferant pathways to the hippocampus, the nerve cells of the hippocampal formation appeared normal. All had severe gliosis and nerve cell loss in the amygdala and the substantia nigra. In the two cases in which the spinal cords were examined there was patchy anterior horn cell loss that varied from level to level and from side to side. In addition, there were swollen anterior horn cells that contained large vacuoles. Similar vacuolar changes have been described in mice engineered to contain a human superoxide dismutase gene with a mutation of amino acid 92 from glycine to alanine (Gurney et al., 1994).

Each of the pathologic features found in DDPAC is nonspecific. When viewed as a group, the six cases have a clear theme. The extent of involvement of particular regions varied from case to case but all share a pattern of involvement (Kormann-Bortolotto et al., 1992).

Clinical-pathologic correlation

Each of the cardinal features in DDPAC correlates with pathologic findings. Disinhibition can be accounted for by the extensive involvement of limbic structures. The spectrum of behavioral changes found in DDPAC overlaps with several frontal-subcortical dysfunction syndromes (Cummings, 1993). These syndromes are most clearly discerned with cortical lesions but can be produced by subcortical lesions. Bilateral anterior cingulate lesions characteristically produce profound abulia and akinetic mutism. Dorsolateral prefrontal cortical lesions result in inattention and reduced executive function. Orbitofrontal lesions produce changes in behavioral disinhibition and emotional lability. The pattern of behavioral changes in patients with DDPAC may correlate at some level with pathologic involvement, but the diffuse pattern of degeneration makes this pattern hard to discern.

The involvement of Limbic structures led AAF Sima to suggest the name Progressive Limbic Lobe Sclerosis as a pathologic description of the syndrome (Simma et al., 1995). In some cases of DDPAC there were Klüver-Bucy-like syndromes with hyper- or hyposexuality, and oral tendencies. This is presumably due to the degeneration of the amygdala and perforant pathway to the hippocampal formation. The cortical deafferentation of the hippocampus is expected to result in disruption of memory. The Parkinsonian features are likely to be secondary to the degeneration of pigmented cells in the substantia nigra. The degeneration of anterior horn cells is presumed to lead to clinical and subclinical amyotrophy.

The relationship of DDPAC to other neurodegenerative diseases

The nonspecific nature of the pathologic findings and the pleiotropic effects of mutations in *wld* on the clinical phenotype raises the possibility that *wld* could be responsible for other neurodegenerative diseases. Numerous reports have described the association of dementia, parkinsonism, amyotrophy, and psychosis [reviewed in Lynch et al. (1994), Wijker et al. (1996)]. These reports also frequently describe pathology that is essentially nonspecific in nature. There is a recurrent description of lobar atrophy with cortical superficial spongiosis and neuronal loss characterized by gliosis without Lewy bodies, neurofibrillary tangles, or senile plaques. Similar pathology leads several investigators to refer to the syndrome as dementia with "nonspecific" pathology. The most extensive pathologic and clinical characterization of patients with dementia and nonspecific pathologic findings have been by the groups in Lund, Sweden and Manchester, UK who have defined the term frontotemporal dementia (FTD) (The Lund and Manchester Groups et al., 1994).

FTD was defined as a behavioral disorder arising from frontotemporal cerebral atrophy and, rarely, from amyotrophy (Brun, 1987, 1993). The Lund and Manchester Groups consensus report (The Lund and Manchester Groups et al., 1994) recognized three pathologic forms of FTD with common behavioral features: Frontal Lobe degeneration type, Pick-type, and motor neuron disease type. The consensus group was careful to state that division into three forms may not reflect differences in etiology. The behavioral disorders, affective symptoms, language and cognitive patterns are very similar to the spectrum found in DDPAC. The Lund and Manchester groups reported that onset is usually before age 65 and there is frequently a positive family history in a first degree relative. The pathologic features found in DDPAC are essentially the same as described in the motor neuron disease type FTD that has the pathologic features of frontal lobe degeneration and anterior horn cell loss.

The clinical and pathologic picture of DDPAC easily fits within the classification of FTD. The genetic localization of *wld*, the locus responsible for DDPAC, is the first locus identified which can cause FTD. The identification of *wld* suggests that a new nomenclature based on genetic etiology, as exists for the hereditary sensory neuropathies (Chance and Fischbeck, 1994), may be possible in the future. FTD segregates with markers on chromosome 17 near *wld* in at least one Swedish family with FTD (Wilhelmsen et al. unpublished data). The author is aware of several additional families with dementia that are also linked to *wld*, including families with the pathologic diagnosis of pallido-ponto-nigral degeneration (Brun, 1993) and subcortical gliosis (Petersen et al., 1995; Lanska et al., 1994). These findings provide independent confirmation that there is a locus on chromosome 17q21-22 that causes atypical dementia.

Discussion

Brun reports that FTD accounts for approximately 8-10% of organic dementia in a large series of demented patients in Lund, Sweden (Brun, 1993). It is surprising that this diagnosis is not made more frequently by other investigators. One possible explanation is that nonspecific features prevent a specific diagnosis. At Columbia University, the cases of DDPAC were reported as dementia with nonspecific neurodegeneration (R. Defendini, personal communication). Without a pathognomonic histologic feature such as in Alzheimer's and Parkinson's disease it is possible that cases with nonspecific neurodegeneration may be under reported or reported under a variety of names.

Table 1 lists names of previously described dementing conditions that show some similarity to FTD and DDPAC. Most of these diagnoses are

Table 1. Familial dementing conditions related to frontotemporal dementia

Dementia lacking distinctive pathology	(Knopman et al., 1990; Kim et al., 1981)
Presenile dementia with motor neuron disease	(Mitsuyama et al., 1985; Horoupian et al., 1984; Pinsky et al., 1975)
Progressive aphasia	(Caselli et al., 1993; Snowden et al., 1992)
Hereditary dysphasic dementia	(Morris et al., 1984)
Progressive subcortical gliosis	(Neumann and Cohn, 1967; Moossy et al., 1987; Lanska and Raff, 1988)
Pallido-ponto-nigral degeneration	(Wszolek et al., 1992)
Familial fatal parkinsonism with alveolar hypoventilation and mental depression	(Purdy et al., 1979; Perry, 1976; Perry et al., 1990)
Hereditary Pick's disease	(Groen and Endtz, 1982; Groen and Hekster, 1982)

differentiated by distinctive clinical features but none of them are made with any regularity in the literature. The trait segregates with *wld* in families with at least three of these conditions. This suggests that there is a group of clinically and pathologically defined entities that may have a common genetic etiology. It is possible that a variety of genetic defects, such as those that occur in Alzheimer's disease, can produce a similar clinical and pathologic picture, presumably by affecting different points in a common pathway.

The identification of a locus on chromosome 17q21-22 that can cause DDPAC raises several questions. What is the frequency of disease caused by mutations in *wld*? What is the range of clinical and pathologic features due to mutations in *wld*? Is the range of clinical and pathologic findings found in association with mutation of *wld* due to differences in the mutations? It is possible that all of the *wld* linked conditions are due to the same mutation whose expression is modified by environmental or other genetic factors? The answers to these questions await the identification of *wld*.

References

Brun A (ed) (1987) Frontal lobe degeneration of non-Alzheimer type. Arch Gerontol Geriatr 6: 193–208

Brun A (1993) Frontal lobe degeneration of non-Alzheimer type revisited. Dementia 4: 126–131

Brun A (ed) (1993) The 2nd international conference on frontal lobe degeneration of non-alzheimer type. Dementia 4: 123–236

Caselli RJ, Windebank AJ, Petersen RC, et al (1993) Rapidly progressive aphasic dementia and motor neuron disease. Ann Neurol 33: 200–207

Chance PF, Fischbeck KH (1994) Molecular genetics of Charcot-Marie-Tooth disease and related neuropathies. Hum Mol Genet 3 [Spec No] : 1503–1507

Cummings JL (1993) Frontal-subcortical circuits and human behavior. Arch Neurol 50: 873–880

Fahn S, Mayeux R, Rowland LP (1995) A new eponym: Wilhelmsen-Lynch disease. Neurology 44: 1980

Groen JJ, Endtz LJ (1982) Hereditary Pick's disease: second re-examination of the large family and discussion of other hereditary cases, with particular reference to electroencephalography, a computerized tomography. Brain 105: 443–459

Groen JJ, Hekster RE (1982) Computed tomography in Pick's disease: findings in a family affected in three consecutive generations. J Comput Assist Tomogr 6(5): 907–911

Gurney ME, Pu H, Chiu AY, et al (1994) Motor neuron degeneration in mice that express a human Cu, Zn superoxide dismutase mutation. Science 264: 1772–1775

Horoupian DS, Thal L, Katzman R, et al (1984) Dementia and motor neuron disease: morphometric, biochemical, and Golgi studies. Ann Neurol 16: 305–313

Kim RC, Collins GH, Parisi JE, Wright AW, Chu YB (1981) Familial dementia of adult onset with pathologic findings of a 'non-specific' nature. Brain 104: 6178

Knopman DS, Mastri AR, Frey WH, Sung JH, Rustan T (1990) Dementia lacking distinctive histologic features: a common non-Alzheimer degenerative dementia. Neurology 40: 251–256

Kormann-Bortolotto MH, Smith M, Toniolo Neto J (1992) Fragile sites, Alzheimer's disease, and aging. Mech Ageing Dev 65: 9–15

Kosik KS (1990) Tau protein and neurodegeneration. Mol Neurobiol 4: 171–179

Lanska DJ, Raff RL (1988) Myokymia in motor neuron disease. J Neurol Neurosurg Psychiatry 51: 1107–1108

Lanska DJ, Currier RD, Cohen M, et al (1994) Familial progressive subcortical gliosis. Neurology 44: 1633–1643

Lynch TS, Sano M, Marder KS, et al (1994) Clinical characteristics of a family with chromosome 17-linked Disinhibition-Dementia-Parkinsonism-Amyotrophy-Complex (DDPAC). Neurology 44: 1878–1884

Mitsuyama Y, Kogoh H, Ata K (1985) Progressive dementia with motor neuron disease. An additional case report and neuropathological review of 20 cases in Japan. Eur Arch Psychiatry Clin Neurosci 235: 1–8

Moossy J, Martinez AJ, Hanin I, Rao G, Yonas H, Boller F (1987) Thalamic and subcortical gliosis with dementia. Arch Neurol 44(5): 510–513

Morris JC, Cole M, Banker BQ, Wright D (1984) Hereditary dysphasic dementia and the Pick-Alzheimer spectrum. Ann Neurol 16: 455–466

Morton NE (1955) Sequential tests for detection of linkage. Am J Hum Genet 7: 277–318

Neumann MA, Cohn R (1967) Progressive subcortical gliosis: a rare form of presenile dementia. Brain 90: 405–418

Perry, TL (1976) Hereditary mental depression with taurine deficiency: further studies, including a therapeutic trial of taurine administration In: Huxtable R, Barbeau A (eds) Taurine. Raven Press, New York, pp 363–374

Perry TL, Wright JM, Berry K, Hansen S, Perry TL Jr (1990) Dominantly inherited apathy, central hypoventilation, and Parkinson's syndrome: clinical, biochemical, and neuropathologic studies of 2 new cases. Neurology 40: 1882–1887

Petersen RB, Tabaton M, Chen SG, et al (1995) Familial progressive subcortical gliosis: presence of prions and linkage to chromosome 17. Neurology 45: 1062–1067

Pinsky L, Finlayson MH, Libman I, Scott BH (1975) Familial amyotrophic lateral sclerosis with dementia: a second canadian family. Clin Genet 7: 186–191

Purdy A, Hahn A, Barnett HJ, et al (1979) Familial fatal parkinsonism with alveolar hypoventilation and mental depression. Ann Neurol 6: 523–531

Simma AAF, Defendini RF, Keohane C, et al (1995) The chromosome 17-linked disinhibition-dementia-parkinsonism-amyotrophy complex is characterized pathologically by progressive limbic lobe sclerosis (unpublished)

Snowden JS, Neary D, Mann DM, Goulding PJ, Testa HJ (1992) Progressive language disorder due to lobar atrophy. Ann Neurol 31: 174–183

The Lund and Manchester Groups, Brun A, Englund B, et al (1994) Clinical and neuropathological criteria for frontotemporal dementia. J Neurol Neurosurg Psychiatry 57: 416–418

Wijker M, Wszolek ZK, Wolters ECH, et al (1996) Localization of the gene for rapidly progressive autosomal dominant parkinsonism and dementia with pallido-ponto-nigral degeneration to 17q21. Hum Mol Genet 5 (1): 151–154

Wilhelmsen KC, Lynch T, Nygaard TG (1994) Localization of disinhibition-dementia-parkinsonism-amyotrophy complex to 17q21-22. Am J Hum Genet 55: 1159–1165

Wszolek ZK, Pfeiffer RF, Bhatt MH, et al (1992) Rapidly progressive autosomal dominant parkinsonism and dementia with pallido-ponto-nigral degeneration. Ann Neurol 32: 312–320

Wszolek ZK, Lynch T, Wilhelmsen KC (1995) Rapidly progressive autosomal dominant parkinsonism and dementia with pallido-ponto-nigral (PPND) and disinhibition-dementia-parkinsonism-amyotrophy complex (DDPAC) are clinically distinct conditions that are both linked to 17q21-22 (unpublished)

Author's address: K. C. Wilhelmsen, M.D., Ph. D., Department of Neurology, University of California, San Francisco, and The Gallo Clinic and Research Center, San Francisco General Hospital, 1001 Potrero Avenue, Bldg. 1, Room 101, San Francisco, California 94110, U.S.A.

Subject Index

Springer Neurology

P. Riederer, D. B. Calne, R. Horowski,
Y. Mizuno, W. Poewe, M. B. H. Youdim (eds.)

Advances in Research on Neurodegeneration
Volume 5

1997. 45 figures. VIII, 215 pages.
Cloth DM 198,–, öS 1386,–, US $ 149.00
ISBN 3-211-82933-4
Special edition of "Journal of Neural Transmission, Supplement 50, 1997"
(Soft cover edition of Supplement 50 only available for subscribers to "Journal of Neural Transmission")

The "International Winter Conferences on Neurodegeneration" have become an established forum to discuss various aspects of basic and clinical topics related to the underlying mechanisms of neurodegenerative disorders. This volume focuses on brain imaging, endogenous and exogenous neurotoxins, programmed cell death, apoptosis and necrosis, and immunoinflammatory mechanisms, infective diseases causing neurological disorders. These topics have been reviewed by invited experts and the articles give an up-to-date reflection of the state of the art in these research fields.

From the contents:

Magnetic Resonance: A multimodal approach to the brain? • Measurement of the dopaminergic degeneration in Parkinson's disease with [^{123}I] β-CIT and SPECT • rCBF SPECT in Parkinson's disease patients with mental dysfunction • IBZM- and β-CIT-SPECT of the dopaminergic system in parkinsonism • Pathophysiology of movement disorders studied using PET • Contributions of Positron Emission Tomography to elucidating the pathogenesis of Idiopathic Parkinsonism and Dopa Responsive Dystonia • Mechanism of 6-Hydroxydopamine neurotoxicity • Induction of mitosis-related genes during dopamine-triggered apoptosis in sympathetic neurons • Neuronal vulnerability in Parkinson's disease • N-Methyl-(R)salsolinol as a dopaminergic neurotoxin: From an animal model to an early marker of Parkinson 's disease • The halogenated tetrahydro-β-carboline "TaClo": A progressively-acting neurotoxin • Developmental and genetic regulation of programmed neuronal death • Apoptosis in neurodegenerative disorders • Mechanisms of cell death in Alzheimer's disease • Assessment of neurotoxicity and "neuroprotection" • Update on management and genetics of multiple sclerosis • Pathogenesis of immune-mediated demyelination in the CNS • Basic mechanisms of brain inflammation • Cell death in prion disease

 Springer Wien New York

Sachsenplatz 4-6, P.O.Box 89, A-1201 Wien, Fax +43-1-330.24.26, e-mail: order@springer.at, Internet: http://www.springer.at
New York, NY 10010, 175 Fifth Avenue • Heidelberger Platz 3, D-14197 Berlin • Tokyo 113, 3-13, Hongo 3-chome, Bunkyo-ku

Springer-Verlag
and the Environment

WE AT SPRINGER-VERLAG FIRMLY BELIEVE THAT AN international science publisher has a special obligation to the environment, and our corporate policies consistently reflect this conviction.

WE ALSO EXPECT OUR BUSINESS PARTNERS – PRINTERS, paper mills, packaging manufacturers, etc. – to commit themselves to using environmentally friendly materials and production processes.

THE PAPER IN THIS BOOK IS MADE FROM NO-CHLORINE pulp and is acid free, in conformance with international standards for paper permanency.